Chemical Biology of the Genome

Chemical Biology of the Genome

Siddhartha Roy
Tapas K. Kundu

ACADEMIC PRESS
An imprint of Elsevier

ELSEVIER

Academic Press is an imprint of Elsevier
125 London Wall, London EC2Y 5AS, United Kingdom
525 B Street, Suite 1650, San Diego, CA 92101, United States
50 Hampshire Street, 5th Floor, Cambridge, MA 02139, United States
The Boulevard, Langford Lane, Kidlington, Oxford OX5 1GB, United Kingdom

Library of Congress Cataloging-in-Publication Data
A catalog record for this book is available from the Library of Congress

British Library Cataloguing-in-Publication Data
A catalogue record for this book is available from the British Library

ISBN 978-0-12-817644-3

For information on all Academic Press publications
visit our website at https://www.elsevier.com/books-and-journals

Publisher: Andre Gerhard Wolff
Acquisitions Editor: Peter B. Linsley
Editorial Project Manager: Samantha Allard
Production Project Manager: Kiruthika Govindaraju
Cover Designer: Greg Harris, Kruthi HT and Authors

Typeset by SPi Global, India

Dedication

We dedicate this book to our parents, who taught us to apply logic and reason and always encouraged and inspired us to go beyond the ordinary to seek the truth.

Contents

Preface

Interdisciplinary sciences have gradually evolved during the last century. One of the pioneers of interdisciplinary sciences was the physicist Jagadish Chandra Bose, to whom many give credit for the discovery of radio communication. In the first decades of the 20th century, he constructed novel instruments and used techniques borrowed from physics to study electrical activities in plants and many other biological phenomena. Even before that, Chemists, notably Emil Fischer, were using chemical tools to elucidate chemical structures of molecules isolated from biological systems. During the second half of the last century, the use of physical and chemical tools in life sciences became widespread. The importance of chemical tools and concepts in the life sciences became overwhelming in the current century. Chemical Biology is now becoming a pivot around which modern biology is developing.

Since the elucidation of DNA structure and the chemical nature of genes, the genome has emerged as the centerpiece of biological processes and function. It is now clear that many of the most important cellular processes are rooted in the genome. The genome has become the central focus of the study of life sciences. Breakthroughs in microscopy, "omics" methods, and DNA sequencing have revolutionized the study of the genome. We now realize that the extraordinary complexity of genome function is underlined by physical and chemical principles, although not yet understood in all its intricacies and nuances.

When we planned the book's content, our aim was to emphasize and underline the importance of chemical principles in the functioning of the genome to senior undergraduate or early career graduate students in Life Sciences. Thus we have recapitulated many chemical principles in the early part of the book at the cost of repeating fundamental chemistry. Chapter I focuses on the integrative nature of biological systems and forms the philosophical backbone of the book. Given our focus on the integrative nature of the living systems, we concentrate on tools widely used in chemical biology to study intact living systems in Chapter II. Chapter III introduces the reader to fundamental aspects of genome organization and function. Chapter IV focuses on the chemistry that underpins the genome function and chemical modifications of the genome used for fine tuning the genome organization and function (epigenome). Chapter V deals with chemical principles of regulation of gene expression. As mentioned above, technologies for large-scale sequencing of DNA were the game changers for studying genome function. Chapter VI describes many of the new generation DNA sequencing technologies and their use in elucidating genome function. By this time, the reader must have become conversant with the idea that genes do not function alone but in conjunction with other genes. Chapter VII focuses on how these networks of genes function and drugs are being developed to intervene in these gene regulatory networks. As genome function is central to much of biology, much attention is now being given to finding drug targets against aberrant genome processes that lead to diseases, with increasing success. The final chapter is devoted to introducing the readers to this aspect of Chemical Biology.

Our aim was to write a book that would lead the uninitiated readers to a comprehensive understanding of the genome function from a chemical perspective and how a chemical understanding gives us a new platform to improve therapeutic outcomes in many disease processes. Naturally, we could not touch on all the aspects of genome function, and examples were only chosen to illustrate some of the major areas of genome science. We hope that students of life sciences would benefit from this book by nudging them toward a better appreciation of inalienable nature of Chemical Sciences and Living systems.

SR acknowledges Prof. Dhananjay Bhattacharyya, Dr. Krishnananda Chattopadhyay, Dr. Debabrata Biswas, and Dr. Abhishek Mazumder for reading parts of the manuscript and giving helpful feedback. SR acknowledges the support he received from his wife during the writing of this book, without which it would have been challenging to finish it.

TKK acknowledges Ms. Kruthi HT for drawing several figures and editorial assistance. He also acknowledges Dr. Anand P Kulkarni for preparing three figures. Last but not least, TKK acknowledges his students and postdoctoral fellows for their fundamental research contributions, used for writing some part of this book.

Siddhartha Roy
Tapas K. Kundu

I

An integrative view of chemical biology

To whatever degree we might imagine our knowledge of the properties of the several ingredients of a living body to be extended and perfected, it is certain that no mere summing up of the separate actions of those elements will ever amount to the action of the living body itself.

John Stuart Mill

I.1 What constitutes Chemical Biology

After its initial appearance in the 1980s, the use of the term Chemical Biology became widespread from the 1990s. What distinguishes "Chemical Biology" from the related area of "Biological Chemistry?" Both encompass scientific inquiry at the interface of Biology and Chemistry. However, the philosophical underpinnings of the two disciplines may be somewhat different. Traditionally, when two words, representing two different fields, are conjoined to define a new discipline at their interface, the last word defines the broader area to which it belongs. Thus Physical Chemistry becomes a part of Chemistry, and generally, Chemical Physics is a part of Physics. Similarly, Mathematical Physics belongs to the domain of Physics.

In the same vein, Chemical Biology tilts more toward Biology than Chemistry. At least the underlying emphasis should be on biological phenomena. What primarily separates Chemistry and Biology is that in Biology, novel properties emerge when molecules come together with a degree of organization not seen in typical chemical systems. The focus here is a collection of molecules that behave as a whole and not merely as the sum of properties of its constituents. The distinguishing feature of biological ensembles is the evolution of new properties, the emergent properties, when compared with the same molecules in the separated state. An integrative approach is not only beneficial but also essential for a proper understanding of Biology. Thus, in our view, Chemical Biology should take more of an integrative approach and, arguably, should be one of its defining characteristics. The mainstay of this book will be on genomes. It is becoming clear that a large part of Biology is underpinned by the regulation of gene expression. Cooperation between various genes and their products is an essential element in the regulation of gene expression. This theme will be important for subsequent chapters, where the variations and nuances of the overall theme will be discussed in detail.

Chemical Biology of the Genome. https://doi.org/10.1016/B978-0-12-817644-3.00004-0

In some circles, Chemical Biology often carries the connotation of the use of small molecules to perturb the cellular systems and derive active drug candidates. Although this is undoubtedly a vital part of Chemical Biology, it is not limited to that. We argue here that obtaining information on functioning ensembles of molecules, cells, or any other related entities, with any chemical tool, should be considered as Chemical Biology. The emphasis here is on ensembles and not individual entities. In particular, elucidating the emergent properties of such functional ensembles may be considered as the most critical component of Chemical Biology. We will adhere to this approach throughout this book.

I.1.1 The cellular interior is highly organized

Molecules are highly organized within the cell (Fig. I.1). The cell is not a bag full of molecules. In simpler systems, such as bacteria, no visible compartments exist. However, if you collect all the molecules and enclose them in a lipid membrane, it is not going to produce a live bacterium. It is increasingly apparent that many molecules inside the cell occur as a part of a larger supramolecular complex. How molecules interact and form defined complexes depend on intermolecular forces and the cellular environment. Intracellular organization is a complex feature of the cell, and its nature is not yet fully understood. An important aspect of this organization is that this is not a static organization; it is an ever-changing one.

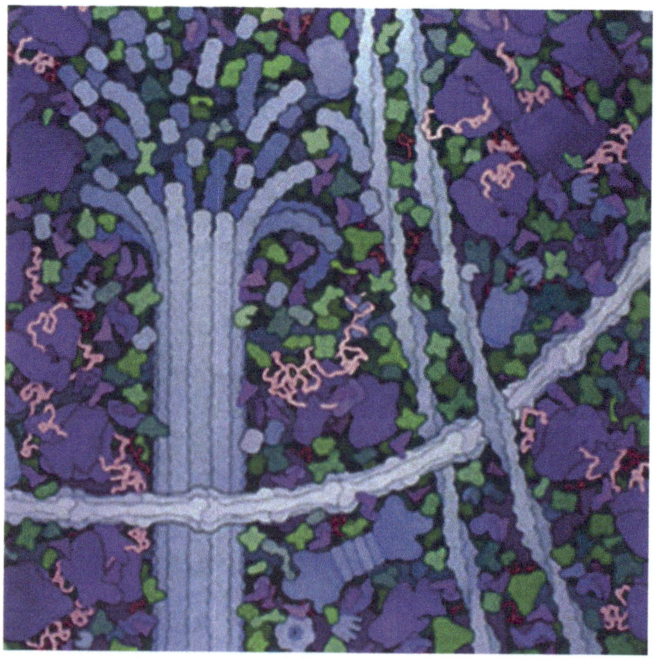

FIG. I.1 An artist's conception of the cellular interior. *Illustration by David S. Goodsell, the Scripps Research Institute obtained from (https://mgl.scripps.edu/people/goodsell/illustration/public/) with permission for reproduction in print as described.*

The organization changes with time, for example, as the cell divides, it changes upon sensing the presence of food or a threat and to many other external signals. This delicate, at the same time, robust organization makes life so different from other organized structures.

I.1.2 Different kinds of organizations

Molecular organizations within a cell are hierarchical. Individual protein molecules may assemble in defined stoichiometries to produce multimeric structures. Multiple types of macromolecules may assemble to produce bigger supramolecular structures. Their formation is better understood in thermodynamic terms. At a higher level, supramolecular structures may be organized into organelles. In a eukaryotic cell, the presence of membrane-bound organelles such as the nucleus and the mitochondrion is well known. Apart from the membrane-bound organelles, there are membraneless organelle-like entities, such as the nucleoli. They belong to a class of structures called liquid-liquid phase-separated condensates. They are perhaps least understood in thermodynamic terms. The full dimensions of the intracellular organization may not have been elucidated.

I.1.2.1 Thermodynamic principles of cellular organization

In the late 19th century, the concept of chemical thermodynamics took shape. The concepts that emerged were applicable to systems at or near equilibrium. However, the principles of equilibrium thermodynamics do not apply to crucial aspects of living systems, as they are not equilibrium structures. Gradually, the principles underlying nonequilibrium thermodynamics became clear. An important concept introduced by Ilya Prigogine is that of the dissipative structure. Dissipative structures are far removed from equilibrium and exist in a steady-state. They are open systems and exchange matter and energy, and their stability is maintained by the expenditure of energy. Many of the organized structures found in living systems are dissipative structures.

A cell is an entity far away from equilibrium and maintained by the expenditure of energy. It is our incomplete understanding of systems far removed from equilibrium that makes it challenging to apply laws of thermodynamics appropriately. According to recent trends, we distinguish between self-assembly and self-organization. Self-assembly is coming together of smaller components forming a supramolecular entity and, in the process, minimizing the free energy (Fig. I.2). An example of self-assembly is the formation of multiprotein complexes. These structures do not require energy to maintain their assembled state. We may apply principles of equilibrium thermodynamics to self-assembled structures, such as multiprotein complexes. Self-organization is coming together of components to create a dynamic structure, maintained by the expenditure of energy, whose composition may well vary. A simple example of such a structure is the self-organized spatio-temporal patterns emanating from Belousov-Zhabotinsky reaction (Fig. I.3). They are in steady-state. Examples of such structures are organelles that are dynamic entities and whose composition may change with time. Thus they are open systems exchanging mass and energy. A self-organized structure that we will intensively deal within this book is the genome.

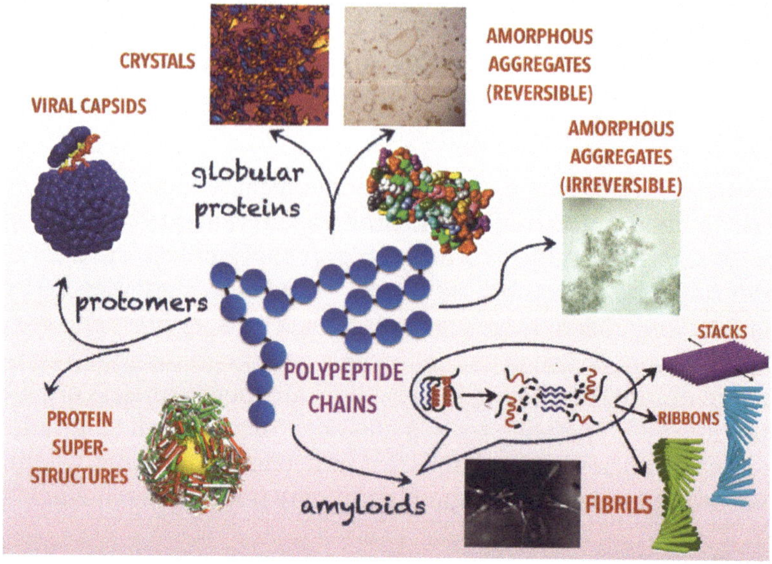

FIG. I.2 Examples of protein self-assembly. A number of different assemblies can be formed by proteins or peptides as illustrated here. Globular proteins can assemble to become either crystals or amorphous aggregates; both types of assembly are reversible *(top)*. Upon a conformational change to a protein or peptide, either amorphous aggregates (that are irreversible) or amyloid fibrils can form *(right)*. For some peptides, a range of different twisted structures have been observed including ribbons, fibrils, and stacks *(bottom right)*. Higher-order assemblies including viral capsids (in the presence of a nucleic acid) and protein superstructures (including bacterial microcompartments) can also form either naturally or in a directed manner, for example by mutagenesis *(left)*. *Reproduced from McManus et al., 2016. Curr. Opin. Colloid Interface Sci. 22, 73–79. Elsevier.*

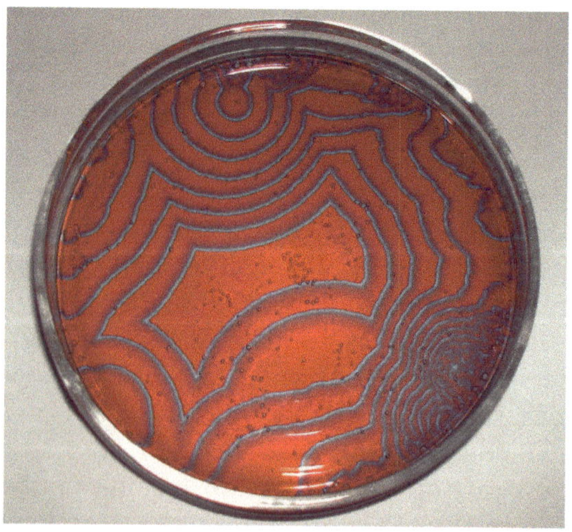

FIG. I.3 An example of self-organized structure. *Authored by Prof. Stephen Morris, University of Toronto, Canada. Reproduced from Wikimedia Commons under CC-BY-2.0.*

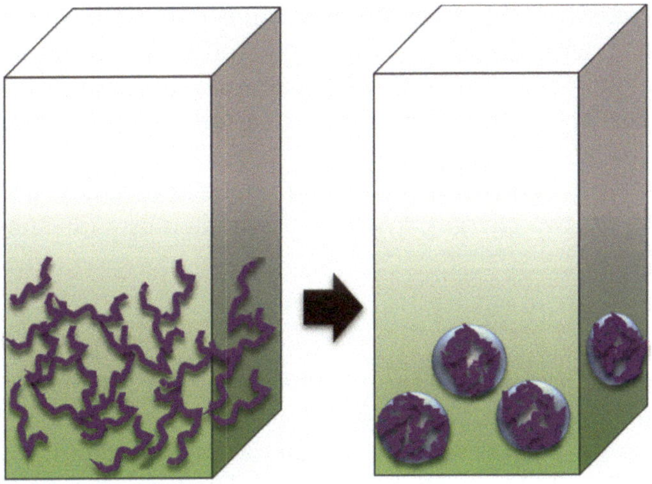

FIG. I.4 A cartoon diagram of liquid-liquid phase separation.

A form of self-assembled structure has drawn much interest in biology recently. This is the liquid-liquid phase separated condensates (Fig. I.4). Although they are thermodynamically stable under the conditions of formation, suggesting a self-assembled structure, unlike macromolecular complexes, they may not have defined stoichiometries.

Liquid-liquid phase separation and formation of condensates

The most commonly observed three phases of matter are solid, liquid, and gas. A key characteristic that differentiates three phases is the short- and long-range order of the molecules. Solids have both short- and long-range orders. In liquids, there is a short-range order but little or no long-range order; whereas, in the gas phase, the molecules lack both short- and long-range orders. In a solution, individual solute molecules are separated and dispersed throughout the solution. Depending on the concentration, temperature, and other variables, solute molecules can separate from the solution phase and coalesce into a new phase. These phase-separated condensates may vary in size. If the solute molecules in the new phase have some short-range order but are dynamic enough to prevent long-range order formation, it is referred to as liquid-liquid phase separation. In biological systems, intrinsically disordered proteins play a crucial role in forming liquid-liquid phase separated condensates.

I.1.2.2 Intermolecular forces are important for self-assembly and self-organization

Intermolecular forces play vital roles in the formation and stability of both self-assembled and self-organized structures. Attractive forces hold complexes of multiple molecules together, overcoming entropy loss due to molecular association. The formation of liquid-liquid phase-separated condensates also depends on intermolecular forces, particularly the attractive forces between macromolecules. Many self-assembled structures are dynamic. Assembled complexes may dissolve and reform upon the change of environmental conditions in situations, where external conditions play a significant role. We will briefly discuss about the different kinds of attractive forces that operate in the cellular organization in the following section.

Ionic interactions are found in many macromolecular complexes. The force (F) along the line connecting the two charges, q_1 and q_2, separated by a distance r, is given by the following equation:

$$F = \frac{q_1 . q_2}{4\pi\varepsilon_o K_r r^2},$$

where ε_o is the permittivity of the vacuum and K_r is the relative permittivity or dielectric constant of the medium. Except for the hydrophobic interactions described below, electrostatic interactions are at the core of other attractive forces.

Hydrogen bonds (Fig. I.5) are weak bonds formed between two electronegative atoms, such as nitrogen and oxygen, with a hydrogen atom sandwiched in between. Typically, the hydrogen atom is attached to one of the electronegative atoms by a covalent bond (the hydrogen bond donor) and noncovalently to the other electronegative atom through its lone pair of electrons (the acceptor). Hydrogen bond strengths vary considerably, from about a few kilocalories per mole to tens of kilocalories per mole. However, their bond strengths are always less than those of the covalent bonds. Because of the strict directional nature of the covalent bond and that of the lone pair of electrons, the hydrogen bonds are linear or nearly so. The requirement for linearity makes the spatial relationship between the atoms critical for bond formation. A precise need for strict spatial relationships makes the hydrogen bond a crucial component of sequence-specific recognition of DNA.

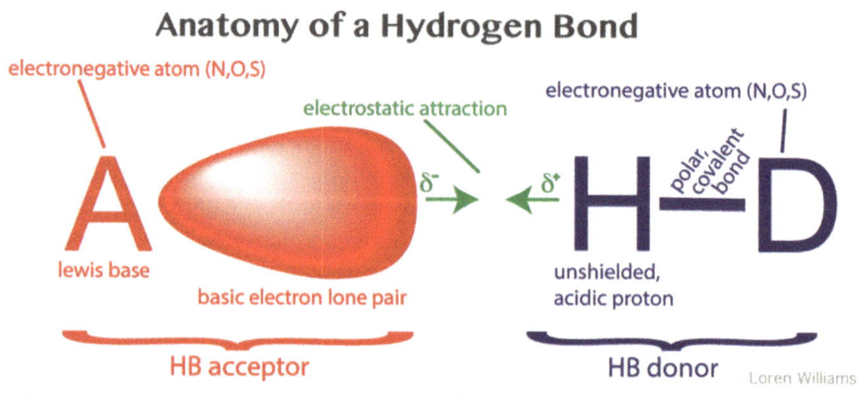

FIG. I.5 Typical arrangement of atoms in a hydrogen bond. The elements of a hydrogen bond include the hydrogen bond acceptor, hydrogen bond donor, the lone pair, and the exposed proton. N, O, and S are the predominant hydrogen bonding atoms (A and D) in biological systems. *Reproduced from the website https://ww2.chemistry.gatech.edu/~lw26/structure/molecular_interactions/mol_int.html with permission from Loren Williams, Professor, Chemistry & Biochemistry, Georgia Tech.*

Van der Waals forces originate from asymmetric electronic charge distributions, either permanent or transient. An electron cloud surrounds an atom or a molecule. For a molecule that is not an ion, the overall charge is zero. However, the molecule may have an asymmetric charge distribution leading to partial charge separation and the presence of a dipole. This partial charge separation leads to a molecule possessing a permanent dipole moment. When two such molecules come close in space, there may be an attractive force, depending upon the orientation of the dipoles, due to electrostatic interactions (Fig. I.6). The second type of van der Waals interaction occurs between a molecule possessing a dipole moment (asymmetric charge distribution) and a molecule having no dipole moment. When a molecule possessing a dipole approaches a molecule that otherwise has no dipole moment, it may induce asymmetry in charge distribution in the latter molecule due to polarizability of the electron clouds. This asymmetry results in an induced dipole and an attractive force as the dipoles are now appropriately aligned (Fig. I.7). The third type of van der Waals force is the London's dispersion force that originates from instantaneous-dipole-induced-dipole interaction. In a molecule possessing no net dipole moment, electrons are moving rapidly, resulting in transient asymmetry in charge distribution. The lack of a net dipole moment is because the time average of such

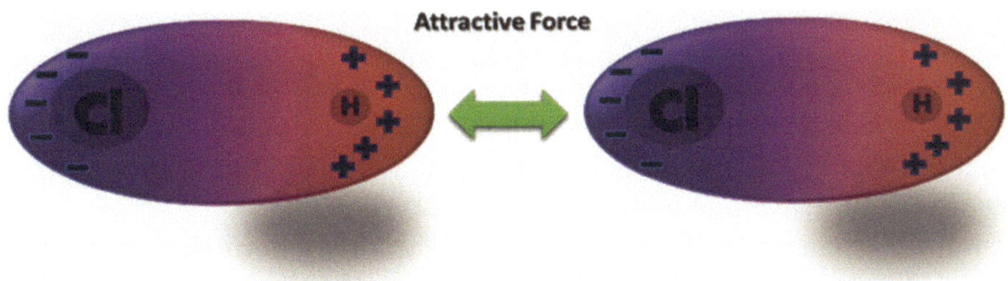

FIG. I.6 Depiction of dipole-dipole interaction between two HCl molecules.

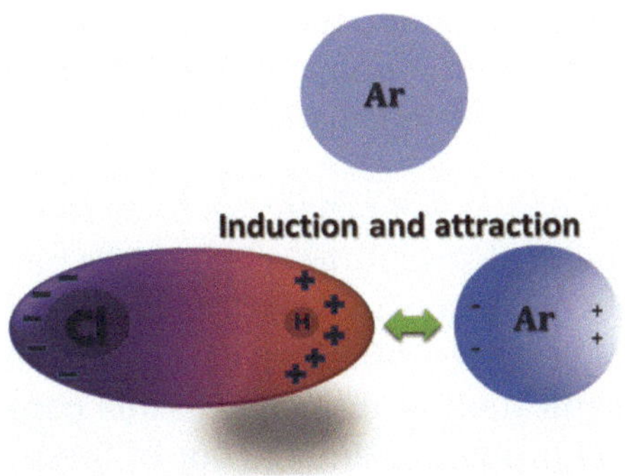

FIG. I.7 Permanent dipole-induced dipole interaction between HCl (a permanent dipole) and argon, an atom possessing no dipole moment, by induction of asymmetric charge distribution leading to transient dipole.

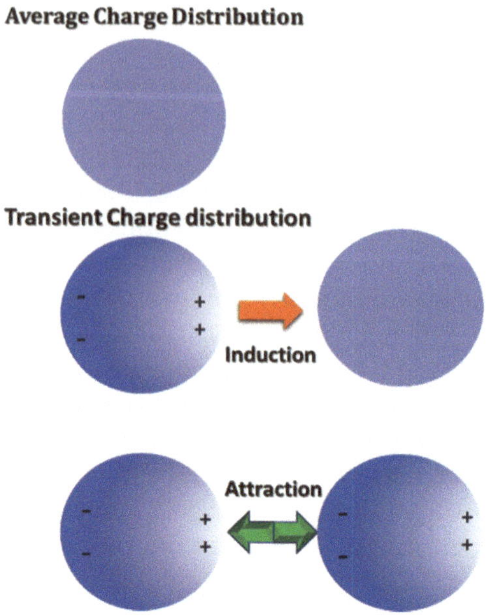

FIG. I.8 Origin of dispersion forces. At any instance, electrons in an atom are asymmetrically distributed, leading to presence of transient dipoles. Such dipoles result in time-averaged attractive forces.

transient dipoles is zero. However, when two such molecules approach each other, the transient dipole induces a dipole moment in the other molecule, which is also transient. The transient dipole moment results in a transient attraction between the two atoms or molecules whose time average is nonzero (Fig. I.8).

Lastly, we will discuss a driver of macromolecular interactions that originate from the hydrophobic effect. Transfer of hydrophobic groups from a polar to a more apolar medium underpins this net attractive effect. In water-exposed hydrophobic surfaces, water molecules in the adjacent layer cannot form hydrogen bonds with surface apolar atoms. Instead, they form hydrogen bond with each other and become fairly ordered. When two hydrophobic surfaces interact, the water in the adjacent layers is released to the bulk water. This released water acquires a more disordered character resulting in entropy increase, hence favoring the association of hydrophobic surfaces (Fig. I.9).

I.1.2.3 Organized interiors imply novel properties of ensembles

The living organisms and their substructures such as cells possess many unique properties. Many of these properties cannot be fully predicted from those of the constituents. The ubiquitous presence of a high degree of molecular organization in the cellular interior appears universal. This organization is an indispensable component of the functioning cell, hinting that some novel properties emerge out of these organizations. This makes it imperative that the assembled molecules must be studied as a whole to elucidate their full functions. A concept of emergent properties is very useful. In the following sections,

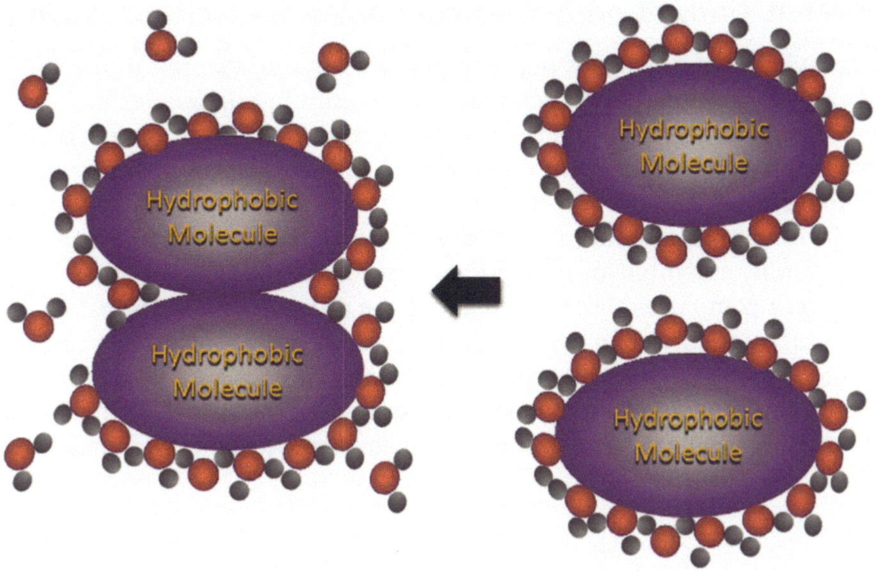

FIG. I.9 The molecular origin of the hydrophobic effect. The *red balls* are oxygen atoms, and the *white balls* are the hydrogen atoms and together they represent a water molecule.

we will describe the nature and properties of the molecular ensembles in more detail, including the emergence of novel properties.

Emergent properties

The phenomenon of emergence spans many disciplines. Its precise meaning has been debated vigorously. For our narrow purpose here, we define it as new properties present in an ensemble of molecules that are not present in their isolated constituent parts.

I.2 Ensembles of molecules

I.2.1 An ensemble of molecules may have novel properties

Cells contain both self-assembled and self-organized structures. Emergent properties are present in both kinds of structures. We will first deal with macromolecular ensembles that are mostly self-assembled structures. Macromolecular ensembles are ubiquitous in eukaryotic cells. In addition to studying the properties of their components, we must study them in a fully assembled state to reveal the complete repertoire of functional properties.

I.2.2 Different kinds of organized ensembles of macromolecules

Many different kinds of molecules populate the interior and exterior of the cell. Cells are not simply a membrane-wrapped collection of functioning molecules but spatially and temporally organized. The genomic DNA is highly organized, as are protein and RNA molecules. Lipid molecules, which form thin layers in the form of membranes, are also organized into domains of distinct compositions. The lipid raft is a well-known example. There are two kinds of supramolecular organizations: transient and permanent. The transient structures generally assemble for a functional purpose and then dismantle. Gene regulatory multiprotein complexes that form near a gene are examples of transient structures. Permanent structures survive many functional cycles. An example of a permanent structure is the ribosome, stable enough to survive numerous functional cycles. Its stability, however, does not imply a lack of dynamism. It undergoes many conformational changes during its functional cycle, and different molecules associate with the ribosome to modify its function.

I.2.2.1 Protein-protein complexes

Multiprotein complexes are ubiquitous in eukaryotic cells. They can be in the form of protein only complexes, such as the complex containing nine aminoacyl-tRNA synthetases and three accessory proteins, the newly discovered purinosome, a complex containing many enzymes that catalyze consecutive steps of the purine biosynthetic pathway, and a complex of helicase, helicase loader, and the primase protein (Fig. I.10). Many other such complexes have been proposed, including the complex involving enzymes of the glycolytic pathway, dubbed the glucosome.

I.2.2.2 Membrane anchored protein-protein complexes

Membranes that envelope the organelles and the cell itself are also highly organized (Fig. I.11). Different lipids and proteins embedded in the membrane undergo phase separation and form localized ensembles such as lipid rafts. Lipid rafts are microdomains within the plasma membrane enriched in cholesterol and sphingolipids. Some signaling proteins, such as the Epidermal Growth Factor Receptor (EGFR), are also abundant in the rafts. The rafts likely function through the concentration of signaling complexes and related proteins.

I.2.2.3 Protein-RNA complexes

Protein-RNA complexes are another class of organized multimacromolecular complexes. Many kinds of protein-RNA complexes are present in the cell, such as the ribosome, spliceosome, and P-bodies. Ribosomes are formed from interactions, well defined in composition and stoichiometry, between component proteins and RNAs. Spliceosomes are dynamic entities that assemble on a precursor mRNA to excise the introns and produce the mRNA, ready for translation. The spliceosome (Fig. I.12) contains several small nuclear RNAs, and many proteins are formed by specific protein-protein and protein-RNA

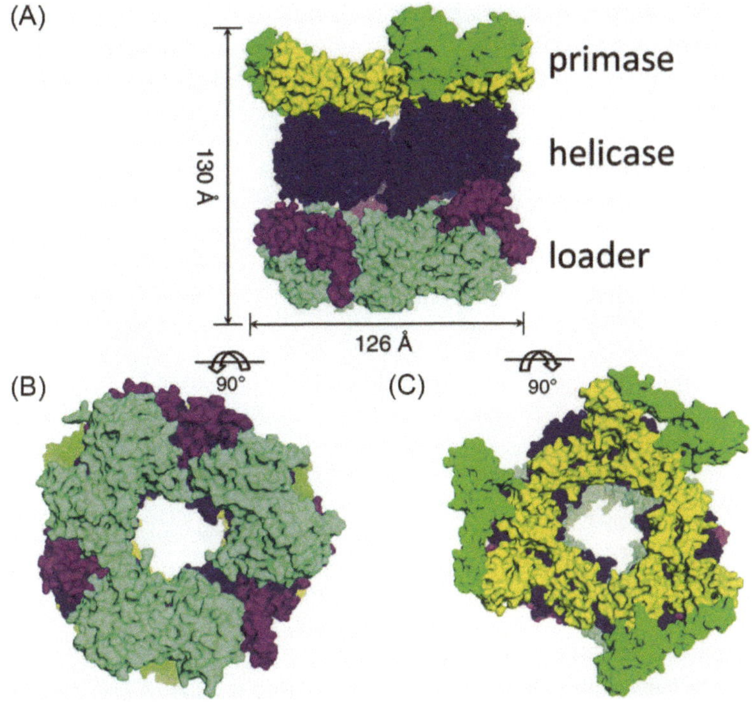

FIG. I.10 Structure of the complex between the helicase, the helicase loader, and the primase proteins: (A) side view, (B) bottom view, (C) top view. *Credit: Pieters et al., 2016. Chem. Soc. Rev. 45, 24–39, licensed by http://creativecommons.org/licenses/by/3.0/.*

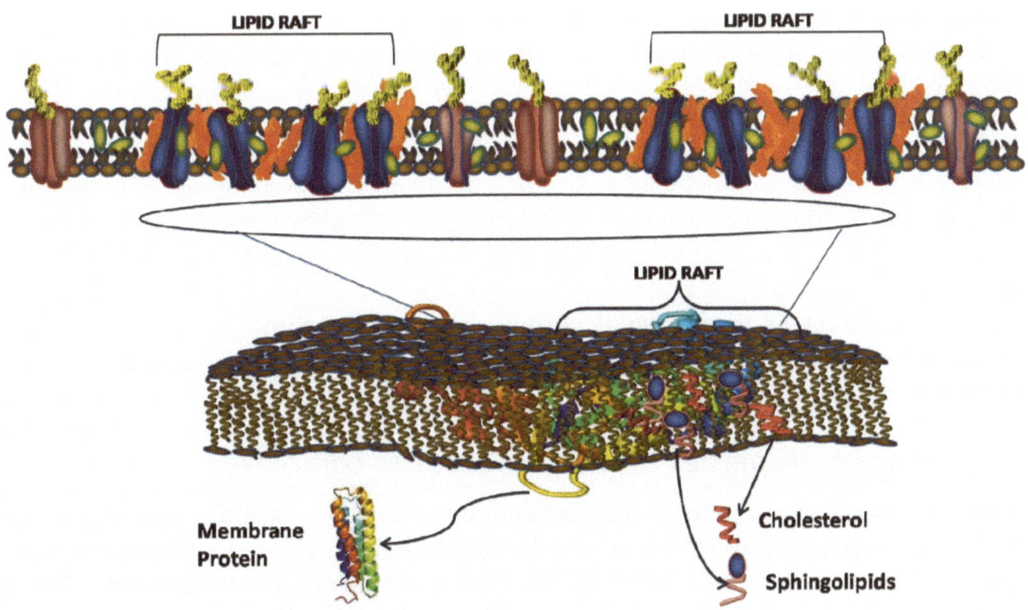

FIG. I.11 A cross-sectional view of the membrane bilayer and lipid rafts.

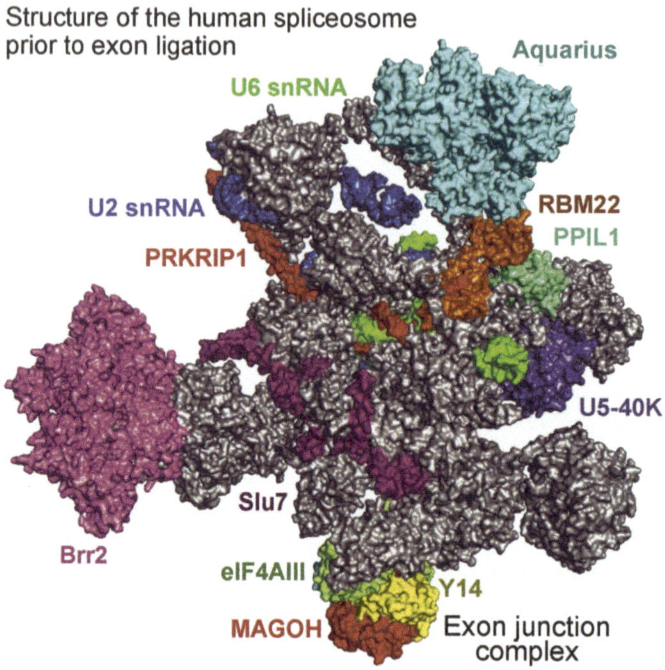

FIG. I.12 Structure of the human spliceosome. *Reproduced from Zhang et al., 2017, Cell 169, 918–929. Elsevier.*

interactions. P-bodies are ribonucleoprotein complexes, where the mRNA processing takes place. They often act as a reservoir of repressed mRNAs that may be released upon appropriate signaling. They are structurally different from ribosomes and Spliceosomes. They are liquid-liquid phase-separated condensates comprising of RNAs and proteins.

I.2.2.4 Protein-DNA complexes

Chromosomes are organized structures of proteins and DNA. Chromosomes, when fully stretched, exceed the dimensions of the nucleus. The average diameter of a nucleus is 5 μm, whereas human chromosome 1 (consisting of 249 million base pairs), when fully stretched, would span about 8 cm. Histones and other chromatin-associated proteins bind the chromosomal DNA and condense it enormously. Recent work suggests that a similar organization of the bacterial chromosome occurs through nonspecific DNA-binding proteins, such as the *E. coli* HU protein. We will provide a detailed view of the genome organization in Section I.3.

I.2.3 Emergent properties of macromolecular complexes

The emergence of new properties is found at many scales of operation. Take the example of a protein. A protein is fully functional when it is folded. The functionality is contingent upon attaining the right spatial relationships among amino acids. Properties of individual

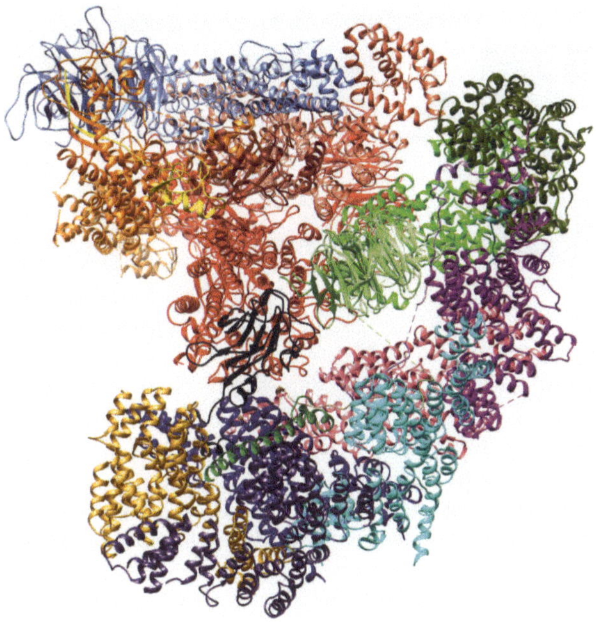

FIG. I.13 The cryoEM structure of the anaphase promoting complex (PDB5KHR). *The image was generated using the freeware chimera.*

amino acids are insufficient for predicting protein function. Another well-known property that emerges upon protein association is that of positive cooperativity of oxygen binding to hemoglobin. The subunit monomers are capable of oxygen binding, but the positive cooperativity is dependent upon the formation of the functional tetramer. Many large protein complexes are known whose function is dependent upon proper assembly of the constituent proteins. Anaphase promoting complex, a huge multiprotein complex, is just one such example (Fig. I.13).

Cooperative binding of ligands to proteins

The term ligand has been used by different authors slightly differently. In this book, we will use the following definition. When a molecule smaller than a protein binds to a protein, we will define the smaller molecule as a ligand. When a protein binds to another protein, we will call it a protein-protein interaction or association. When multiple identical ligands bind to a protein, they generally bind to different subunits of the same protein. If the binding of the first ligand enhances the binding of subsequent ligands, it is defined as positive-cooperative binding. If the binding of the first ligand weakens the binding of the subsequent ligands, it is defined as negative cooperativity.

The cooperativity often arises in multisubunit proteins, with each subunit possessing one ligand-binding site. Two models have been proposed to account for binding cooperativity, the concerted or Monod-Wyman-Changeux model and the stepwise or Koshland-Nemethy-Filmer

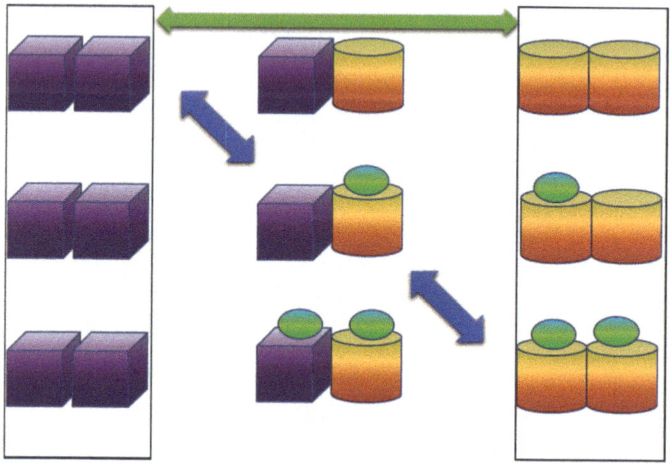

FIG. I.14 Comparison between Monod-Wyman-Changeux and Koshland-Nemethy-Filmer models. The states in the diagonal connected by *blue arrows* represent the Koshland-Nemethy-Filmer model. In the Monod-Wyman-Changeux model, the two states are at equilibrium with one state binding the ligand with high affinity (the cylinder).

(KNF) model. In both the models, the subunits can assume two different conformations having different affinities for the ligand. In the former, only two conformations of the proteins can exist, all subunits in one or the other conformation, whereas in the latter, the change takes place in a stepwise fashion (Fig. I.14).

I.2.3.1 Interfaces as novel functional sites

A majority of proteins occur as multimers. They may either contain multiple identical or nonidentical subunits. In some multimeric proteins, the functional region occurs at the subunit interface. In these proteins, both subunits may contribute to the function. Enzyme active sites at the subunit interfaces are abundant in nature. One example is 5-aminolevulinic acid synthase (ALAS). ALAS uses pyridoxal phosphate (PLP) as the cofactor. Like many PLP-containing enzymes, the PLP here is covalently linked to a lysine residue (K337 from subunit B). Other residues also help in positioning the PLP in the proper orientation. At least two residues in the active site come from subunit A (T367 and T366) (Fig. I.15). Thus the association of two subunits in the correct orientation is essential for the formation of the active site, and the catalytic activity may be called an emergent property of the dimeric structure.

I.2.3.2 Entropic advantages in multienzyme complexes catalyzing consecutive reactions

The association of enzymes catalyzing consecutive steps of a biochemical pathway is known. Previously, we have cited the example of the purinosome in which several enzymes of the purine biosynthetic pathway come together, forming a multienzyme complex. The purinosome appears to form when the flux through the pathway needs to

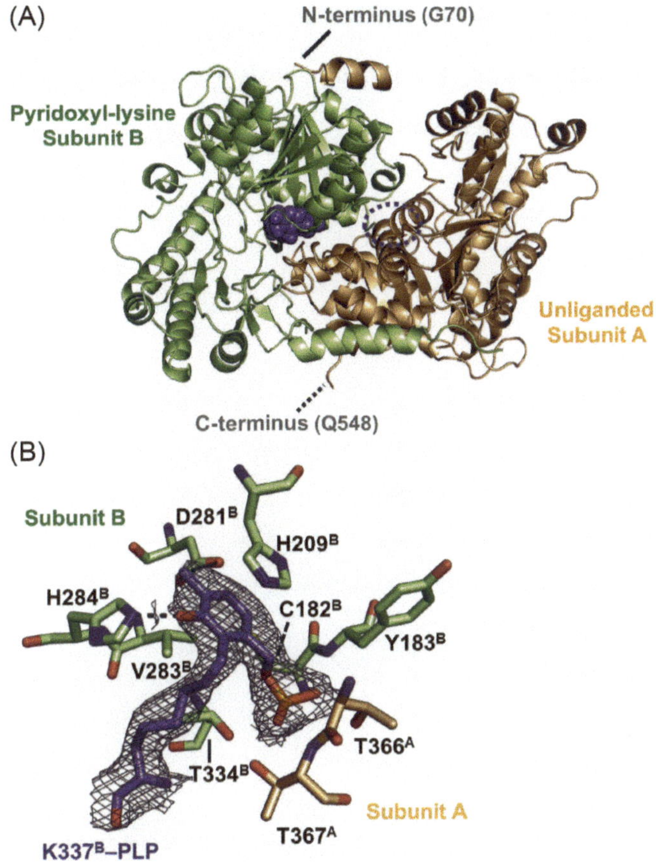

FIG. I.15 Structure of mitochondrial aminolevulinic acid synthase from *S. cerevisiae*. (A) Cartoon representation of the aminolevulinic acid synthase dimer with subunits A (unliganded) and B (containing covalently bound lysine-pyridoxal phosphate) colored tan and green, respectively. Lysine-pyridoxal phosphate is shown as blue spheres. The unoccupied cofactor-binding pocket is outlined with a blue dotted line. The structured N and C termini from subunit A are labeled with name and residue number. (B) Pyridoxal phosphate-binding pocket containing the covalently bound cofactor with 2mFo-Fc electron density for lysine-pyridoxal phosphate ligand is contoured at 1.0σ and shown as a gray mesh. Residues that directly interact with pyridoxal phosphate are shown as sticks and are labeled according to their subunit with a superscript notation. *Reproduced from Brown et al., 2018. Structure 26 (4), 580–589.e4. Elsevier.*

increase. By bringing the enzymes of a biosynthetic pathway in proximity, diffusion of intermediate metabolites may be prevented. This may lead to an increase in flux through the pathway. The purinosome may be a liquid-liquid phase-separated structure, rather than a defined multiprotein complex.

I.2.3.3 Ribosomes as an exemplar of protein-nucleic complexes with novel emergent properties

Ribosomes are arguably the best examples of many protein and RNA molecules coming together and forming a defined complex that performs the complex function of protein

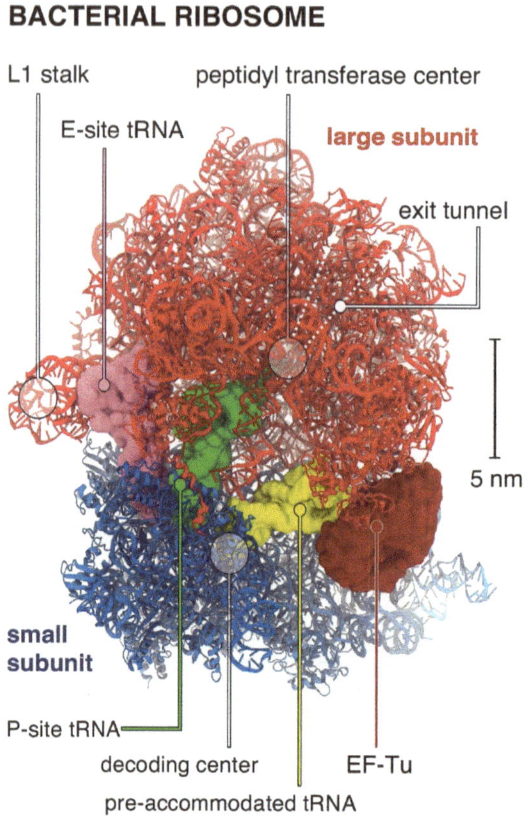

FIG. I.16 Structure of the bacterial ribosome in a complex with EF-Tu (PDB 5AFI). *Adapted from Bock et al., 2018. Curr. Opin. Struct. Biol. 49, 27–35. Elsevier.*

synthesis (Fig. I.16). Although the ribosomes are present in all known organisms, their compositions vary somewhat. The complexity of the ribosome can be gauged by the fact that it consists of at least 2 very large RNAs and more than 50 proteins. Although the large RNAs by themselves have some catalytic activity, they require the assembly of the full structure to carry out the sophisticated functions required for a faithful translation of the mRNA.

I.3 Genome: Complexity of organization

I.3.1 Genome organization in the three domains of life

Genomes—the focal point of this book—are sophisticated organized structures. We will dig deeper into the genome organization in later chapters but will provide a glimpse of its complexity here. The genomic DNA is compact in all organisms. Even simpler prokaryotic genomes are highly organized and compact. Different compaction mechanisms prevail in different domains of life, but DNA bending induced by DNA-binding proteins plays a pivotal role everywhere. These DNA-binding proteins generally lack the ability to recognize specific sequences. Multiple chromosomes in eukaryotes are spatially organized inside the nucleus in a nonrandom manner. This complex system plays a crucial role in regulating the expression of the information inherent in the nucleotide sequence. Understanding the emergent properties of the genome requires elucidation of its complex organization.

Three domains of life

Domains are the highest levels of taxonomic rank. All the extant and extinct living beings belong to the three domains of life. Devised by Carl Woese, the three domains are Archaea, Eubacteria, and Eukarya (Fig. I.17). The eukarya have the nucleus and other membrane-bound cellular compartments, whereas the other two are prokaryotes, lacking membrane-bound cellular compartments. All three domains of life have been thought to arise from an ancestral life form, the Last Universal Common Ancestor (LUCA). In many quarters, it is believed that eukaryotes evolved from archaea. A subgroup of archaea has remarkably similar genome organization as that of the eukaryotes, suggesting their ancestral relationship with the eukaryotes.

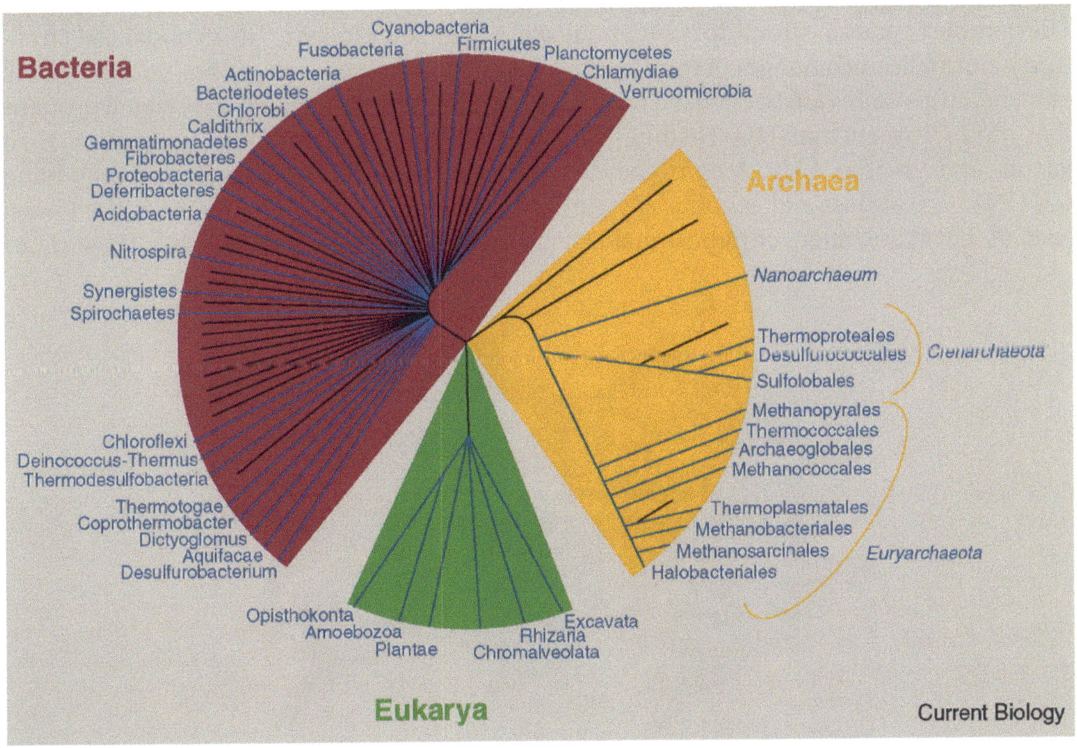

FIG. I.17 Diagrammatic representation of the organization of life into three domains based on the similarity of small subunit rRNA gene sequences. *Blue branches* represent those groups with cultured representatives; *black branches* represent groups only known from culture-independent environmental studies. *Reproduced from Walsh, Doolittle, 2005. Curr. Biol. 15 (7), PR237–R240, Elsevier.*

I.3.1.1 Some common features of genome organization

Reactions that take place on the genome, often localize precisely on a segment of DNA. An example is the assembly of protein complexes to initiate DNA-based transaction reactions. Apart from physical compaction, the complex organization of the genome help localize chemical reactions, thus, evolving new properties. We may cite the trans-cellular inheritance of specific chemical modifications as an example. In many ways, preservation of patterns of chemical modifications upon replication of the DNA is a unique phenomenon. Preservation of modification sites, after DNA replication and cell division, is the emergence of a new property. Neither the protein nor the DNA by itself is capable of transmitting positional information over cell generations. Many such examples are known in Biology, and more are being brought to light. Thus, we must look at the genome from an integrative viewpoint, starting from its organization first, to fully comprehend its function in all its subtlety and beauty.

I.3.1.2 Genome organization in Eubacteria

Eubacteria do not have a nucleus. Notwithstanding the lack of a cellular compartment in which to reside, genomes still occupy a fraction of the cell volume through compaction. The compact genome structure is hierarchically organized. However, this condensed structure is not static and changes in response to cellular state and environmental conditions. All eubacterial genomes are condensed through supercoiling and nucleoid-associated proteins that bind DNA, such as HU, Histone-like nucleoid-structuring protein, Integration Host Factor, and Fis (Fig. I.18). A common characteristic of these proteins is that they bend the DNA. The eubacterial genome is organized at different scales (Fig. I.19). At the lowest level of the organization, chromosomal DNA is organized into loops (Fig. I.20). These loops

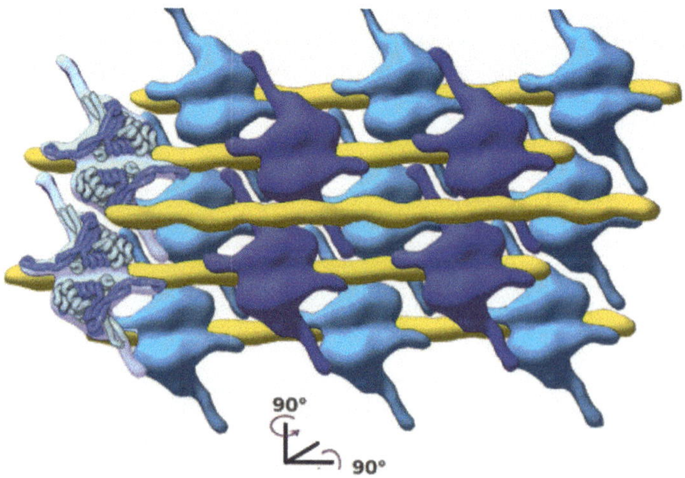

FIG. I.18 A view of HUαα DNA nucleoprotein network that induces genome compaction in eubacteria. *Reproduced from Hammel et al., 2016. Sci. Adv. 2, e1600650. a Creative Commons Attribution Non-Commercial License 4.0 (CC BY-NC). https://doi.org/10.1126/sciadv.1600650. Received permission from the corresponding author Michal Hammel by email.*

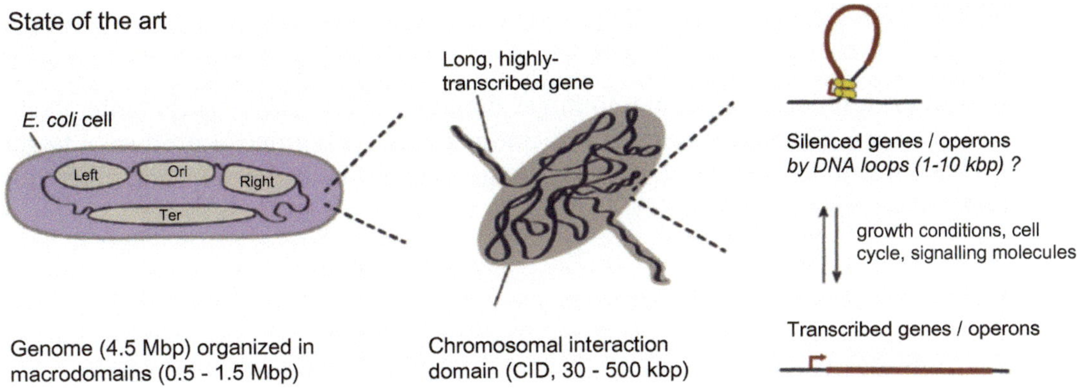

FIG. I.19 Organization of eubacterial genome at different scales. *Reproduced from Brocken et al., 2018. Curr. Opin. Sys. Biol. 8, 137–143. Elsevier.*

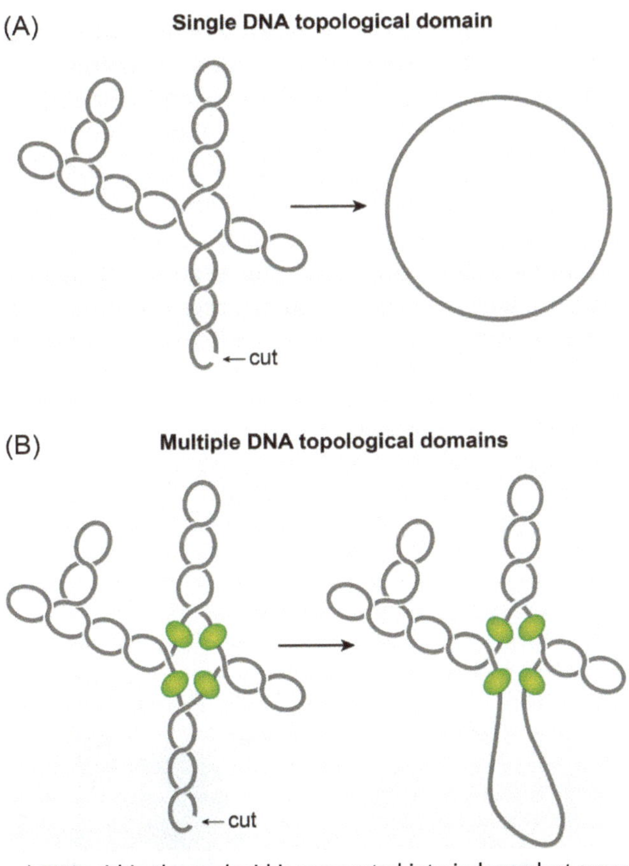

FIG. I.20 The chromosomal DNA within the nucleoid is segregated into independent supercoiled topological domains. (A) An illustration of a single topological domain of a supercoiled DNA. A single double-stranded cut anywhere would be sufficient to relax the supercoiling tension of the entire domain. (B) An illustration of multiple topological domains in a supercoiled DNA molecule. The presence of supercoiling-diffusion barriers segregates a supercoiled DNA molecule into multiple topological domains. Hypothetical supercoiling-diffusion barriers are represented as green spheres. As a result, a single double-stranded cut will only relax one topological domain and not the others. Plectonemic supercoils of DNA within the *E. coli* nucleoid are organized into several topological domains, but only four domains with a different number of supercoils are shown for simplicity. *Reproduced from Verma SC, Qian Z, Adhya SL (2019) Architecture of the* Escherichia coli *nucleoid. PLoS Genet 15(12): e1008456. doi:10.1371/journal.pgen.1008456. Open access, free to use CC license.*

interact with each other, coalescing into larger domains called chromosomal interaction domain (CID). At a higher level, some CIDs interact with each other to produce a larger domain called macrodomain (Fig. I.19). CIDs range from 20 Kbp to 400 Kbp, and macrodomains may be as large as 1.5 Mbp.

I.3.1.3 Genome organization in Eukaryotes

Eukaryotes possess much bigger genomes than prokaryotes and are encased by the nuclear membrane. Their genome consists of multiple linear chromosomes. Histone proteins are an integral feature of the eukaryotic chromosome. At the first level of compaction, the linear DNA is wrapped around a histone octamer core, forming the nucleosome (Fig. I.21). Nucleosomes are not static entities. They can be formed and removed with the help of a cellular machinery. Linker histones bind to nucleosomes and assemble them to higher-order structures (Fig. I.22). These higher-order structures may be liquid–liquid phase-separated condensates. They are also dynamic and respond to regulatory signals, giving access to the transcription machinery. Thus the whole chromosome acts like a highly interconnected system. An even higher-order structure exists in the form of packing of multiple chromosomes in the nucleus. Their distribution within the nucleus is nonrandom, and the position of a chromatin may change depending upon the cellular state. Because of this nonrandom distribution, certain regions of individual chromosomes are packed against their neighbors. These contact regions appear to play a crucial role in regulating gene expression.

I.3.1.4 Genome organization in Archaea

Archaeal genome is outwardly similar to the eubacterial genome in having a single copy of a circular chromosome. However, unlike eubacteria, most archaeal species contain

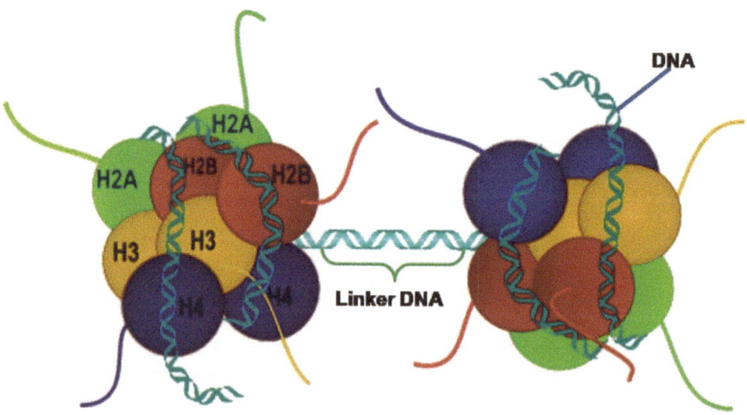

FIG. I.21 Cartoon diagram of nucleosomes.

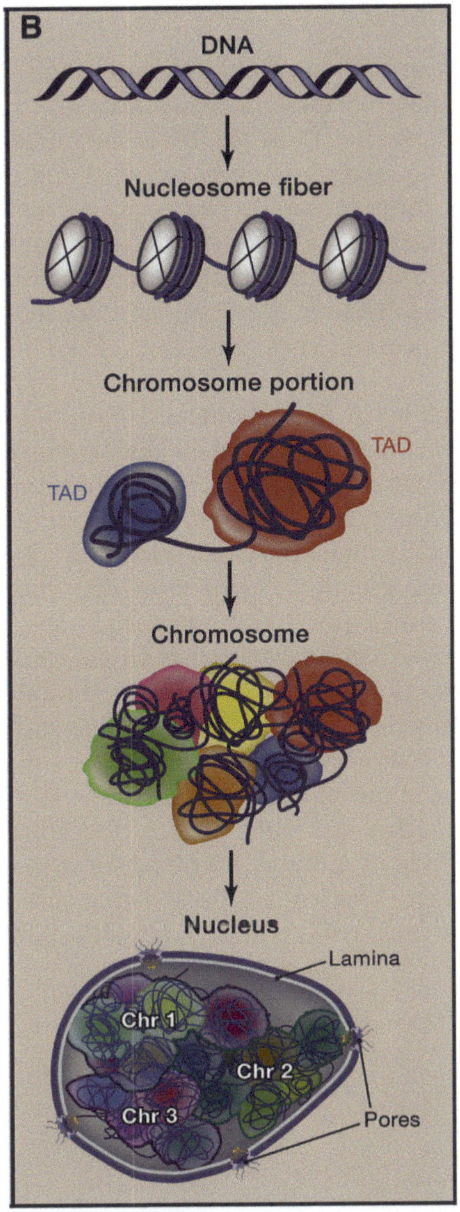

FIG. I.22 Hierarchical organization of genomes in higher organisms. *Reproduced from Sexton et al., 2015. Cell 160 (6), 1049–1059. Elsevier.*

histones that pack and compact the genome. Those that do not have histones, have similar DNA-binding proteins, for example, HU, as that of the eubacteria. Another DNA-binding protein that plays a role in organizing and condensing archaeal chromosomes is Alba (Acetylation lowers binding affinity). Alba is involved in the compaction and organization of some archaeal chromosome in an acetylation-dependent manner.

I.3.2 Genomes are dynamic

One of the key properties of biological ensembles is their dynamic nature. Proteins, nucleic acids, and other biomolecules that constitute the supramolecular complexes undergo coordinated motions that form the basis for functions of the ensembles. As described in the earlier sections, genomes are packaged and condensed with many proteins. In eukaryotes, they form multiple chromosomes that are condensed dynamic assemblies of DNA, protein, and to a lesser extent, other molecules such as RNA. They change their structural organization and chemical modifications in a very precise manner. Their base sequence may also change randomly as a consequence of DNA replication, repair, and transposition processes. These processes make the genome as a continually changing entity.

DNA itself is highly mobile. In the canonical B-form, all the bases are paired with complementary bases from the other strand in the Watson-Crick geometry. However, the base-paired state does not mean a static and invariant structure. The bases move, break interbase hydrogen bonds, and occasionally flip out of the helix. These transient base flips have been utilized by many enzymes that modify DNA (Fig. I.23). The motion and flipping of the bases are greatly influenced by distortion of the helix induced by protein binding, packaging, and general conformational states of the DNA in the chromosome. The proteins are also highly dynamic entities, undergoing constant thermal motions. An important class of motions is the correlated motions spanning many bonds and residues. It is generally believed that such long-range correlated motions give rise to allosteric effects that are important for the recognition of specific base sequences and other postbinding events. The protein motions are, in general, substantially affected by association with DNA or other proteins. Hence, the formation of supramolecular complexes leads to new kinds of motions that may form the basis for some emergent functions. The picture that emerges is that of a dynamic chromosome, and the dynamic

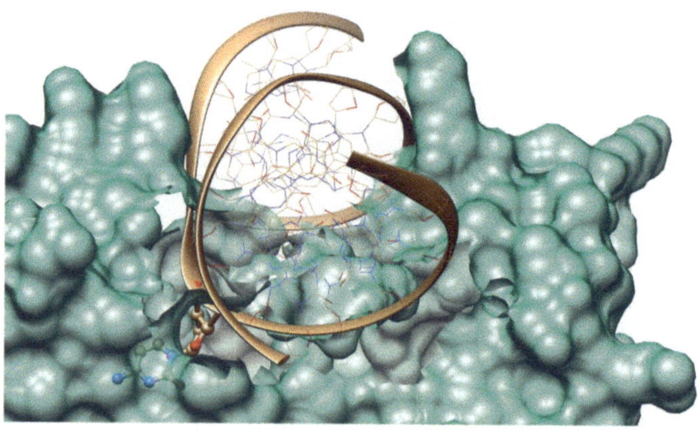

FIG. I.23 The structure of M.*Hha*I C5-Cytosine DNA methyltransferase with base-flipped unmodified DNA. The protein is shown in a light semitransparent surface. The DNA is shown in ribbon and wire representation, with the flipped cytosine inserted into the protein interior (ball and stick model). *This figure was generated from Pdb3MHT.*

character is vital for many of the emergent properties of the chromosome that do not occur in its components.

I.3.2.1 Chromatin organization and territories change with cellular states

In eukaryotic cells, in general, the genome consists of multiple chromosomes. In the interphase, the chromatins are located in definite spatial positions within the nucleus, known as chromatin territories. These positions may not be absolutely invariant. The chromatin folds into domains called topologically associated domains (TADs) and lamina-associated domains (LADs) (Fig. I.24). TADs promote more frequent intraDNA contact such as promoter-enhancer interactions, and LADs tend to promote gene repression. However, the boundaries of TADs and LADs vary and may change with cellular states, altering gene expression. In general, the highly transcribed genes tend to be positioned away from the nuclear membrane, more toward the center. TADs are dynamic entities in which promoters and enhancers can come nearby in space to promote interactions. Thus chromosomes are dynamic entities, and their dynamic character is crucial for the regulation of gene expression.

I.3.2.2 Nucleosomes can form and disassemble dynamically

In nucleosomes, the double-stranded DNA tightly wraps around a histone octamer core. The tightly wrapped DNA cannot be readily accessed by proteins that carry out DNA-based transactions. The histone-DNA interactions must be weakened for the proteins to access the wrapped DNA. Despite tight wrapping, nucleosomes are not static.

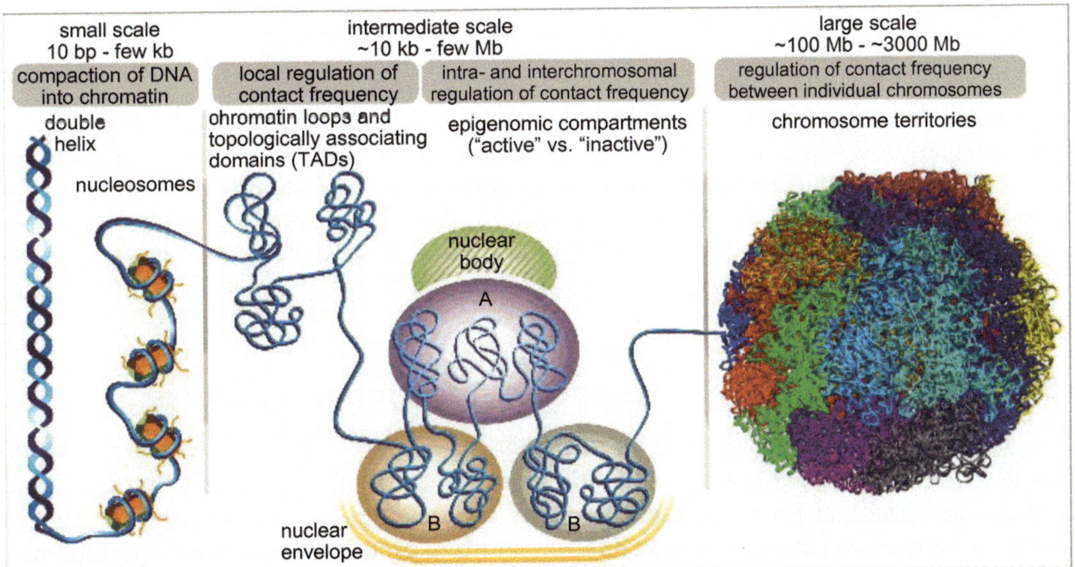

FIG. I.24 A cartoon diagram of topologically associated domains, lamina-associated domains, and chromatin territories. *Reproduced from Hansen et al., 2018. Nucleus 9 (1), 20–32, http://creativecommons.org/licenses/by/4.0/.*

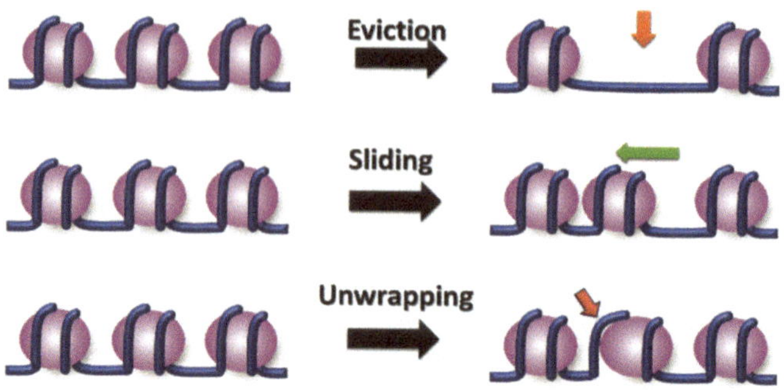

FIG. I.25 Nucleosome dynamics. Examples of eviction, sliding, and unwrapping are shown.

They can slide along the DNA and move from the original site to a different one (Fig. I.25). The nucleosomes can be ejected as well, making the wrapped DNA accessible to other proteins. These dynamic changes are crucial for the regulation of gene expression in eukaryotes.

I.3.2.3 Transposon insertions can move genes

Transposable elements are a source of change in the genome. Transposable elements are short DNA sequences that can jump from one position of the genome to another. These elements are present in most life forms, and in humans, these elements may consist of as much as 50% of the genome. By duplicating themselves and inserting at different positions of the genome, these elements can significantly alter the DNA sequence and may introduce functionally significant mutations. We will discuss transposons in another chapter.

I.3.2.4 Replication errors and damage can alter the DNA sequence

During cell division, the genomic DNA replicates with high fidelity. Thus the error rate in DNA replication is low, and the repair systems can repair many errors. Notwithstanding the high fidelity of the copying process and subsequent corrections, some errors are passed on to progeny cells. This gives rise to mutations that accumulate over many cell generations. Oxidative damage and spontaneous deamination of cytosine also introduce errors that accumulate over cell generations. Thus the genome changes as organisms grow and become aged.

I.3.3 Genomes are complex systems with emergent properties

The genome is constantly changing its composition, conformation, and spatial positions. It is also subject to change in the nucleotide sequence over cell generations. Some of these changes are essential for evolution, but some are deleterious. Many properties of the genome emerge from the assembly of different kinds of molecules that make up a genome. The emergence of these properties makes it essential that genomes are studied in a functional state, and reductive approaches will not be able to resolve some outstanding questions. It is also an interesting question, whether the genome is a self-assembled or a self-organized structure. However, given the precise packing order inside the nucleus, change of its composition, and constant energy input that goes in for it to be functional, a strong argument can be made that it belongs to the class of self-organized structure.

Chapter summary

In this chapter, we have seen that a cell is a highly organized system. This organized character begins at the bottom of the scale, where proteins, DNA, and other molecules associate to form supramolecular assemblies. At a higher level, the genome is a far more intricately organized system. However, notwithstanding its intricacies, the genome is an immaculately functioning system. Importantly, we have seen that such organizations give rise to emergent properties not found in their constituting parts. Thus the emphasis on studying biology using chemical tools and concepts must focus on the integrative character of the biological systems.

Further reading

Ackers, G.K., Holt, J.M., 2006. Asymmetric cooperativity in a symmetric tetramer: human hemoglobin. J. Biol. Chem. 281, 11441–11443.

Alberti, S., Dormann, D., 2019. Liquid–liquid phase separation in disease. Annu. Rev. Genet. 53, 171–194.

Bae, T., Tomasini, L., Mariani, J., Zhou, B., Roychowdhury, T., Franjic, D., Pletikos, M., Pattni, R., Chen, B.-J., Venturini, E., 2018. Different mutational rates and mechanisms in human cells at pregastrulation and neurogenesis. Science 359, 550–555.

Ben-Shem, A., Jenner, L., Yusupova, G., Yusupov, M., 2010. Crystal structure of the eukaryotic ribosome. Science 330, 1203–1209.

Blokzijl, F., De Ligt, J., Jager, M., Sasselli, V., Roerink, S., Sasaki, N., Huch, M., Boymans, S., Kuijk, E., Prins, P., 2016. Tissue-specific mutation accumulation in human adult stem cells during life. Nature 538, 260–264.

Bonev, B., Cavalli, G., 2016. Organization and function of the 3D genome. Nat. Rev. Genet. 17, 661.

Brocken, D.J., Tark-Dame, M., Dame, R.T., 2018. The organization of bacterial genomes: towards understanding the interplay between structure and function. Curr. Opin. Syst. Biol. 8, 137–143.

Do Young Hyeon, J.H.K., Ahn, T.J., Cho, Y., Hwang, D., Kim, S., 2019. Thematic series: tRNAs and aminoacyl-tRNA synthetases in human disease: evolution of the multi-tRNA synthetase complex and its role in cancer. J. Biol. Chem. 294, 5340.

Doolittle, W.F., 2020. Evolution: two domains of life or three? Curr. Biol. 30, R177–R179.

Epstein, I.R., Pojman, J.A., Steinbock, O., 2006. Introduction: self-organization in nonequilibrium chemical systems. Chaos. 16, 037101.

Falahati, H., Haji-Akbari, A., 2019. Thermodynamically driven assemblies and liquid–liquid phase separations in biology. Soft Matter 15, 1135–1154.

Feng, X., Huang, Q., She, Q., 2019. Crenarchaeal 3D genome: a prototypical chromosome architecture for eukaryotes. Cell 179, 56–58.

Grishin, N.V., Phillips, M.A., 1994. The subunit interfaces of oligomeric enzymes are conserved to a similar extent to the overall protein sequences. Protein Sci. 3, 2455–2458.

Koshland, D.E., Hamadani, K., 2002. Proteomics and models for enzyme cooperativity. J. Biol. Chem. 277, 46841–46844.

Lingwood, D., Simons, K., 2010. Lipid rafts as a membrane-organizing principle. Science 327, 46–50.

Liu, B., Eliason, W.K., Steitz, T.A., 2013. Structure of a helicase–helicase loader complex reveals insights into the mechanism of bacterial primosome assembly. Nat. Commun. 4, 1–8.

Misteli, T., 2001. The concept of self-organization in cellular architecture. J. Cell Biol. 155, 181–186.

Misteli, T., 2007. Beyond the sequence: cellular organization of genome function. Cell 128, 787–800.

Misteli, T., 2009. Self-organization in the genome. Proc. Natl. Acad. Sci. 106, 6885–6886.

O'Connor, T., 1994. Emergent properties. Am. Philos. Q. 31, 91–104.

Pedley, A.M., Benkovic, S.J., 2017. A new view into the regulation of purine metabolism: the purinosome. Trends Biochem. Sci. 42, 141–154.

Peeters, E., Driessen, R.P., Werner, F., Dame, R.T., 2015. The interplay between nucleoid organization and transcription in archaeal genomes. Nat. Rev. Microbiol. 13, 333–341.

Peters, J.-M., 2006. The anaphase promoting complex/cyclosome: a machine designed to destroy. Nat. Rev. Mol. Cell Biol. 7, 644–656.

Schmitt, D.L., An, S., 2017. Spatial organization of metabolic enzyme complexes in cells. Biochemistry 56, 3184–3196.

Schuwirth, B.S., Borovinskaya, M.A., Hau, C.W., Zhang, W., Vila-Sanjurjo, A., Holton, J.M., Cate, J.H.D., 2005. Structures of the bacterial ribosome at 3.5Å resolution. Science 310, 827–834.

Shin, Y., Brangwynne, C.P., 2017. Liquid phase condensation in cell physiology and disease. Science. 357, eaaf4382.

Verma, S.C., Qian, Z., Adhya, S.L., 2019. Architecture of the *Escherichia coli* nucleoid. PLoS Genet. 15.

Werner, B., Sottoriva, A., 2018. Variation of mutational burden in healthy human tissues suggests non-random strand segregation and allows measuring somatic mutation rates. PLoS Comput. Biol.. 14, e1006233.

Will, C.L., Lührmann, R., 2011. Spliceosome structure and function. Cold Spring Harb. Perspect. Biol. 3, a003707.

II

Methods: Tools for studying genomes and gene functions in intact cells

II.1 Tools for studying genomes in intact cells

An integrative approach requires tools capable of revealing properties of genomes and their associated molecular ensembles within the natural milieu and in the functional state. A cell is the lowest level of an autonomously functioning live entity. Cells contain enormous numbers and types of molecules. Due to this diversity, tools designed for studying purified molecules are often not adequate for investigating whole cells. Several techniques have been developed for simultaneously analyzing cellular ensemble of molecules at an instant (Fig. II.1). These techniques generally sample the state of the ensemble by quenching the content of the cell at a given time point and have a suffix of "omics." For example, when we measure the expression of the whole repertoire of genes present in an organism into respective transcripts, we call it transcriptomics. We will briefly describe some "omics" techniques that are capable of providing snapshots of genome modifications and functions using high-throughput methodologies.

Another class of important techniques that can elucidate processes and mechanisms operating in a cell involves inhibiting a particular target and observing changes in molecular compositions, often using high-throughput techniques. Small molecules are generally used to inhibit specific target proteins. We call this approach, chemical genetics. Another method of inhibition is to knock down the expression of a gene or introduce specific changes in the genomic DNA (genome editing) allowing selective modification of the function of a single gene (called the reverse genetics approach). We will describe some representative techniques belonging to these classes as well.

Physiologists and microbiologists have used microscopy for many centuries. Different forms of microscopy have emerged as an essential component of the tool kit of chemical biologists. The resolution of standard optical microscopy has a lower limit, the so-called Abbe limit, that depends upon the observation wavelength. For visible wavelengths, this limit is in the range of hundreds of nanometers. It is now possible to bypass this limit and study live cells at resolutions down to low nanometer levels. These techniques are called super-resolution optical microscopy. We will also describe optical microscopic techniques, including a representative resolution-enhancement technique, widely used for studying both fixed and live cells. Our focus in this chapter will be illustrative and not

Chemical Biology of the Genome. https://doi.org/10.1016/B978-0-12-817644-3.00002-7

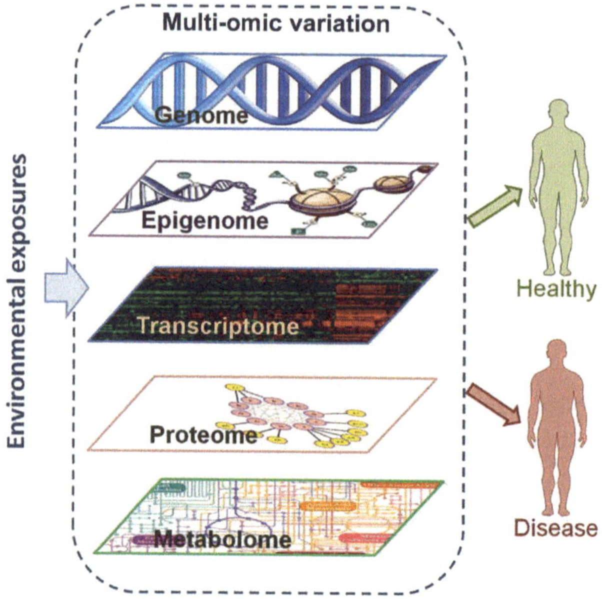

FIG. II.1 Integration of various "omics" approaches. *Reproduced from Sun, Hu, 2016. Adv. Genet. 93, 147–190. Elsevier.*

exhaustive. Thus only some techniques useful for studying chromatin structure and function will be discussed to give the readers a flavor of the kind of techniques used to study molecular ensembles from an integrative viewpoint.

II.1.1 Fluorescence: A powerful approach to study genome organization and function

We use different methodologies and approaches to study the genome. A reductive approach of purifying genome-associated proteins and studying them in vitro has its limitations. Investigating the genome in its functional state is essential for fully comprehending its function. Fluorescence-based methods have emerged as techniques of choice for studying isolated biomolecules and whole cells. Its versatility, ease of use, rapid response, and environment sensitivity makes fluorescence a method of choice for many genome-focused investigations. We will discuss the basic concepts of fluorescence-based methodologies while focusing on techniques important for investigating genomes.

II.1.1.1 Basic principles of fluorescence

The phenomenon of fluorescence is emission from an excited state produced by the absorption of electromagnetic radiation. When a molecule absorbs a photon of appropriate wavelength, an electron moves from the ground state to a higher energy state. Each electronic state is associated with multiple vibrational states of modestly different energies (Fig. II.2). From the excited state, the electron decays rapidly to the lowest vibrational

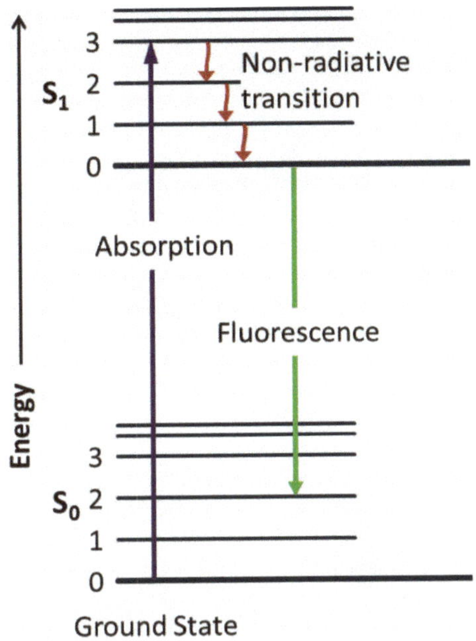

FIG. II.2 Molecular energy levels and fluorescence emission occur (Jablonski's diagram). Numbers 0, 1, 2, etc. signify different vibrational states associated with an electronic state. *Reproduced from Wikimedia Commons. Acknowledgment: By Jacobkhed—Own work, CC0, https://commons.wikimedia.org/w/index.php?curid=19180813.*

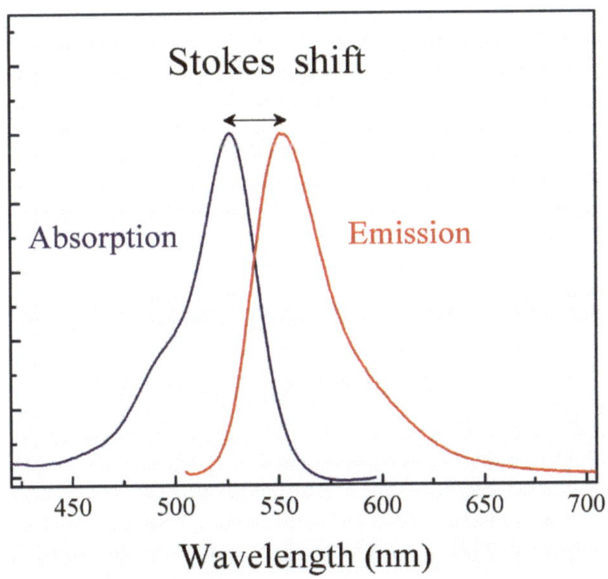

FIG. II.3 Normalized absorption and emission spectra of Rh6G. *Reproduced from Wikimedia. The original contribution by Sobarwiki—Own work, Public Domain, https://commons.wikimedia.org/w/index.php?curid=29474504.*

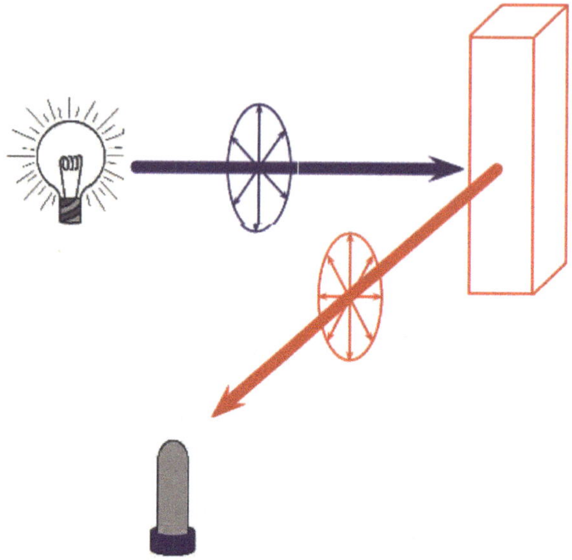

FIG. II.4 A schematic drawing of measurement of fluorescence. *The original file from Wikimedia was modified. This file is made available under the Creative Commons CC0 1.0 Universal Public Domain Dedication.*

state associated with the excited electronic state without emitting a photon. The electron can then return to the ground state with the emission of a photon. Because the emission takes place from a state of lower vibrational energy, the energy of the emitted photon is less than that of the absorbed one. Thus the emission wavelength is longer than the absorption wavelength. The difference between wavelength maxima of the absorption and emission is called the Stokes shift (Fig. II.3). Commonly, fluorescence is measured by sending a beam of light of a particular wavelength through a dilute solution of the fluorophore and detecting the emitted light at right angles to the incident beam (Fig. II.4). Measurement at a right angle causes minimum interference from the incident light.

Quantum yield

As shown in Fig. II.2, upon absorption of a photon, an electron is promoted to a higher energy state as a singlet. It quickly returns to the lowest excited electronic state and its associated lowest vibrational state. From this state, return to the ground state can take different paths. It can emit light as fluorescence (as shown). Like a chemical reaction, we can consider this process to be characterized by a rate constant (k_f). However, this is not the only process by which the excited state can be depleted. The excited state can decay without emitting light (internal conversion; rate constant k_{ic}), quenched due to various processes (rate constant k_q), and can get converted to a triplet state (intersystem crossing; rate constant k_{is}). A fraction of an excited singlet that gets deexcited through the fluorescence emission pathway is the quantum yield (ϕ).

$$\phi = k_f / (k_f + k_{ic} + k_{is} + k_q)$$

Fluorescence quenching or enhancement (dequenching) is essentially an alteration of the quantum yield of a fluorophore, which may occur upon processes such as macromolecular interactions, energy transfer, collisional deactivation of the excited state, or other processes.

The average time an electron spends in the excited state before decaying is an important parameter for many experiments. The fluorescence lifetime is a measure of the average time the molecule spends in the excited state. A common method of measuring fluorescence lifetime is to excite the fluorophore by a light pulse of very short duration at wavelengths, where the fluorophore absorbs light and then to count the emitted photons at different time windows. The measured decay is then fitted to one or more exponential functions, and the fluorescence lifetime is extracted (Fig. II.5).

Fluorescence resonance energy transfer (FRET) is a very useful technique for the measurement of molecular proximity. FRET may occur between two fluorophores that are close in space (typically $<10\,\text{nm}$). One of the fluorophores acts as the donor and the other acts as

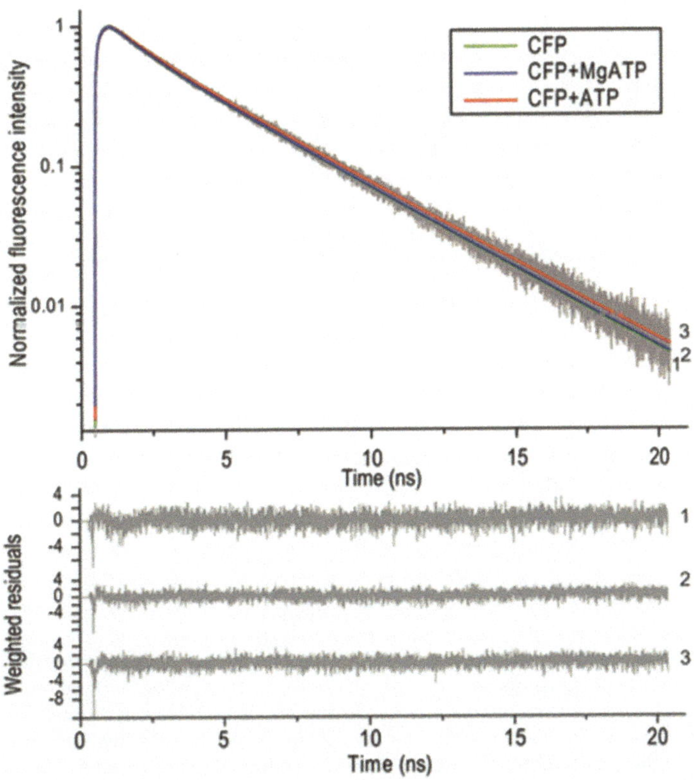

FIG. II.5 Example of a time-resolved fluorescence decay of cyan fluorescent protein *(top panel)*. The *lower panel* is the residuals of the fit. *Borst et al., PLoS ONE 5(11): e13862. https://journals.plos.org/plosone/article?id=10.1371/journal.pone.0013862 (Reproduced under CC-BY license).*

the acceptor. A necessary condition for FRET is that the donor emission spectrum must significantly overlap with the excitation spectrum of the acceptor (as shown in Fig. II.6). When the donor is selectively excited, a radiation-less energy transfer may occur (not a photon emission from the donor and absorption by the acceptor) to the acceptor. As a consequence, the acceptor goes to the excited state (for an energy level explanation, see Fig. II.7). The acceptor then emits the energy at its fluorescence emission wavelength. The amount of energy transferred (FRET efficiency is a measure of this) is strongly dependent on the distance between the two fluorophores and, hence, can be used as a molecular ruler. The relationship between the FRET efficiency (E) and the interfluorophore distance (R) can be written as follows:

$$E = \frac{1}{\left\{1 + \left(\frac{R}{R_0}\right)^6\right\}},$$

where R_0 is the interfluorophore distance at which FRET efficiency is 50%. R_0 is dependent on the spectral characteristics of the donor-acceptor pair and orientation between the two fluorophores.

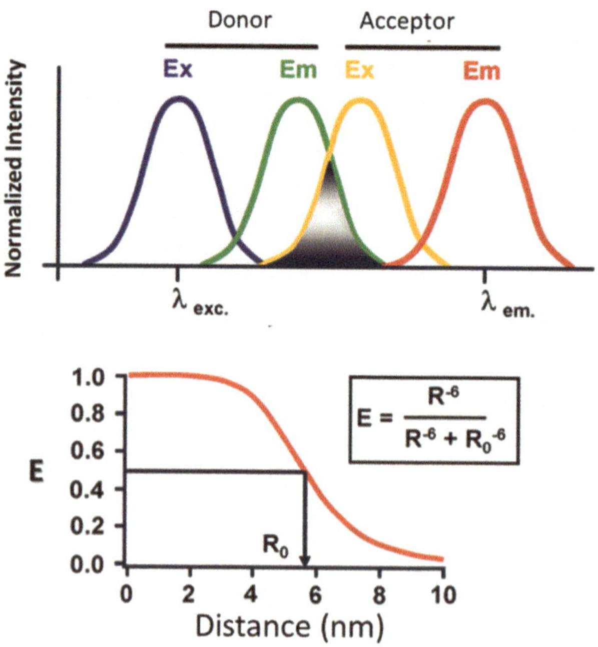

FIG. II.6 A typical excitation and emission spectra of a fluorescence resonance energy transfer donor-acceptor pair *(top panel).* The blue and the green colors are those of the donor, and the *yellow* and the *red* colors are those of the acceptor. *Bottom panel,* fluorescence resonance energy transfer efficiency (E) versus distance between the donor and the acceptor. *From Wikimedia commons modified by the author. https://commons.wikimedia.org/w/index.php?search=FRET+overlap&title=Special%3ASearch&go=Go&ns0=1&ns6=1&ns12=1&ns14=1&ns100=1&ns106=1#/media/File:FRET_cz_zavislost_ucinnosti_na_ruznych_faktorech.tif. Reproduced under GNU Free Documentation License, Version 1.2.*

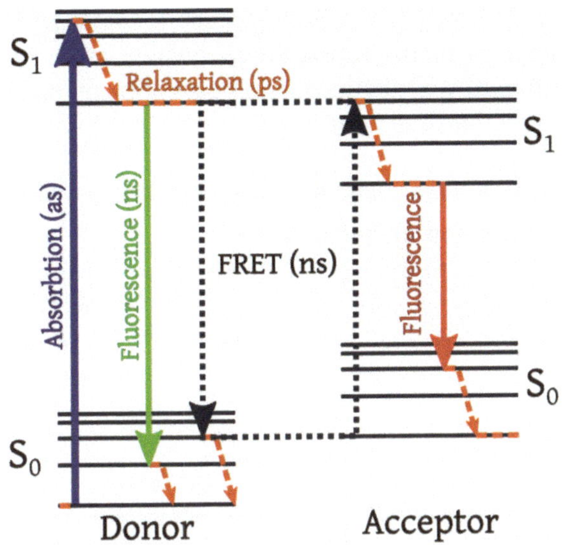

FIG. II.7 Energy level diagram of fluorescence resonance energy transfer. *By Alex M Mooney—Own work, CC BY-SA 3.0. https://commons.wikimedia.org/w/index.php?curid=23197114.*

Anisotropic fluorescence emission is unequal fluorescence emission intensities in different directions. Anisotropic emission may occur when excited by plane-polarized light. Light is a form of electromagnetic radiation with electric and magnetic components (vector quantities) vibrating in perpendicular planes and propagating in the third direction, perpendicular to the other two components. We will define the plane of polarization as the plane that contains the electric vector and the direction of propagation. The light coming out of a common source is unpolarized, that is, all the planes of polarization are equally probable. Thus the intensity of light is the same in all the possible planes of polarization. The unpolarized light, when passed through a polarizer, becomes plane polarized. The plane of polarization depends on which way the polarizer is oriented. Plane-polarized light (or linearly polarized light), when passed through a solution of a fluorophore, excites only those fluorophores whose transition moments are parallel to the electric vector of the light. This orientation-selective excitation results in a population of excited molecules whose transition moments are oriented parallel to the direction of the original plane of polarization. If the molecule has a very short excited-state lifetime, the fluorescence emission will primarily take place from the oriented molecules in that particular direction. The emission will be polarized in the same direction of the exciting light. However, if the excited state lifetime is comparable to the timescale of molecular reorientation, the orientations of the excited molecules will gradually randomize due to molecular motion, resulting in the loss of emission of polarization.

A fluorescence probe attached to a macromolecule reorients as the macromolecule tumbles. Fluorescence anisotropy is a measure of this reorientation, hence the tumbling of the molecule. Because the tumbling rate is dependent on the molecular size and shape, so is the fluorescence anisotropy. Thus fluorescence anisotropy is useful for estimating molecular size and shape. In many interactions between molecules, the molecular size and shape change significantly. Thus fluorescence anisotropy has emerged as an important technique for deriving binding isotherms.

Fluorescence polarization and anisotropy

Fluorescence polarization and anisotropy are different ways of measuring polarized emission. Polarizers are often used to measure fluorescence polarization, as shown in Fig. II.8A and B. Unpolarized light is first passed through a polarizer to polarize the emerging light in the vertical

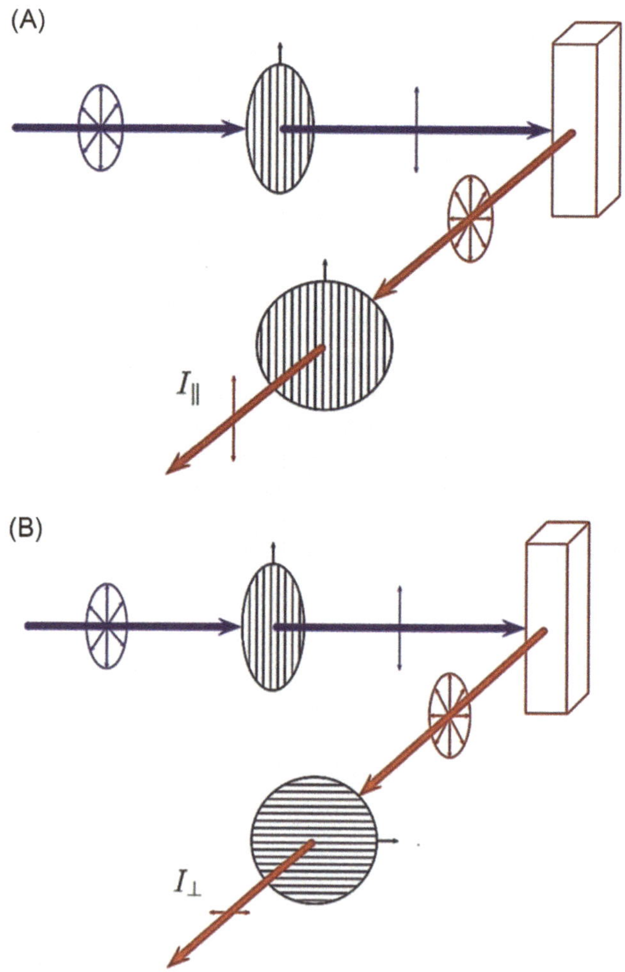

FIG. II.8 Measurement of fluorescence anisotropy. (A) Measurement of $I_{\parallel}$ component of fluorescence anisotropy. (B) Measurement of $I_{\perp}$ component of fluorescence anisotropy. *From Wikimedia https://commons.m.wikimedia.org/wiki/File:FP-Abb.png. This file is made available under the Creative Commons CC0 1.0 Universal Public Domain Dedication.*

plane. Upon excitation of the fluorophore, the emitted light is passed again through a polarizer whose orientation is either in parallel orientation to the excitation polarizer (Fig. II.8A) or perpendicular (Fig. II.8B). In the former case, the parallel component of the emitted light ($I_{\parallel}$) reaches the detector, and in the latter case, the perpendicular component of the emitted light ($I_{\perp}$) reaches the detector.
Fluorescence polarization, *P*, is defined as:

$$P = \frac{I_{\parallel} - I_{\perp}}{I_{\parallel} + I_{\perp}}$$

Fluorescence anisotropy, *A*, is defined as:

$$A = \frac{I_{\parallel} - I_{\perp}}{I_{\parallel} + 2I_{\perp}}$$

II.1.1.2 Fluorescence microscopy including enhanced-resolution techniques

Fluorescence-based techniques have proven their utility for investigating macromolecules. Whole cells, fixed or live, have proven to be particularly amenable to fluorescence microscopy-based techniques. Specific labeling techniques, which stain a particular molecule or molecules with a fluorescent dye, are in widespread use. Some dyes bind to specific cellular substructures, such as mitochondria, and some, such as fluorescent proteins, can be expressed as a fusion to make individual proteins fluorescent. Real-time tracking of cellular processes in living cells or organisms by fluorescence-based techniques is widely performed. Because of continuous methodological improvements in fluorescence microscopy over the last several decades, such as confocal microscopy and super-resolution microscopy, fluorescence microscopy-based techniques are gradually emerging as essential techniques for studying the genome.

Using super-resolution techniques, it is now possible to resolve the fluorescence of molecules present in the cell at a lateral (X and Y direction) resolution close to 10 nm. Coupled with optical sectioning in the *Z*-direction, a detailed three-dimensional view of chromatins may be obtained, opening up new avenues for studying the chromatin structure in the nucleus.

In fluorescence microscopy, the sample contains one or more fluorophores that emit light at longer wavelengths than the excitation wavelengths. The most common setup of the fluorescence microscope is that the incoming light beam and the emitted fluorescent light are both transmitted and focused through the same objective lens. In this setup, a mirror, called the dichroic mirror, that has wavelength-dependent transmission or reflection property plays a crucial role in selecting the excitation and emission light paths. The dichroic mirrors, in general, reflect specific wavelengths but transmit others. Dichroic mirrors used in fluorescence microscopy (Fig. II.9) reflect shorter wavelengths and transmit longer wavelengths. The mirror reflects the excitation beam toward the objective or

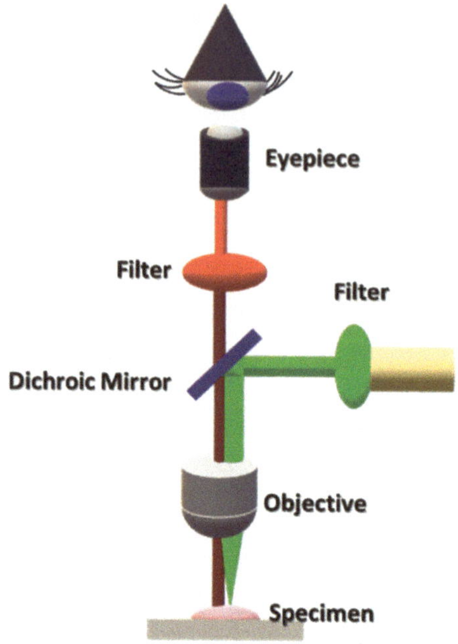

FIG. II.9 Schematic of a fluorescence microscope.

sample. The returning emitted light of longer wavelength from the sample is transmitted to the camera or the eyepiece. In this setup, the image of the uniformly illuminated sample forms as a whole. In samples of finite thickness, the light from all vertical sections goes to the eyepiece to form an image. Images from above and below the focal plane will be out of focus. This phenomenon would cause overlap of images from different sections; thus, the sharper image from the focal plane will become fuzzy.

The purpose of confocal microscopy is to preferentially select the image from the in-focus plane. In confocal microscopy, the incoming excitation light beam is focused on the sample using a pinhole (blue rays in Fig. II.10). The beam also excites adjoining areas and planes. The dichroic mirror transmits the fluorescence emission from the sample while cutting off the reflected and scattered light. Another pinhole in the conjugate plane preferentially selects the image coming from the focal point of illumination (green rays). The fluorescence emanating from out of focus planes is largely cut off by the pinhole in the detector side producing a sharper image of the in-focus plane.

Another significant variation of the confocal microscope is to scan the sample illumination point throughout the sample and collect the images of objects lying in that plane. Images from different planes can also be collected, and a three-dimensional image of the sample is reconstructed by image reconstruction techniques.

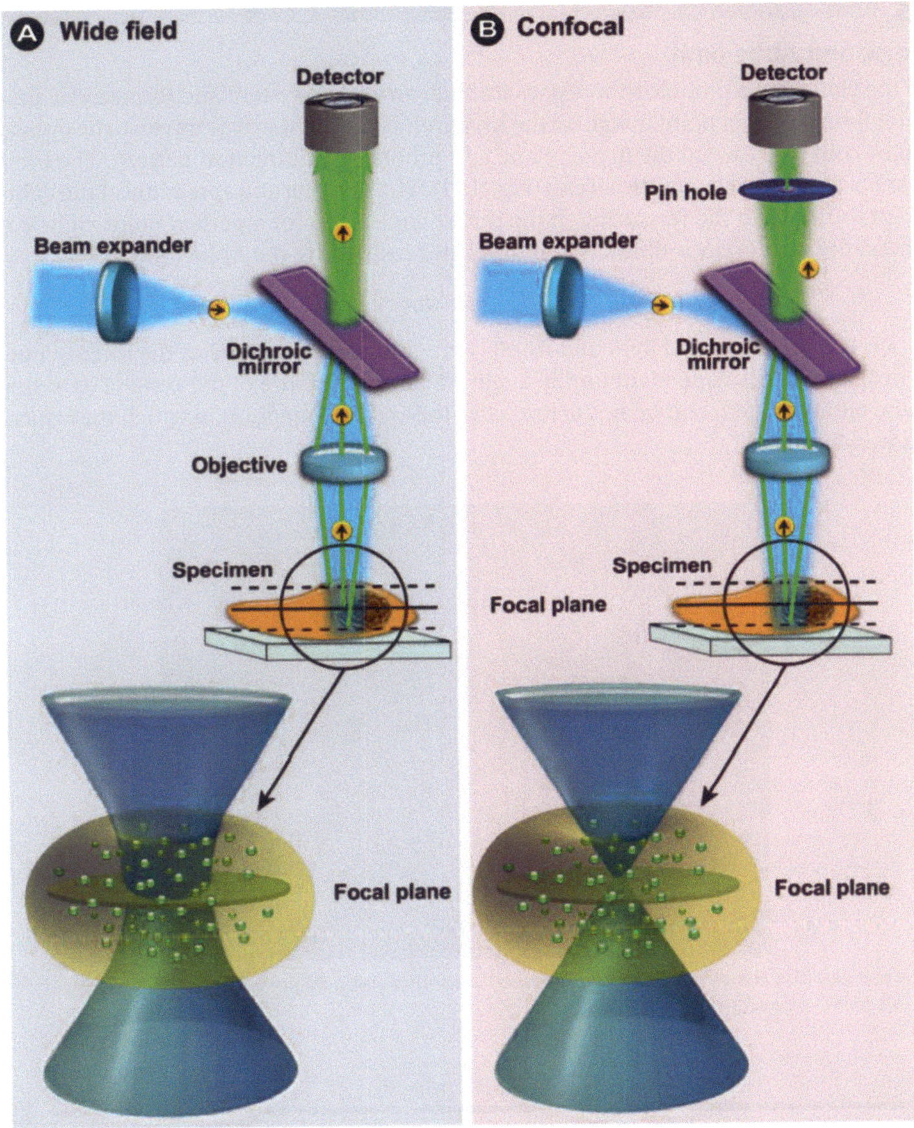

FIG. II.10 Comparison of principles of a confocal microscope with a wide-field fluorescence microscope. In confocal microscopy (B), the concept of illuminating the sample with excitation light (e.g., *blue light*) and the sample emitting light with a longer wavelength (e.g., *green light*) is identical to the general principles of fluorescence in wide-field microscopy (A). The differences in wide field microscopy are (i) the excitation laser light is scanned over the sample, and the emitted light originates from this area; (ii) on the detection beam path, a pinhole aperture in front of the detector prevents light emitted from above or below the focal plane *(dotted lines)* from reaching the detector; and (iii) because only the light from the focal plane *(solid line)* reaches the detector, an optical section is generated. *Reproduced from Ishikawa-Ankerhold et al., 2012. Molecules 17 (4), 4047–4132.*

Airy disk and Abbe limit

When the light from a pinhole source goes through an optical system and forms a real image, it does not form a perfect point image on the image plane. As light waves traverse the optical system, they converge toward the image plane and produces a diffraction pattern. The image on the plane is a series of concentric circles (Fig. II.11). The blurring and spreading of the light from the point source limit the resolution of the microscope. Even for a perfect point emitter in the specimen, the resolution is diffraction limited and defined by the Abbe limit:

$$d = \lambda / 2nsin\theta,$$

where λ is the wavelength of light and $nsin\theta$ is the numerical aperture of the microscope. The numerical aperture is dependent on the angle θ that is a measure of the cone of light that can enter the optical system and on n, the refractive index of the medium in which the optical system is working.

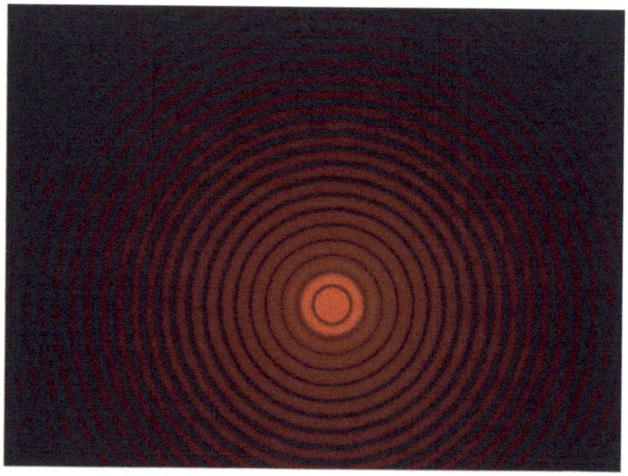

FIG. II.11 A real airy disk created by passing a red laser beam through a 90-μm pinhole aperture. *Reproduced from https://en.wikipedia.org/wiki/Airy_disk. CC0 license.*

The nucleus of a mammalian cell typically has a diameter of about 5000 nm. Chromatin territories span approximately 1000 nm. Substructures of folded chromatins are even smaller. In confocal microscopy, we can realistically expect to reach a resolution of approximately 500 nm in the lateral dimension. Thus resolution of finer structural details of the nucleus and chromatins is impossible with simple confocal microscopes. Abbe's limit can be bypassed through methodologies developed in recent times, and this has proved to be extremely useful in probing finer features of chromatin structure and function. Many methods have been developed to bypass the Abbe limit. We will describe the stimulated emission depletion (STED) method as an illustrative example.

Stimulated and spontaneous emission

In optical spectroscopy, electronic transitions occur when a photon interacts with an atom or a molecule. A necessary condition is that the energy difference between the ground and the excited state (ΔE) must match the energy of the photon ($h\nu$).

$$\Delta E = h\nu,$$

where h is the Planck's constant, and ν is the frequency of the incoming light. It is essential to know that in the presence of a photon of such an energy, transitions from ground-to-excited and excited-to-ground states are equally probable. When the atoms are primarily in the ground state, the net effect will be excitation as more atoms are in the ground state. However, if the atoms are primarily in the excited state, the net effect will be deexcitation. The latter is called stimulated emission (Fig. II.12). The excited state can also return to the ground state spontaneously, the probability of which is dependent on the magnitude of ΔE. This deexcitation process is called spontaneous emission.

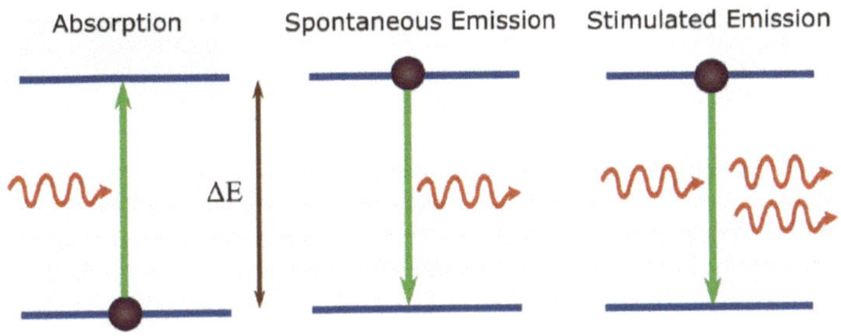

FIG. II.12 Schematic diagram of the processes of spontaneous and stimulated emission.

In the confocal fluorescence microscopy, the excitation beam is focused on a relatively narrow spot. The focused excitation beam creates a circular area of excitation on the sample, where the fluorophores are excited to the higher energy levels (Fig. II.13, left panel). The dimension of the excitation spot is diffraction limited, setting a lower limit to the final observed resolution. STED microscopy reduces the dimension of this spot by causing stimulated emission of fluorophores present in the outer annulus of the original spot by a second pulse of annular shape (Fig. II.13, upper-middle left panel). This leads to a much smaller remaining excitation area (left panel, rightmost spot) that is scanned to yield an image of higher resolution (lower right panel). Resolution improvement is visible when compared with an image obtained by confocal microscope.

II.1.1.3 Fluorescence probes for study of the genome

Visible light causes minimal damage to cells and organisms and is preferred for probing living systems. Thus fluorescent probes used in biological experiments typically have molecular energy levels whose energy gap falls into the visible wavelengths. Such energy gaps are usually found in aromatic molecules or molecules having extensive conjugated

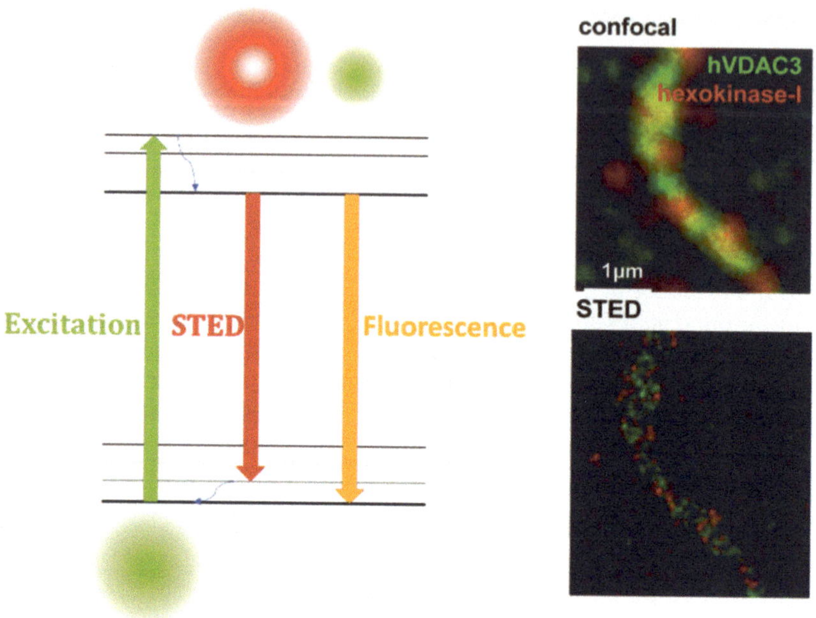

FIG. II.13 Stimulated emission depletion nanoscopy. *Left panel*: *(green)*. An annular-shaped laser beam of longer wavelength *(red)* spatially confines the central signal, thereby reducing the effective PSF (point spread function) size below the diffraction limit. *Right panel*: dual-colour stimulated emission depletion resolved the separation of the indicated mitochondrial proteins in cultured cells. *Adapted from Kohl et al., 2013. J. Mol. Cell. Cardiol. 58, 13–21 (Elsevier).*

double bonds. The brightness of a probe is related to how efficiently it can absorb light, reflected in its molar extinction coefficient, and what fraction it can reemit as fluorescence, reflected in its quantum yield. Fluorescent probes have become essential tools for sensing intracellular environments and imaging. Despite much work done in the past two or three decades, development and introduction of newer probes are continuously taking place to probe live cells and organisms.

We will divide the fluorescent probes into two major classes based on their molecular nature: (i) small molecules and (ii) proteins. Each class has found its unique areas of application. Small molecule probes are the oldest class of probes. They and their designed derivatives have been widely used, for example, in molecular beacons for sensitive nucleic acid detection. Small molecule probes are generally small aromatic molecules that can either bind to macromolecules, ions, or react with reactive species, small or macromolecules, to produce highly fluorescent molecules.

Small molecule DNA-binding probes: The best-known DNA-binding fluorescence probe is ethidium bromide (Fig. II.14). Ethidium bromide intercalates into double-stranded DNA and becomes intensely fluorescent, making it useful for DNA staining. However, lack of absorption bands in visible wavelengths prevents its use in cellular experiments. Propidium iodide, a derivative of ethidium bromide, has excitation in visible

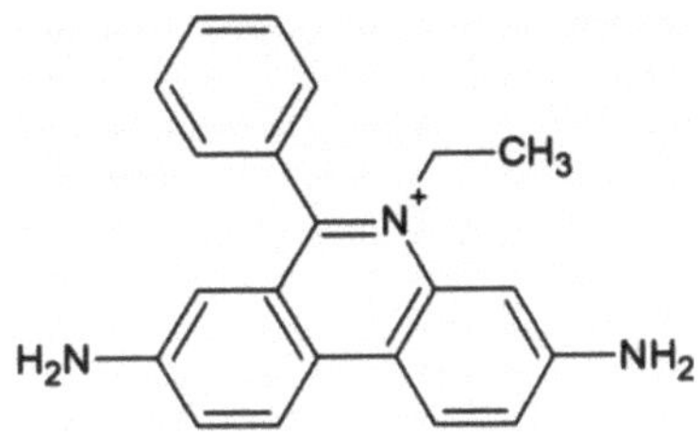

FIG. II.14 Chemical structure of ethidium bromide.

wavelengths. This makes propidium iodide suitable for use in flow cytometry for detecting cells with a compromised plasma membrane (dead or apoptotic cells).

There are a few critical points to consider regarding the use of DNA-binding fluorescence probes. Probe emission in longer wavelengths is generally preferred due to less autofluorescence from cellular materials. Plasma membrane permeability is another vital consideration. The other important issue is the ability to discriminate between DNA and RNA. The two probes referred above are cell impermeable and bind to RNA. 4′,6-diamidino-2-phenylindole (DAPI) (Fig. II.15) is a minor groove-binding dye widely used for nuclear staining. It emits blue fluorescence upon binding to DNA, and its plasma membrane permeability is a bit higher than the other two dyes mentioned above. However, its excitation at blue wavelengths is not convenient, and it is generally not used in live-cell staining. Hoechst stains are similar to DAPI but offer better membrane permeability making it more convenient to use for live cell staining. Recently some new dyes have come into use in live-cell nuclear staining. We will give examples of two such dyes: Draq5 and SiR-Hoechst.

Draq5 is an anthraquinone-based DNA-specific dye with far-red excitation (~650 nm), emission (~680 nm), and excellent cell permeability (Fig. II.16). Its mode of action is through intercalation into double-stranded DNA, and its binding appears to be selective for DNA. Its particular advantage is that its excitation and emission spectra are far removed toward the red when compared with other fluorophores typically used as cellular probes. So, it can be used simultaneously with fluorescein, green fluorescent protein (GFP), and other widely used dyes. It has found many applications in flow cytometry, live-cell imaging, and cell-based assays. It has some significant disadvantages in terms of affecting cellular processes, thus producing some cellular toxicity (Fig. II.17).

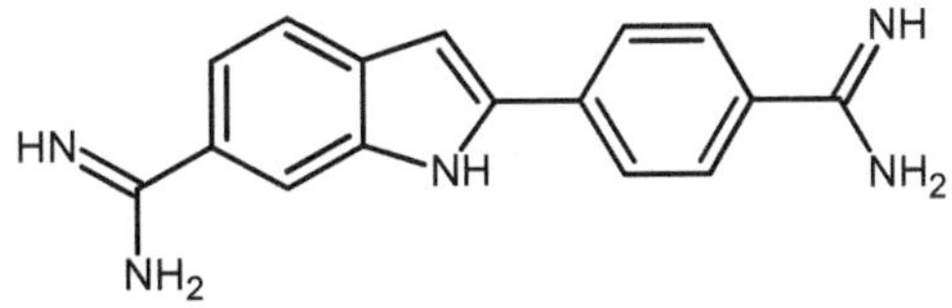

FIG. II.15 Chemical structure of 4′,6-diamidino-2-phenylindole.

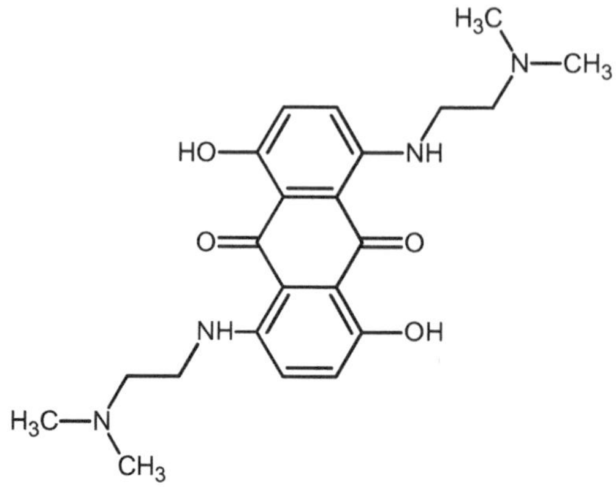

FIG. II.16 Chemical structure of Draq5.

Another dye with high cell permeability, DNA selectivity, and far-red excitation is SiR-Hoechst. It is structurally related to the Hoechst group of dyes, which has significant applications in microscopy. Hoechst dyes, however, are fluorescent in green regions of the electromagnetic spectrum. SiR-Hoechst is a derivative of Hoechst 33342, conjugated to a carboxylated silicon-rhodamine moiety. This conjugation produces a dye with far-red absorption and high DNA selectivity, low toxicity, and compatibility with STED super-resolution microscopy.

Protein-specific labeling: Small-molecule fluorophores for site-specific labeling of proteins are currently available. Some such probes are derivatives of the SIR group described above. Selective labeling of individual proteins often utilizes engineering the target protein at the level of its gene. A widely employed technique uses GFP or its derivatives. We will focus here on the GFP and its derivatives, extensively used to selectively tag an individual protein in live cells. GFP was first isolated from the jellyfish *Aequorea victoria* by Osamu Shimomura. GFP has a very stable β-barrel structure with 11 strands (Fig. II.18). The fluorophore, 4-(*p*-hydroxybenzylidene) imidazolidin-5-one, present in GFP forms entirely from amino acids present inside the barrel without requiring any cofactor. This makes it particularly useful for labeling proteins in live cells without external manipulation. In conventional applications, the GFP gene is fused in frame with the targeted gene, and the targeted protein is expressed as a GFP fusion in the cell. The fluorescent fusion protein makes it possible to selectively observe and track a single type of protein within live cells.

II.1.1.4 Fluorescence recovery after photobleaching: A useful fluorescence technique for studying genome dynamics

Proteins are highly mobile entities. They move from one cellular compartment to another and diffuse within a compartment. Fluorescence recovery after photobleaching (FRAP) can measure the translational diffusion rate of a molecule. In FRAP, a high-intensity laser

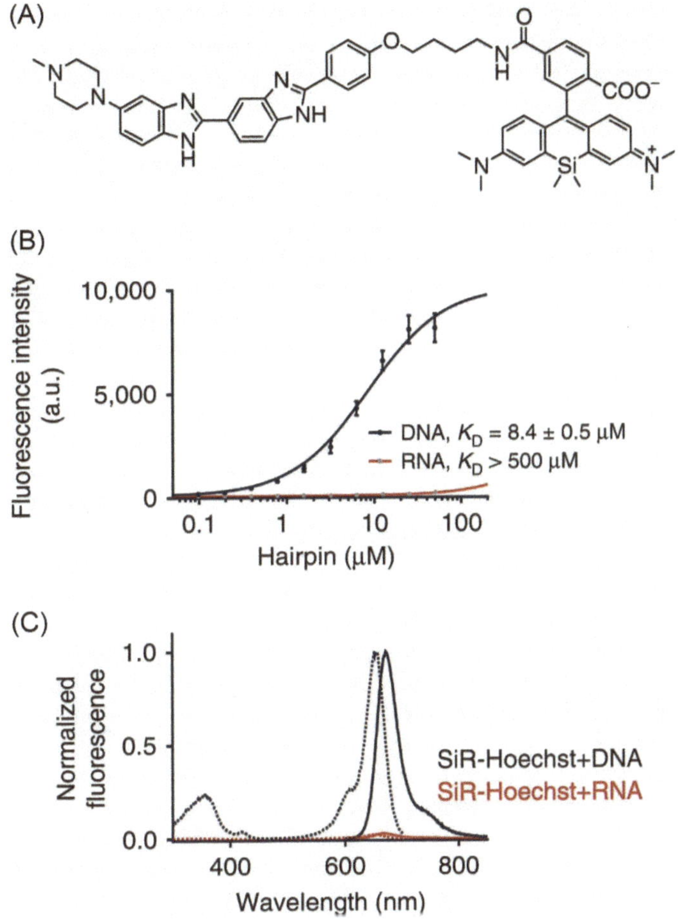

FIG. II.17 Properties of SiR-Hoechst. (A) Structure of the SIR-Hoechst. (B) Titration of 0.1 μM SiR-Hoechst probe with varying concentrations of hairpin DNA or hairpin RNA. Data points represent mean ± s.d. of three independent replicates measured in triplicates. (C) Excitation spectra and absorbance spectra of 1 μM SiR-Hoechst in the presence of a 50-μM hairpin DNA or RNA. Determined excitation maximum is at 652 nm and emission maximum is at 672 nm. *Reproduced from Lukinavičius, G., Blaukopf, C., Pershagen, E., et al. 2015. SiR–Hoechst is a far-red DNA stain for live-cell nanoscopy. Nat Commun 6, 8497. https://doi.org/10.1038/ncomms9497. License: http://creativecommons.org/licenses/by/4.0/.*

pulse (a bleaching pulse) is applied to a small area containing a fluorescent protein. Upon excitation, an excited-state reaction occurs leading to irreversible loss of fluorescence. A high-intensity pulse causes rapid cycles of excitation and emission, hastening the excited-state reaction. Fluorescence recovery in the bleached area occurs exclusively by translational diffusion of existing fluorescent molecules from adjacent areas (Fig. II.19). Thus translational diffusion coefficient of fluorescent molecules can be obtained from the rate of fluorescence recovery. The translational diffusion coefficient is environment dependent, and its magnitude may reveal the nature of the molecular environment.

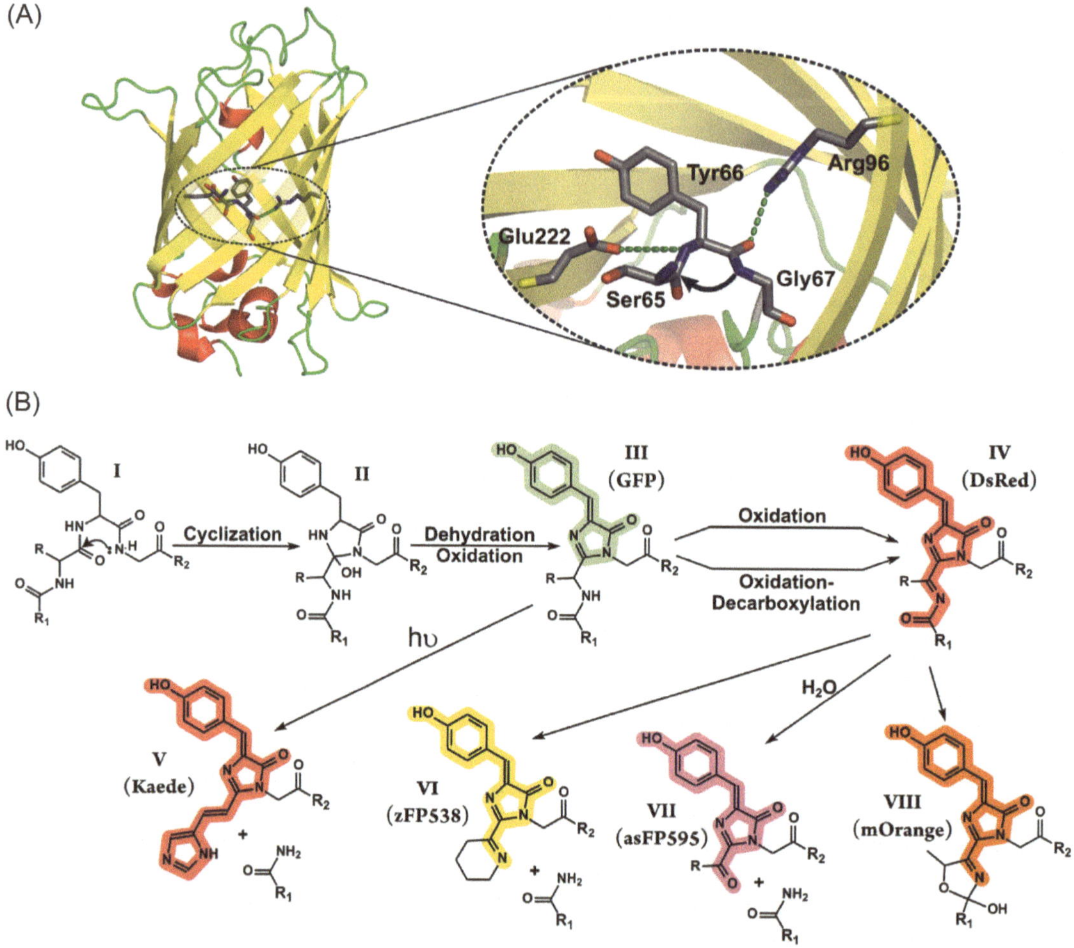

FIG. II.18 Posttranslational reactions and representative chromophore structures of green fluorescent protein-like proteins (A) The β barrel structure of green fluorescent protein-like proteins, precyclization state, and catalytic Arg96 and Glu222 in the chromophore surroundings of avGFP (magnified). (B) Chromophore biosynthesis and the diversity of chromophore chemical structures of green fluorescent protein-like proteins (π-electron conjugation is indicated in color). The figure is not intended to show ionization states of the chromophore upon its biosynthesis. *Credit: Pakhomov, Martynov, 2008. GFP family: structural insights into spectral tuning. Chem. Biol. 15 (8), 755–764. Elsevier.*

II.1.2 Chemical modifications and proximity analysis

We will discuss methodologies for global profiling techniques, such as DNA and histone methylation profiling. These techniques have emerged as crucial elements for studying genome structure and function. Two methodologies, formaldehyde cross-linking and massively parallel DNA sequencing, play critical roles. Principles of parallel DNA sequencing will be described in a later section. We describe the chemistry of formaldehyde cross-linking before describing the main content of this section.

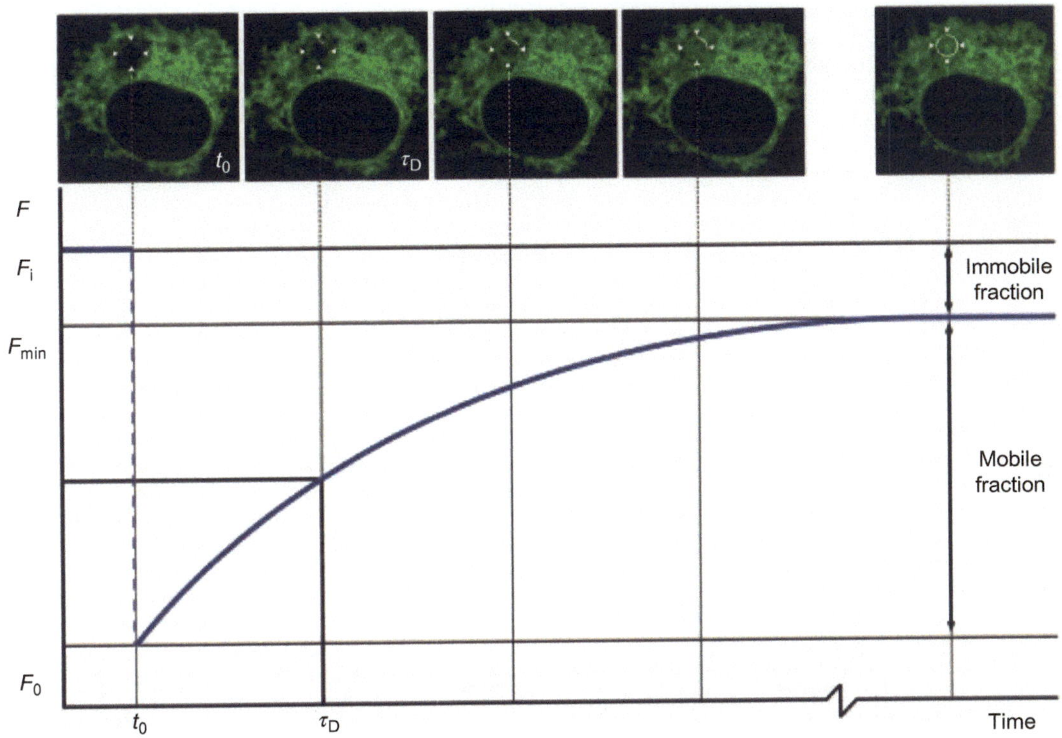

FIG. II.19 Principles of fluorescence recovery after photobleaching. When a region in the fluorescent area (here the endoplasmic reticulum) is bleached at time t_0, the fluorescence decreases from the initial fluorescence F_i to F_0. The fluorescence recovers over time by diffusion until it has fully recovered (F_∞). The characteristic diffusion time τ_D indicates the time at which half of the fluorescence has recovered. The mobile fraction can be calculated by comparing the fluorescence in the bleached region after full recovery (F_∞) with that before bleaching (F_i) and just after bleaching (F_0). *Reproduced from Reits, E., Neefjes, J., 2001. From fixed to FRAP: measuring protein mobility and activity in living cells. Nat Cell Biol 3, E145–E147. https://doi.org/10.1038/35078615https://doi.org/10.1038/35078615https://doi.org/10.1038/35078615.*

Proteins and DNA contain many nucleophilic amino groups. Prominent among them are the ε-amino groups of lysine and the 2-amino group of guanine in DNA. The former group, being polar and charged at neutral pH, generally resides on the protein surface and is solvent exposed. The latter lies in the minor groove of the B-DNA and is generally solvent exposed. The carbonyl carbon of the formaldehyde is highly electrophilic due to the electron withdrawal by the carbonyl oxygen. Protein or DNA nucleophilic groups may undertake a nucleophilic addition reaction to create methylol, which then eliminates a water molecule to yield a Schiff's base. If another nucleophile from the protein or DNA is nearby, the Schiff's base may undergo another nucleophilic addition reaction, creating a methylene bridged cross-link (Fig. II.20). Formaldehyde, as a cross-linking agent, has two additional advantages. Being small, it penetrates the cellular milieu quite efficiently, and the cross-linked product can be reversed under certain conditions.

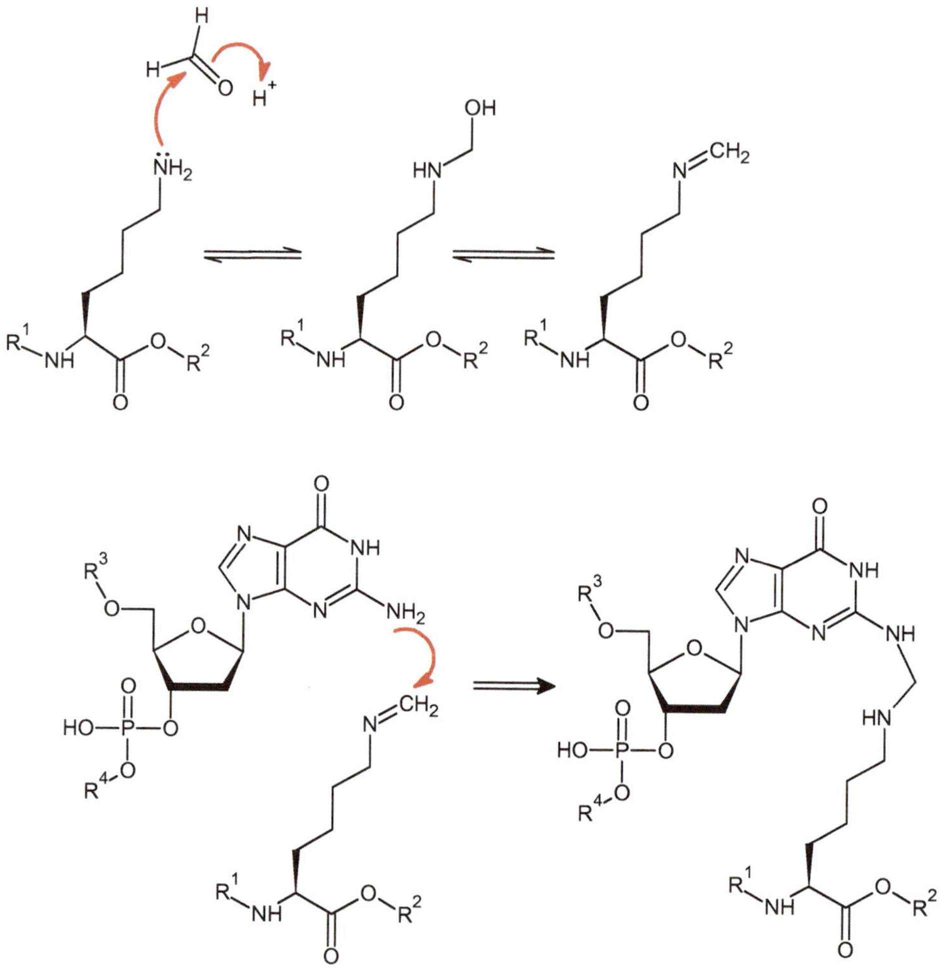

FIG. II.20 Chemical reactions occurring during formaldehyde cross-linking of protein and DNA. The top panel shows the mechanism of formation of the Schiff's base. A specific example of a protein-DNA cross-link is shown in the bottom panel.

II.1.2.1 Chromatin immunoprecipitation

Chromatin immunoprecipitation (ChIP) is a technique to determine the location of a DNA-binding protein on the genome. In ChIP, proteins and genomic DNA are cross-linked by formaldehyde at a predetermined time point and pulled down with an antibody specific to the desired protein, along with the cross-linked DNA. The cross-linked protein-DNA complex is then separated by reversing the cross-link. If the goal is to determine the occupancy of a target site, the precipitated DNA is amplified by primers specific for that site and analyzed (Fig. II.21). In a later chapter, we will show that the eukaryotic transcription factors bind to many sites on the genome. For the global profiling of binding sites, deep sequencing of the precipitated DNA yields all the occupied sites on the genome

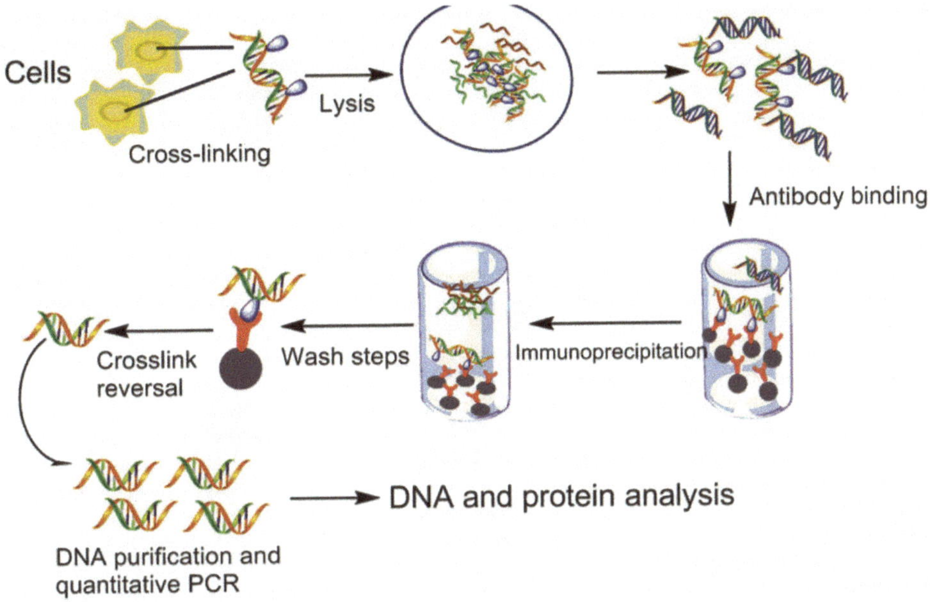

FIG. II.21 Chromatin immunoprecipitation. Protein and associated chromatin in living cells or tissues are temporarily bonded, and the DNA-protein complexes (chromatin-protein) are then sheared into ~500 bp DNA fragments using either enzymatic digestion or physical shearing by sonication. Cross-linked DNA fragments associated with the protein(s) of interest using formaldehyde are selectively immunoprecipitated from the cell debris using appropriate protein-specific antibody. After the cross-links are reversed, the associated DNA fragments are purified and their sequence is determined. These DNA sequences are supposed to be associated with the protein of interest in vivo. The DNA undergoes polymerase chain reaction amplification using primers targeting a particular genomic locus. These DNA sequences can be subjected to a number of downstream analysis techniques, including targeted approaches, such as semiquantitative polymerase chain reaction, quantitative polymerase chain reaction, and genome-wide analyses using microarrays (ChIP-chip) and deep sequencing (ChIP-seq), ChIP-on-chip (Shah, 2009; Vinckevicius and Chakravarti, 2012). *Reproduced from Song et al., Front. Microbiol. 6. https://doi.org/10.3389/fmicb.2015.01049. Under Creative Commons Attribution 4.0 International.*

(ChIP-seq). The ChIP provides a methodology to determine the occupancy of a transcription factor at a target sequence and its occupied sites on a genome-wide scale.

II.1.2.2 Elucidation of DNA and histone methylation at global scale

Site-directed chemical modification of DNA and its associated proteins, particularly histones, has emerged as a significant theme of gene regulation. The most common chemically modified protein in the genome is the histone. Among modifications that occur on the DNA, the 5-methylation of the cytosine base is the most common. We will describe methods to determine genome-wide location of 5-methyl-cytosine and histone modifications. In one method, the genomic DNA is first fragmented by mechanical force or enzymatic digestion, and 5-methyl cytosine containing fragments are pulled down by specific antibodies or affinity columns (Fig. II.22). Deep sequencing of the methylated DNA fragments yields the methylation sites. In another method, sodium bisulfite is first used to specifically deaminate unmethylated cytosines to uracil (Fig. II.23). The bisulfite-treated

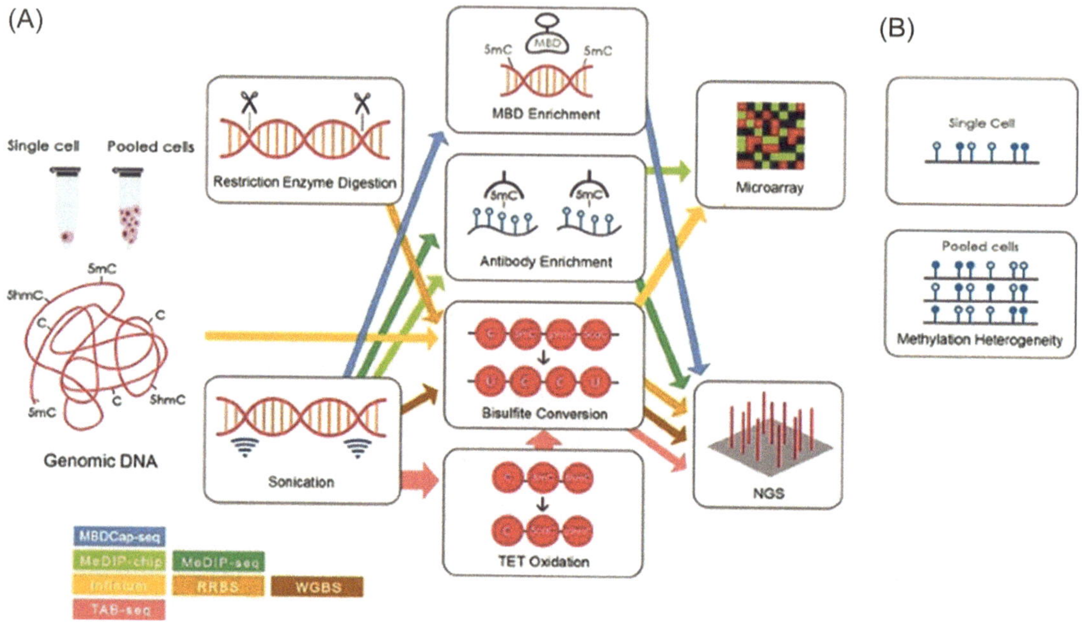

FIG. II.22 Commonly used methods for genome-wide DNA methylation analysis. (A) The procedures may involve fragmentation of genomic DNA by restriction enzyme digestion or sonication. The genomic DNA can be subjected to methyl-CpG-binding domain or MBD enrichment, antibody enrichment, bisulfite conversion, or ten-eleven translocation (TET) oxidation before analyzing by microarray or next-generation sequencing platform. (B) Single-cell DNA methylation analysis that involves the isolation of single cells allows the assessment of methylation heterogeneity in cell populations, whereas other genome-wide DNA methylation profiling methods using pooled heterogeneous cell populations are not capable to dissect the methylation heterogeneity. *Blue* concrete dots represent 5mC, and hollowed ones represent C. Each track represents one read. *Reproduced from Yong, W., Hsu, F., Chen, P., 2016. Profiling genome-wide DNA methylation. Epigenet. Chromatin 9, 26. https://doi.org/10.1186/s13072-016-0075-3.*

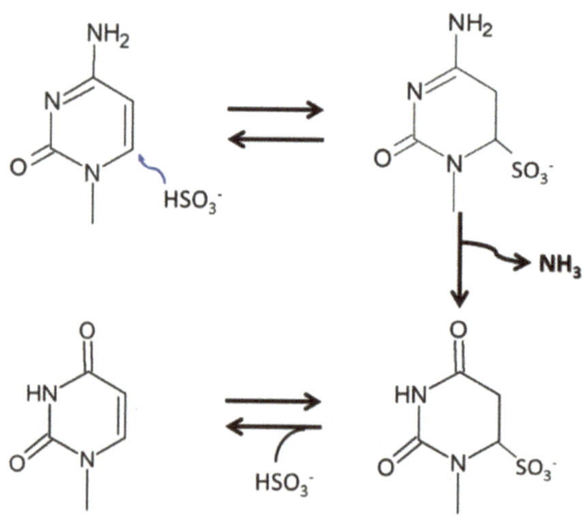

FIG. II.23 Reaction mechanism of bisulfite catalyzed deamination of cytidine. Similar reactions do not occur with 5-methyl cytidine.

DNA is sequenced using deep sequencing methods to determine the global methylation pattern at single-nucleotide resolution. For profiling of histone modifications, ChIP-seq is carried out with an antibody specific for the desired modification.

II.1.2.3 Chromosome conformation capture techniques

Chromatins pack inside the nucleus in a spatial arrangement that is specific for a cellular state. Individual chromatins also fold compactly with some sections attached to the nuclear lamina (Fig. II.24). DNA segments within chromatins often associate with nearby regions of the same chromosome creating modules called topologically associating domains (Fig. II.25). Thus folding and packing of chromatins lead to extensive intra and interchromatin contacts. An important method to determine the proximity of chromatin segments is called chromosome conformation capture (3C) or its high-throughput version, Hi-C. In the Hi-C version, the chromatins are cross-linked and fragmented, typically by a restriction endonuclease. The fragments are then biotin labeled and ligated. The ligated fragments are again fragmented and pulled down using biotin as a handle after removing biotin from unligated fragments. Paired-end sequencing of the isolated fragments establishes the proximity of different fragments at a global scale (Fig. II.26).

II.2 Tools for studying gene functions

II.2.1 Forward and reverse chemical genetics

Forward and reverse genetics

A significant part of genetics deals with the roles genes play in an organism. A favorite method in the early days was to create mutations by ionizing radiation or chemical mutagens and select a particular phenotype. Once a mutant organism with the desired phenotype was isolated, an attempt was made to map the mutation to a gene and establish the gene responsible for the phenotype. This approach dominated early genetics research, particularly in the study of unicellular organisms. We now call this the forward genetics approach.

With the development of targeted gene knockouts and knockdown techniques, we can now selectively annul or modify genes and study their effect on cellular or organismal phenotype. We call this the reverse genetics approach. The reverse genetics approach has made it easier to gain knowledge about gene function and connectivity.

II.2.1.1 Basic concepts of chemical genetics

The use of externally delivered small molecules or RNA to inhibit functions of a gene product gives a powerful handle on elucidating functions and interconnectedness of genes. In addition to revealing information about functional roles, a bonus may be to obtain lead molecules for future drug discovery. The chemical genetics approach attempts to inhibit a

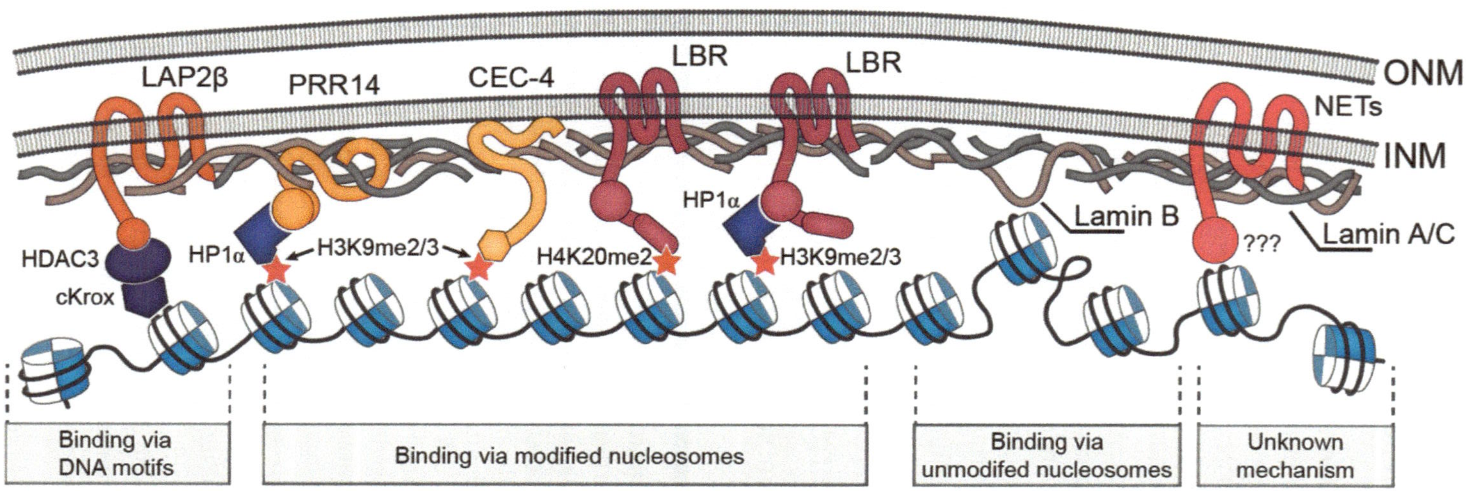

FIG. II.24 Schematic representation of the main NL-chromatin tethering mechanisms. *Reproduced from Shevelyov et al., 2019. Cells 8 (2), 136; https://doi.org/10.3390/cells8020136. This is an open access article distributed under the Creative Commons Attribution License.*

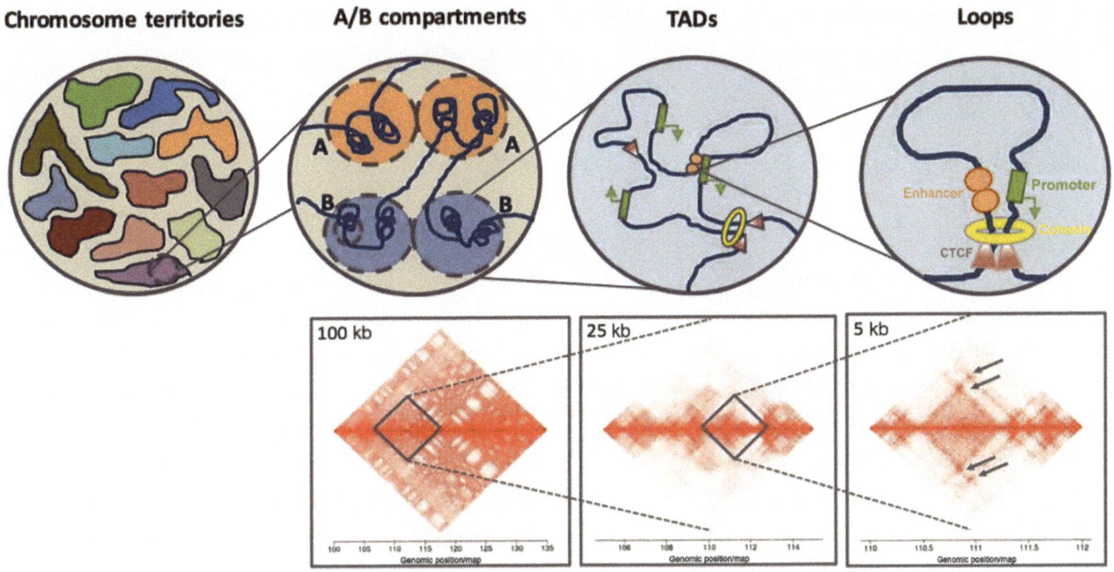

FIG. II.25 Hierarchical organization of the higher eukaryotic genome. Individual chromosomes occupy independent chromosome territories that are divided into active (A) and inactive (B) chromatin compartments, yielding a chessboard-like pattern of interaction frequencies in the contact matrix 2 and 12. Each compartment region encompasses several topologically associating domains that are visible in the contact matrix (below) as red triangles (highlighted by the *gray squares*) containing high interaction frequencies. The higher the intensity of red coloration, the higher is the frequency of interaction detected between the two regions. Within the topologically associating domains, subregions can be distinguished with even higher interaction frequencies. Often the borders of the topologically associating domains or the subtopologically associating domains exhibit punctual and very high interaction density, corresponding to the anchoring of the loops *(marked by arrows). Reproduced from Wright, R.H.G., Le Dily, F., Beato, M., 2019. ATP, Mg2+, nuclear phase separation, and genome accessibility. Trends Biochem. Sci. 44 (7), 565–574. Elsevier.*

single protein inside a cell, often by small-molecule ligands. The overarching goal is to determine the function of the targeted protein inside a functioning cell. Like genetics, chemical genetics can also be approached by either randomly targeting proteins by a library of small molecules and making a phenotype-based selection or inhibiting a predetermined protein by a ligand, known or designed, and observing the phenotype. We call the former approach as forward chemical genetics and the latter as the reverse chemical genetics.

II.2.1.2 Forward and reverse chemical genetics

In the forward chemical genetics approach, a library of small molecules is screened against one or more cell types to obtain a particular phenotype. The phenotype can be anything as long as it can be assayed, preferably in a high-throughput fashion. Once we know the molecule that produces the desired phenotype, the target needs to be identified. Identification of the target, in many cases a protein, is a challenging task and may require novel approaches. Upon identification of the target, it is important to establish a

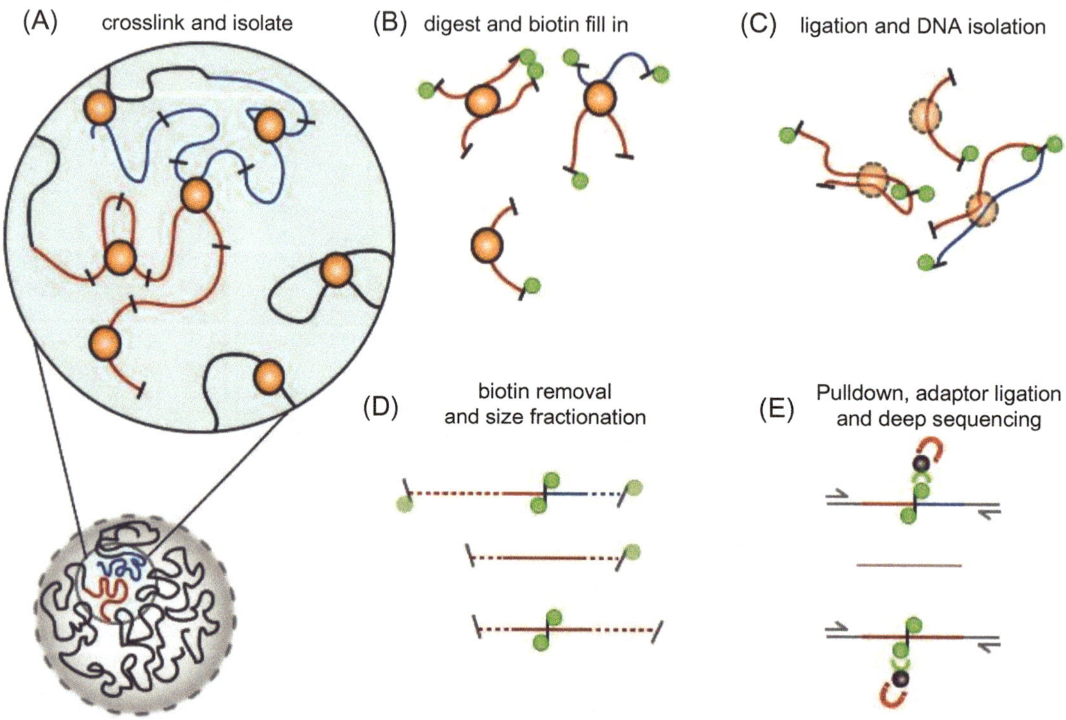

FIG. II.26 Overview of Hi-C technology. (A) Hi-C detects chromatin interaction both within and between chromosomes by covalently cross-linking protein or DNA complexes with formaldehyde. (B) The chromatin is digested with a restriction enzyme, and the ends are marked with a biotinylated nucleotide. (C) The DNA in the cross-linked complexes is ligated to form chimeric DNA molecules. (D) Biotin is removed from the ends of linear fragments, and the molecules are fragmented to reduce their overall size. (E) Molecules with internal biotin incorporation are pulled down with streptavidin-coated magnetic beads and modified for deep sequencing. Quantitation of chromatin interactions is achieved through massively parallel deep sequencing. *Reproduced from Belton et al., 2012. Methods 58 (3), 268–276. Elsevier.*

causal connection with the phenotype (Fig. II.27). In the reverse chemical genetics approach, the investigator knows the target and wants to elucidate its function. The most important step is to find a selective cell-permeable inhibitor of this target. A cell is treated with the selective inhibitor, and the changes in cellular functions is determined (Fig. II.28). The advantage here, in contrast to reverse genetics, is that the inhibition is reversible and tunable. A potentially significant problem in the chemical genetics approach is the selectivity of the ligand molecule. The off-target binding of the ligand may intertwine many effects, thus, making it difficult to deconvolute the real cause and effect relationship.

II.2.2 Small molecules as tools for chemical genetics

II.2.2.1 Chemical libraries

One of the critical requirements of a successful forward chemical genetic approach is to have a large and diverse collection of small molecules. Ligands with larger surface areas provide a

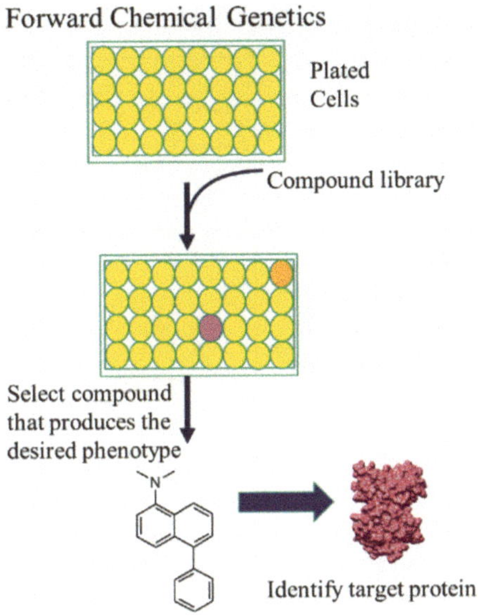

FIG. II.27 The forward chemical genetics approach.

larger interaction surface with more bond formation possibilities; thus, they have a higher probability of having greater interaction energy and affinity. The specificity of the ligand for the target protein is a product of many factors but having multiple functional groups and chiral centers may allow the molecule to display unique atomic patterns that may

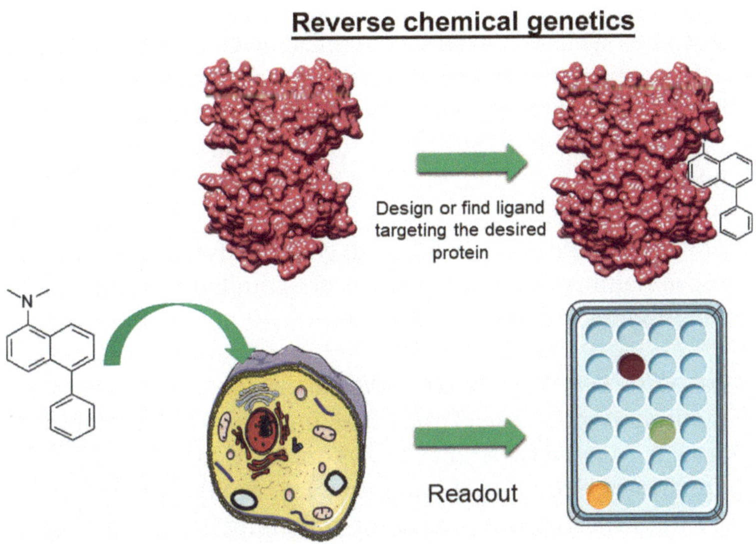

FIG. II.28 The reverse chemical genetics approach.

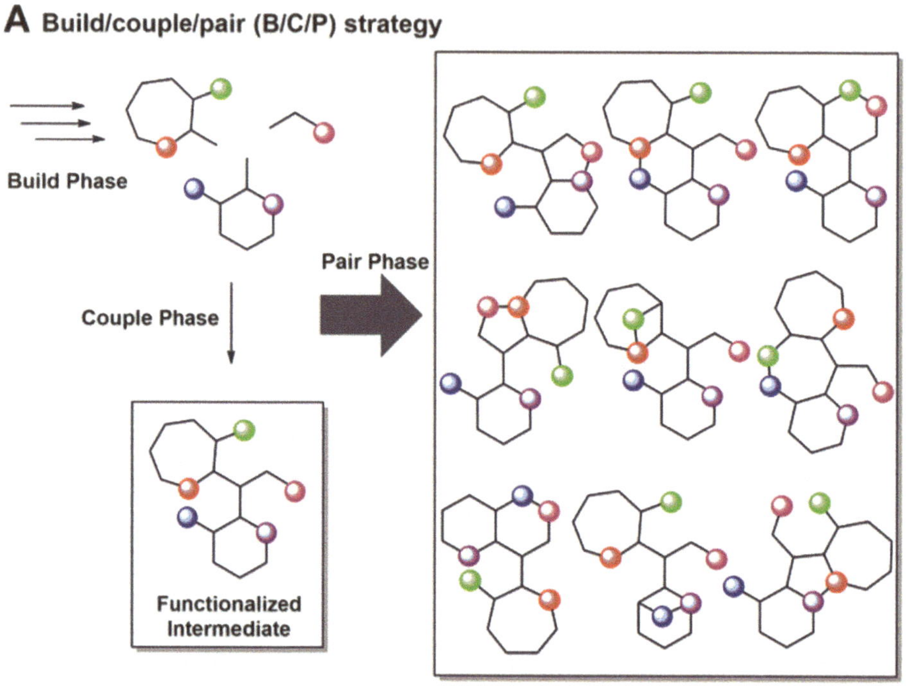

FIG. II.29 A schematic representation of the build or couple or pair strategy. A brief overview of two major strategies in diversity-oriented synthesis: Build or couple or pair and ring distortion. *Reproduced from Yi, S., Varun, B.V., Choi, Y., Park, S.B., 2018. Front. Chem. 6, 507. Copyright © 2018 Yi, Varun, Choi and Park. This is an open-access article distributed under the terms of the Creative Commons Attribution License (CC BY).*

enhance specificity. Thus skeletal and stereochemical diversities in the chemical library are likely to enhance the probability of discovering the desired ligand. Many methods are available for generating a library of chemicals having skeletal and stereochemical diversity. Fig. II.29 shows one such strategy called the build-couple-pair strategy of diversity-oriented synthesis. In the build phase, several chiral building blocks are synthesized with multiple functional groups, followed by the couple phase in which the building blocks are coupled intermolecularly in all possible combinations to generate functionalized intermediates. In the pair phase, the remaining functional groups are coupled intramolecularly to generate a diverse class of compounds.

II.2.2.2 Target-based screening of small molecule

An appropriate method of screening of chemical libraries is pivotal for the forward chemical genetics approach. Massively parallel assays are ideal for screening a large number of compounds. If the library size is small, one may use rapid sequential assays as well. In recent times, very high-throughput assays for screening are becoming common. In all high-throughput methods, the rapidity of the assay is critical, and a parallel readout is ideal. Many types of methods can be adapted for high-throughput screening, but fluorescence-based methods are relatively simple and popular. We will mainly describe the fluorescence-based assays here as they are sensitive, quantitative, rapid, and can easily be automated. In biochemical screening assays, one is usually assaying for either an enzymatic activity or association of two molecules. The two molecules can be either a macromolecule and a smaller molecule or two macromolecules. In both cases, many fluorescence properties can be utilized to develop an assay.

We have seen before that change in the environment may change the quantum yield of a fluorophore and, hence, its emission intensity. Changing the environment may reduce the quantum yield due to enhanced depletion of the excited state through radiationless processes. In fluorescence-based enzyme assays, a fluorescent substrate is often used. Enzymatic conversion of the substrate leads to a change in the environment of the fluorophore and, hence, altered quantum yield and intensity. Similarly, when two macromolecules associate, a fluorescence probe, internal or external, may change its environment resulting in altered quantum yield. The alteration in quantum yield may be read out as enhanced or diminished fluorescence emission intensity. The major problem with the fluorescence quenching-based assays is that the quantum yield, in general, is very sensitive to impurities, both fluorescent and nonfluorescent, and interference could be significant. The fluorescence intensity is also dependent on the intensity of the exciting light; any fluctuation of light intensity may introduce drift and bias in the measurement.

The fluorescence anisotropy assay relies on the fact that the anisotropy of a fluorescence probe is dependent upon its tumbling rate. When a fluorescent molecule associates with another molecule of significant mass, its tumbling slows down, increasing its fluorescence anisotropy. This principle forms the basis for the fluorescence anisotropy-based assays. We have previously seen that fluorescence anisotropy is a ratio of fluorescence intensities and, hence, is less prone to errors due to intensity variations.

FRET-based assays depend on the change of spatial positions of two fluorophores, one donor and the other acceptor. If during enzymatic reactions or molecular interactions, the distance between two fluorophores changes, the FRET efficiency will also change. Altered FRET efficiency is read out as a change in acceptor emission intensity. A number of donor-acceptor pairs are available for use in FRET-based assays, having different quantum yields, excitation, and emission wavelengths, and R_0 the distance at which 50% of energy transfer occurs.

II.2.2.3 Phenotypic screening of small molecule and target identification

In contrast to the target-based drug screens, the phenotypic screen reads out a prechosen phenotype of a biological entity, often a cell. The phenotypic screens may be superior to target-based screens when the targeted disease is not well understood in terms of the underlying biochemistry. Target-based screens that assay a particular molecular property identify molecules that hit a single target. In some situations, better results may be obtained from hitting multiple targets. In such situations, phenotypic screens are most useful. A significant drawback in the phenotype-based approach is the elucidation of the mechanism of action and identification of the target(s) involved. However, target identification is becoming increasingly amenable to newer methods, widening its applicability.

A case study will illustrate the power of phenotypic screens. The goal of this study was to identify a small molecule that will promote cartilage repair. For this, the target cell was bone marrow-derived mesenchymal stem cells. The appearance of chondrogenic nodules containing cartilage-specific matrix components allowed the staining of the cells by rhodamine B and formed the basis for an image-based assay (Fig. II.30). The positive stain with rhodamine B was the phenotypic end point, which was the surrogate marker for differentiation to chondrocytes. The phenotypic screen produced a molecule, kartogenin (KGN), having the desired property. Affinity purification of the KGN cross-linked proteins identified filamin A (FLNA) as the target. Once the target was known, further exploration indicated that KGN disrupted protein-protein interaction between FLNA and CBFβ thus

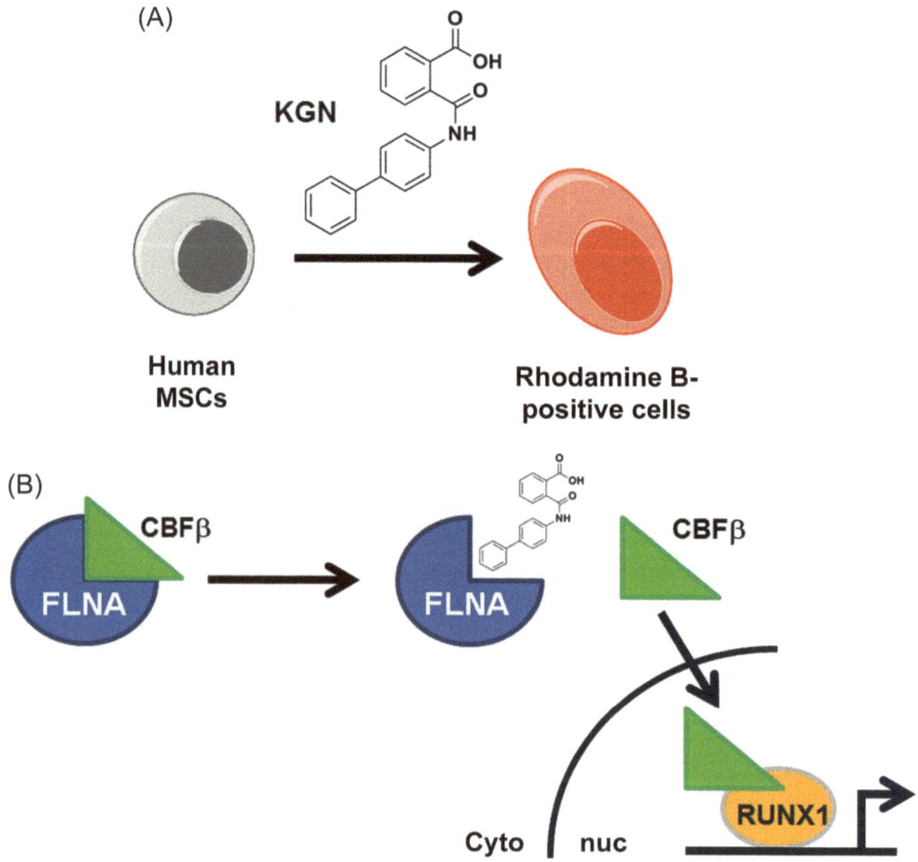

FIG. II.30 Example of a phenotypic assay-based drug discovery. (A) Schematic of the phenotypic assay for rhodamine-positive chondrocyte-like cells derived from human mesenchymal stem cells. (B) Affinity-based pull-down experiments revealed filamin A as a cellular binder of kartogenin. Kartogenin disrupts the cytoplasmic interaction of filamin A and CBFβ, enabling CBFβ nuclear translocation and RUNX1-dependent transcription of cartilage-specific genes. *Wagner, B.K., Schreiber, S.L., 2016. The power of sophisticated phenotypic screening and modern mechanism-of-action methods. Cell Chem. Biol. 23 (1), 3–9. Elsevier.*

allowing CBFβ (core-binding factor beta) to relocate to the nucleus and upregulate target genes, initiating chondrocyte differentiation.

High-content screening (HCS) also relies on observing phenotypes of cells. Fluorescence microscopy is the technique most used for observation. In HCS, we observe and measure a large number of cellular phenotypic characters and parameters simultaneously. Under favorable conditions, we may be able to extract meaningful conclusions about the underlying processes affected by a perturbation. Fig. II.31 shows a typical workflow of HCS. Initially, the cells are plated on a microwell plate, followed by perturbing

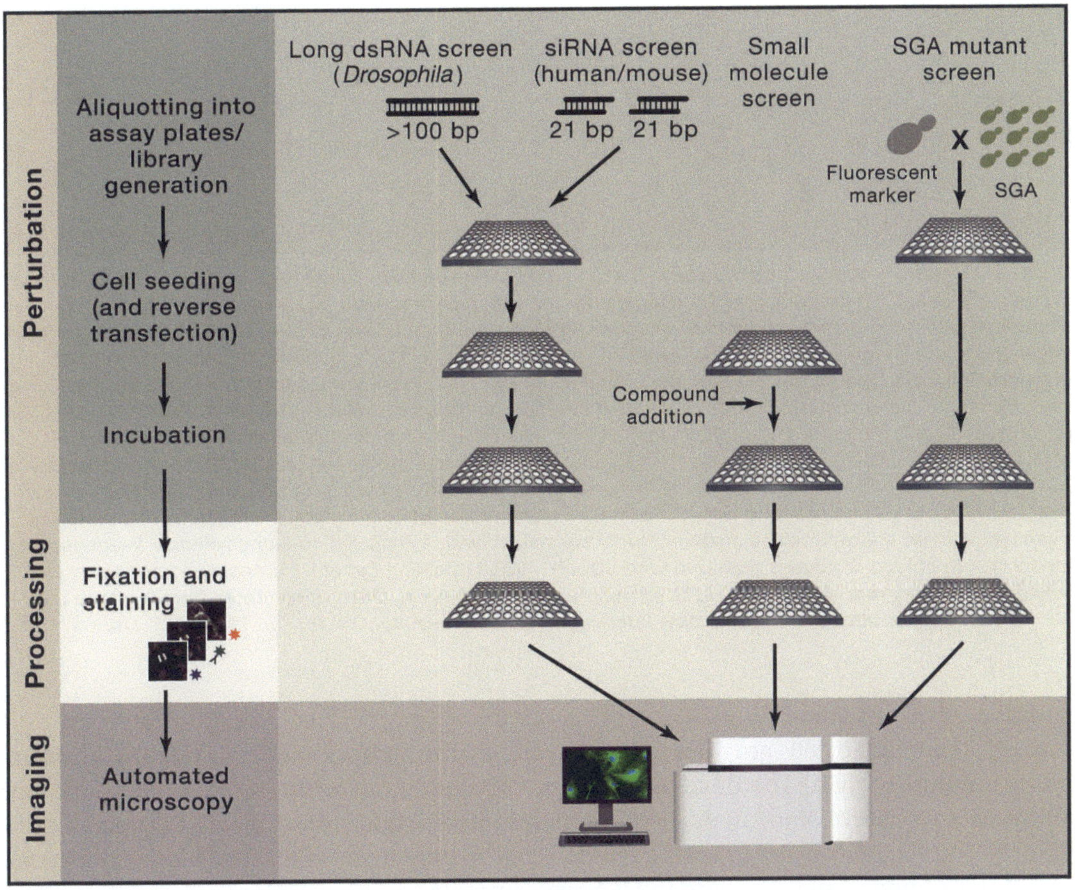

FIG. II.31 Workflows of image-based screens using different perturbation reagents. Despite differences in perturbation agent, cellular model, or experimental scope, the main steps of an image-based screen are similar. The library of perturbation reagents is distributed onto assay plates and then cells are seeded into the wells (and, if applicable, reversely transfected with RNAi reagents). In microscopic screens in yeast, fluorescently tagged reporters can be crossed into library mutant strains. Following an incubation time of typically 24h–96h, the cells are fixed and stained with fluorescent probes that label the relevant cellular features. Subsequently, the images are captured by automated microscopy and passed on to the image-analysis pipeline. *SGA*, synthetic genetic arrays. *Reproduced from Boutros, M., et al., 2015. Microscopy-based high-content screening. Cell 163 (6), 1314–1325. Elsevier.*

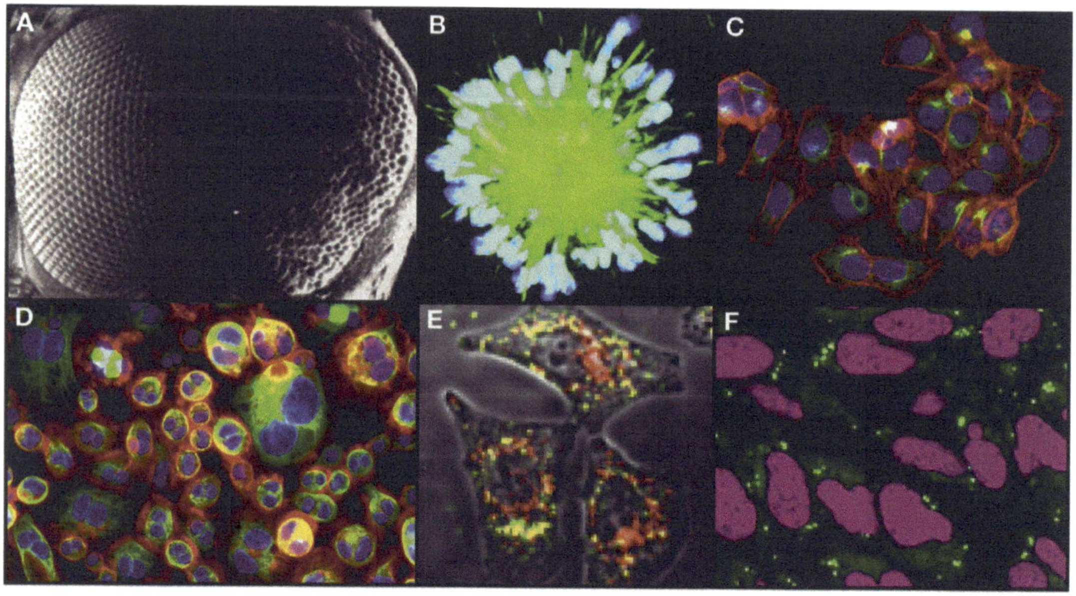

FIG. II.32 An example of high-content screens illustrating the range of image-based read-outs. (A) Scanning electron micrographs of adult Drosophila eyes. Compared with the wild type, a sev-Ras1V12 transgene led to a rough eye phenotype with misarranged and fused ommatidia. (B) High-resolution images of BS-C-1 cells are stained for DNA *(blue)* and α-tubulin *(green)*, showing the effect of monastrol treatment on spindle formation. In mitotic cells, monastrol induces a mono-astral microtubule formation. (C) HCT116 cells are stained for DNA, α-tubulin, and actin to extract quantitative phenotypic features describing cell morphology as described in Laufer et al. (2013). (D) Drosophila cells are stained as in (C), imaged at 20× magnification for detailed analysis of morphological phenotypes. RNAi-mediated knockdown of Rho1 leads to enlarged, multinucleated cells. (E) Confocal micrographs of A549 cells infected with influenza A/WSN/33 virus and stained for influenza virus *(green)* and CD63 *(red)*. Yellow indicates virus particles colocalizing with CD63-labeled late endosomes. (F) Confocal micrographs of L cells after treatment with Wnt3a-conditioned medium. Nuclei are stained with 4′,6-diamidino-2-phenylindole *(magenta)*, and the formation of lipid droplets is visualized with BODIPY 493/503 staining *(green)*. *(A) Adapted from Karim et al. (1996). (B) Adapted from Mayer et al. (1999). (E) Adapted from Karlas et al. (2010). (F) Adapted from Scott et al., 2015. Reproduced from Boutros et al., Microscopy-based high-content screening. Cell 163 (6), 1314–1325. Elsevier.*

additions, such as small molecules from a chemical library. The cells are then fixed and stained. The stained cells are then imaged in a high-throughput mode generating an enormous amount of data. The developments in cellular engineering and big data handling allow us to observe and analyze more parameters, thus, offering an unprecedented glimpse into cellular processes affected by a perturbing agent. Fig. II.32 shows the range of different phenotypic readouts in HCS experiments.

II.2.2.4 Peptides as tools for chemical genetics

Small molecules have emerged as a leading tool for interrogating important targets and processes. Significant advantages of using small molecules lie in many different domains. There is a large number of chemical libraries available, making it easier to find an appropriate molecule. Many small molecules are membrane permeable and diffuse to different cellular compartments easily. Their serum stability is generally excellent. In addition to that, their pharmacokinetic and pharmacodynamic properties are either favorable or can be modified to make them favorable for developing them as drug candidates. Despite having these major advantages, they often suffer from substantial off-target effects. Off-target effects can be understood as follows. A small molecule derived from a synthetic chemical library has, in general, not coevolved with the proteins in an organism. The millions of proteins and protein variants present a very diverse array of atomic and electronic patterns, among which it has to match only one ideally. This ideal condition is often not met, and it has to go through several cycles of optimization to obtain a molecule having a high degree of target specificity. Addressing these off-target effects may also require extensive redesigning, and sometimes it is impossible to address the problem adequately. The off-target effects create profound complexity in interpreting the data of a chemical genetics experiment and are difficult to deconvolute.

The peptide-based chemical genetics approach offers a somewhat different perspective. Peptides tend to have a higher target specificity but suffer from delivery problems. Proteins in the cell interact with a variety of molecules, including other proteins and nucleic acids. Protein segments and motifs used for recognition of other macromolecules can be used as a template for developing peptide mimics. This class of mimics may inhibit targets with adequate specificity. One can understand this specificity from an evolutionary perspective. Because many proteins have coevolved inside a cell, they have optimized their interactions to attain the desired levels of specificity. Thus any molecule mimicking the interacting surfaces may attain similar specificity. These considerations have led to the development of peptide mimics as suitable tools for chemical genetics.

Peptides and peptide-based molecules suffer from two major drawbacks as tools for chemical genetics. One is the instability of unmodified peptides in the cellular milieu, and the other is the lack of cell penetration ability. One can overcome the former drawback by constraining the conformation of the peptide by chemical modifications. Cellular proteases frequently interact with peptides in disordered conformation. Constraining the peptide to a more ordered conformation makes it less prone to proteolysis, thus, improving its intracellular stability. The cell penetration problem has now been overcome by attaching a cell penetration tag, usually consisting of several positively charged amino acids, to the peptide. Many protein-protein and protein-DNA interactions use helix as a motif. Thus helix has emerged as a template for constraining peptides to an ordered conformation for probing drug targets. We will discuss this theme in the next chapter, where we go into design of synthetic transcription factors.

II.2.3 Global profiling of gene expression

Typically, several thousand types of RNAs are present in a cell at any moment. Many of these RNAs are translated into proteins as well. Individual RNA levels change throughout the cellular lifecycle as its synthesis and degradation rates change. Analysis of RNA levels is a popular method, albeit an indirect one, for inferring the functional state of the genome. We will describe three methods for quantitation individual RNA levels: reverse transcription polymerase chain reaction (RT-PCR), DNA microarray, and RNA sequencing (RNA-seq).

II.2.3.1 Quantitative reverse transcription polymerase chain reaction (RT-qPCR)

A widely used method for measuring the number of a particular mRNA present in the cell utilizes amplification by PCR. PCR is one of the most widely used techniques in biology laboratories around the world. We will summarize the basic principles first. PCR is a nucleic acid amplification method in which two primers in head-to-head direction amplify a nucleic acid template; the two primers hybridize to opposite strands. If the target of amplification is DNA, it can be used directly as the template and amplified. If the targeted nucleic acid is an RNA, it first needs to be converted into DNA and then amplified. An enzyme called reverse transcriptase copies RNA into its complementary DNA (cDNA). This enzyme uses an RNA as a template and an oligodeoxynucleotide primer and 2′-deoxynucleoside triphosphates to synthesize the cDNA. The cDNA can then act as the template for PCR amplification. PCR is recursive in nature in which the complementary strand synthesized in one cycle is used as a template in the next, thus, doubling the concentration of the template (only the sequences between the two primers are amplified; see Fig. II.33) in each cycle. This sequential amplification leads to a chain reaction and massive amplification of the targeted sequence. The amplification process ends after the starting materials become depleted. Traditionally, the amplified DNA is then analyzed by gel electrophoresis or related techniques.

RT-PCR, as described above, suffers from two drawbacks in quantitating the amount of template present. At the end point, the amount of amplified DNA present is often not proportional to the input template. This is due to the decline of amplification efficiency as the amplification proceeds. Second, continuous observation of the amplification is not possible. Real-time PCR overcomes these deficiencies by recording the early amplification cycles directly. In real-time PCR, a fluorescent dye, whose quantum yield increases significantly upon binding to double-stranded DNA, is added to the reaction mixture. Because dye binding is proportional to the double-stranded DNA present, the progress of amplification can be monitored in real time by recording the fluorescence.

In the first few cycles of real-time PCR, the double-stranded DNA concentration is too low for the fluorescence to rise significantly above the background. As the concentration of the double-stranded DNA builds up, the fluorescence becomes sufficiently intense for detection above the background. The point at which the observed fluorescence crosses a preset "threshold" value is known as the C_T value (Fig. II.34). The C_T value is inversely proportional to the concentration of the template present. RT-qPCR is widely used for quantitating RNA, and in multiwell format, hundreds of types of RNAs can be quantitated in parallel. However, it cannot rival the massive throughput of the techniques described below.

II.2.3.2 DNA microarray

The basis for the DNA microarray methodology is the ability of complementary DNA sequences to form double-stranded DNA quickly, with high sequence-specificity. Upon

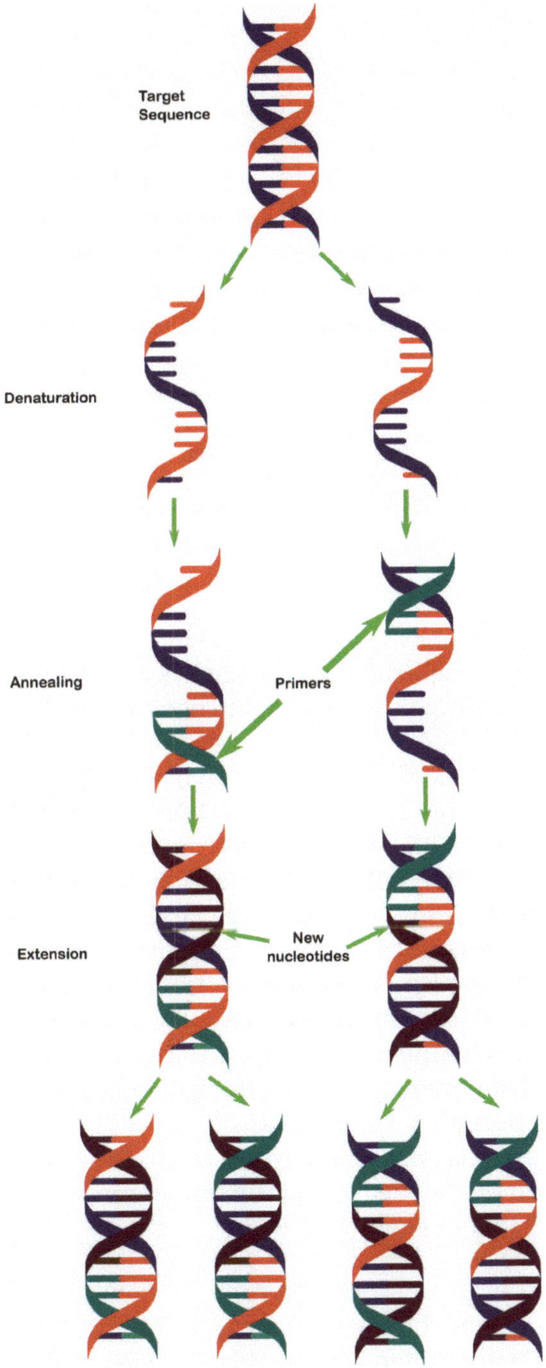

FIG. II.33 Steps of polymerase chain reaction amplification.

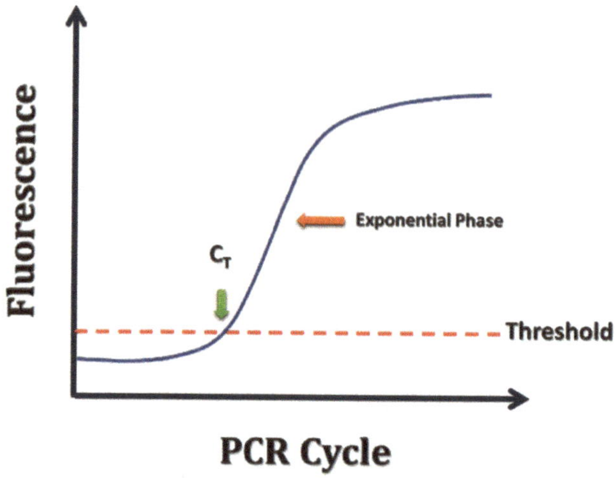

FIG. II.34 A real-time polymerase chain reaction profile.

mixing of two complementary DNA sequences, a double-stranded DNA forms whose stability depends on the base composition and sequence. Mismatched base pairs destabilize the duplex DNA, substantially. Even one mismatched base pair can destabilize the DNA duplex by 1–2 kcal/mole.

The first step in the analysis of gene expression by DNA microarrays is to prepare the microarray. The DNA microarray consists of many DNA sequences immobilized on a solid surface in an ordered array. In one method, the array consists of 20–100 nucleotide-long oligonucleotides. Each chosen oligonucleotide matches uniquely with one mRNA sequence and with none other in the genes of the targeted organism. Such unique oligonucleotides are usually generated through computational analysis of the genome sequence of the targeted organism. The total RNA from the cells is then extracted, copied to DNA by reverse transcription, and labeled (often by a fluorescent probe). The fluorescently labeled collection of complementary DNAs is then hybridized to the array, washed, and scanned to read the fluorescence signal. The signal intensity of a spot is proportional to the concentration of the mRNA. For comparison, two different samples are labeled with two different fluorescent dyes, mixed, and hybridized to the array. The hybridized array is then read at two different wavelengths, and the ratio represents the differential effect of treatment on the expression of genes (Fig. II.35).

II.2.3.3 Global profiling of gene expression through RNA sequencing

Transcription from the genome produces many types of RNAs. As mentioned before, two competing processes of transcription and degradation determine the RNA levels. Thus the concentration of an RNA species is the steady-state concentration derived from the kinetics of these two processes. Sequencing the RNA pool in a cell and quantifying each species

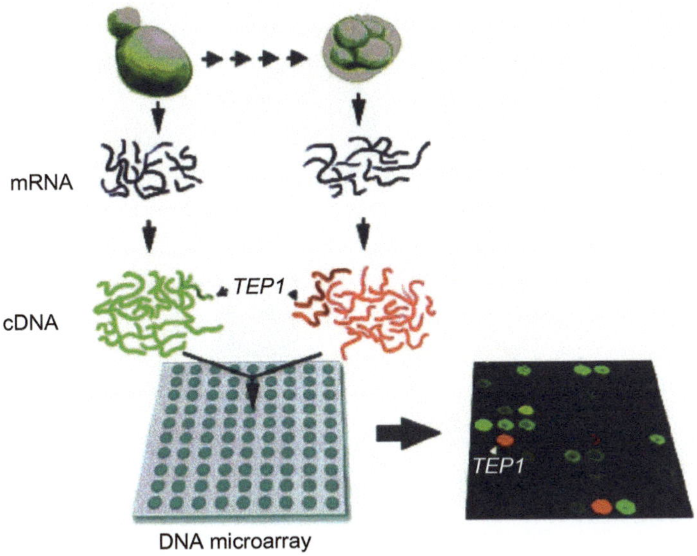

FIG. II.35 Methodology of DNA microarray analysis. In this illustration, mRNA samples from vegetative and sporulating yeast cells are compared. The total pool of messenger RNA from each cell population is used to prepare fluorescently labeled complementary DNA by reverse transcription in the presence of fluorescently labeled nucleotide precursors. To allow direct comparison of the abundance of each gene in the two samples, the two samples are labeled with different fluors—in this example, a *red* fluor for the mRNA from sporulating yeast and a *green* fluor for the mRNA from the vegetative yeast cells. The two fluorescently labeled complementary DNAs are then mixed and hybridized with a DNA microarray in which each yeast gene is represented as a distinct spot of DNA. Irrespective of their fluorescent labels, the complementary DNA sequences representing each individual transcript hybridize specifically with the corresponding gene sequence in the array. Thus the relative abundance in sporulating when compared with vegetative yeast cells of the transcripts from each gene is reflected by the ratio of "*red*" to "*green*" fluorescence measured at the array element representing that gene. For example, the greater relative abundance of the TEP1 mRNA in the sporulating cells results in a high ratio of *red*-labeled to *green*-labeled copies of the corresponding complementary DNA, and an equivalent ratio of red to green signal hybridized at the array element composed of DNA from TEP1. *Brown, P., Botstein, D., 1999. Exploring the new world of the genome with DNA microarrays. Nat. Genet. 21, 33–37. https://doi.org/10.1038/4462.*

directly would give us some idea about the functional state of the genome. With the development of next-gen DNA sequencing methods, it is now possible to sequence the RNA pool and quantify each species. In the current literature, this methodology is called the RNA-seq.

High-throughput parallel DNA sequencing

Many techniques that we discuss in this chapter stand on massively parallel DNA sequencing methods. Maxam, Gilbert, and Sanger initially developed DNA-sequencing methodologies. Both are sequential methods, requiring pure fragments of DNA. Although these sequencing methods revolutionized molecular biology, these techniques are not suitable for large-scale sequencing needed for determining whole-genome sequences. The invention of massively parallel sequencing methodologies was the game-changer. Many methods exist for the sequencing of vast stretches of DNA, including whole genomes. In early parallel sequencing methods, long

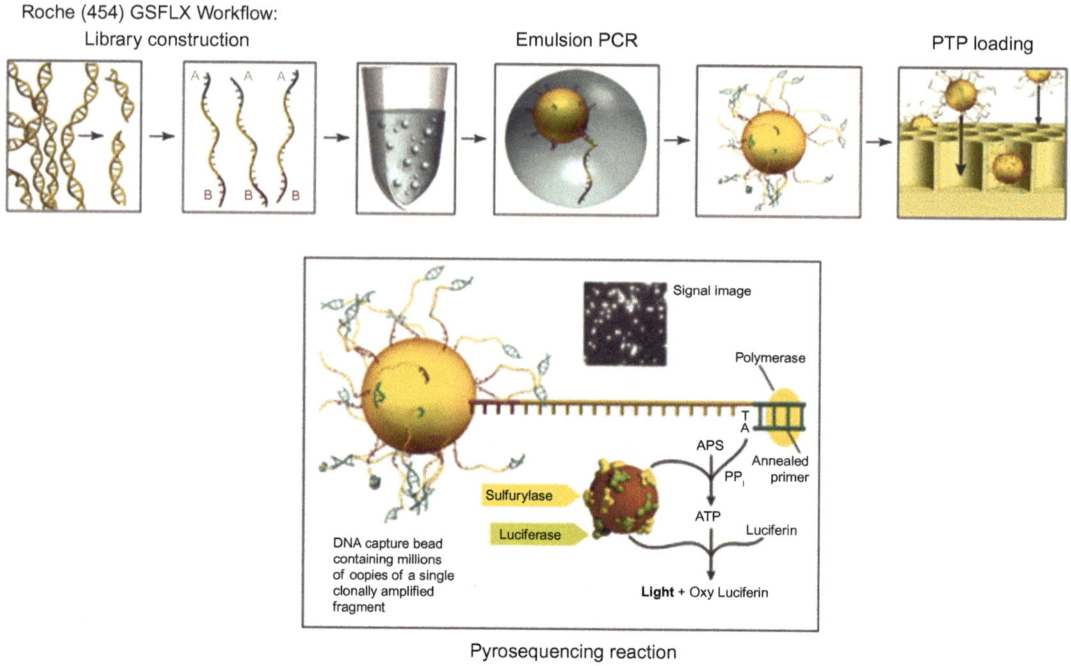

FIG. II.36 Workflow of a next-gen method. Library construction ligates 454 specific adapters to DNA fragments and couples amplification beads with DNA in an emulsion polymerase chain reaction to amplify fragments before sequencing. The beads are loaded into the picotiter plate. The *bottom panel* illustrates the pyrosequencing reaction that occurs on nucleotide incorporation to report sequencing by synthesis. *Reproduced from Mardis, E.R., 2008. Trends Genet. 24, 133–141. Elsevier.*

stretches of DNA are fragmented, by mechanical force or enzyme digestion, into much smaller pieces (Fig. II.36). These fragments are then attached to a solid surface through an anchor and sequenced in parallel. Many types of sequencing reactions exist; a popular early one detected fluorescence emission upon the addition of a nucleotide to a growing chain. In one of the first commercial sequencers, the four nucleoside triphosphates flowed in and out sequentially. Incorporation of the correct nucleotide-released pyrophosphate is then coupled to a reaction that produced luminescence. A camera, in a position-dependent manner, detected the coordinates of the luminescence signal. The coordinates, in turn, identified which chain is incorporating that nucleotide. Newer methodologies have now increased the read length (length of a nucleotide sequence that is determined in one go) and improved read errors.

An RNA-seq experiment can be conducted in several ways depending upon the goal. We will focus on the sequencing of poly-adenylated RNAs, the most common application of the RNA-seq technique. First, the total cellular RNA pool is isolated, followed by separation of poly-A containing RNAs using oligo-dT beads or columns. In one method, the RNA pool is broken down to smaller fragments, copied to cDNA by reverse transcription, and ligated with adapters to generate an RNA-seq library. Deep sequencing of the library yields a large number of sequence reads. The mapping of sequence reads onto the known genome sequence identifies the parent genes. The number of sequence reads that map onto a gene is proportional to the corresponding mRNA concentration (Fig. II.37). Thus RNA-seq can provide a quantitative profile of the mRNAs present at that instant.

II.2.4 Reduction or abrogation of gene expression

Reducing or abolishing the function of a protein is an important tool for understanding the function of a protein. This reduction and abrogation can be done at the level of protein, mRNA, or DNA. Gene knockdown by RNA interference is a very popular method for reducing the level of a protein in a cell. Gene knockout, that is complete abolition of a functional protein, is usually carried out by editing the gene at the level of the DNA. With the recent introduction of proteolysis-targeting chimera (PROTAC), one can now reduce the protein function at the protein level as well. We will discuss all three techniques in this section.

II.2.4.1 Methods for gene knockdown

Knocking down a gene ultimately involves lowering the concentration of its protein product. Lowering the concentration of a protein has a similar biological effect to that of inhibiting its activity by an inhibitor. Thus gene knockdown techniques have broad applications in drug target validation. In eukaryotic cells, a single gene may produce multiple functional entities through RNA splicing and posttranslational chemical modifications. Inhibiting protein modifications or altering splicing patterns may be effective ways to selectively knock down one or more functional entities produced from a gene. However, some of these techniques are still evolving, and we will focus on two other techniques of reducing the overall concentration of the protein product, RNA interference and targeted protein degradation.

Fig. II.38 depicts the cellular RNA interference pathway that naturally exists in many cells. Double-stranded RNA or hairpin RNA is processed into short interfering RNA (siRNA), approximately 21 bp, by a cellular enzyme called Dicer. One of these strands becomes part of the RNA-induced silencing complex (RISC). The siRNA-loaded RISC complex binds to the targeted mRNA and induces its degradation, thus, lowering the rate of synthesis of the corresponding protein. A lower rate of protein synthesis without a corresponding reduction in the degradation rate will result in lower protein levels. For knockdown, a short double-stranded RNA that matches the sequence of a particular mRNA is transfected into the cell. This RNA enters the extant cellular RNA interference pathway and downregulates the level of the targeted protein.

In the targeted protein degradation approach, a PROTAC is first synthesized. PROTAC consists of a ligand that binds specifically to a protein and a ligand that recruits an E3 ubiquitin ligase, connected by a linker. The E3 ubiquitin ligase recruiters, in many cases, are peptides. The ligand binds the target protein, and the recruiter brings in the E3 ligase, forming a ternary complex. This ternary complex attracts the ubiquitination machinery, resulting in protein degradation (Fig. II.39). Enhancement of degradation rate, without corresponding enhancement of synthesis rate, results in lower protein levels in the cell.

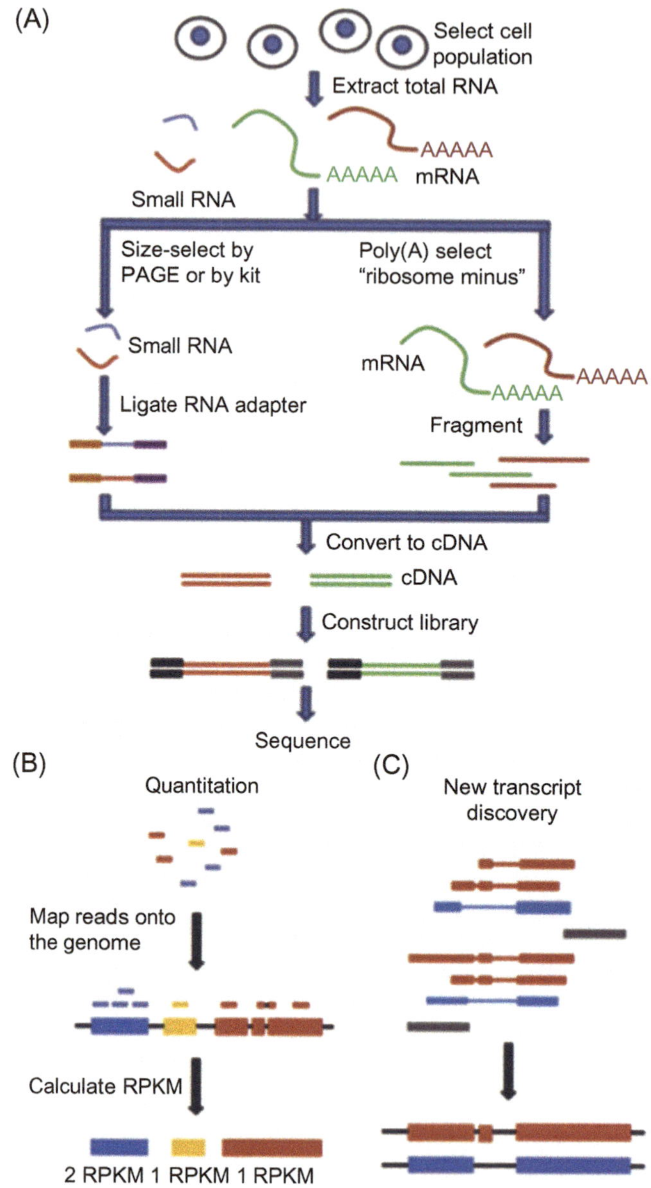

FIG. II.37 *RNA sequencing* work flow. (A) Schematic diagram of *RNA sequencing* library construction. The total RNA is extracted from cells, and a small aliquot is used to measure the integrity of the RNA. rRNA is then depleted through one of the several methods to enrich subpopulations of RNA molecules, such as mRNA or small RNA. mRNA is fragmented into a uniform size distribution, and the fragment size can be monitored by RNA gel electrophoresis or a microfluidics-based bioanalyzer such as the Agilent 2100 Bioanalyzer (Agilent Technologies, Santa Clara, CA). The complementary DNA is then made into a library. (B) Mapping programs align reads to the reference genome and map splice junctions. Gene expression can be quantified as absolute read counts or normalized values such as RPKM (reads per kilobase of transcript per million mapped reads). (C) If *RNA sequencing* data sets are deep enough and the reads are long enough to map enough splice junctions, the mapped reads can be assembled into transcripts. *Reproduced from Whitley et al., 2016. J. Investig. Dermatol. 136 (8), e77–e82. Elsevier.*

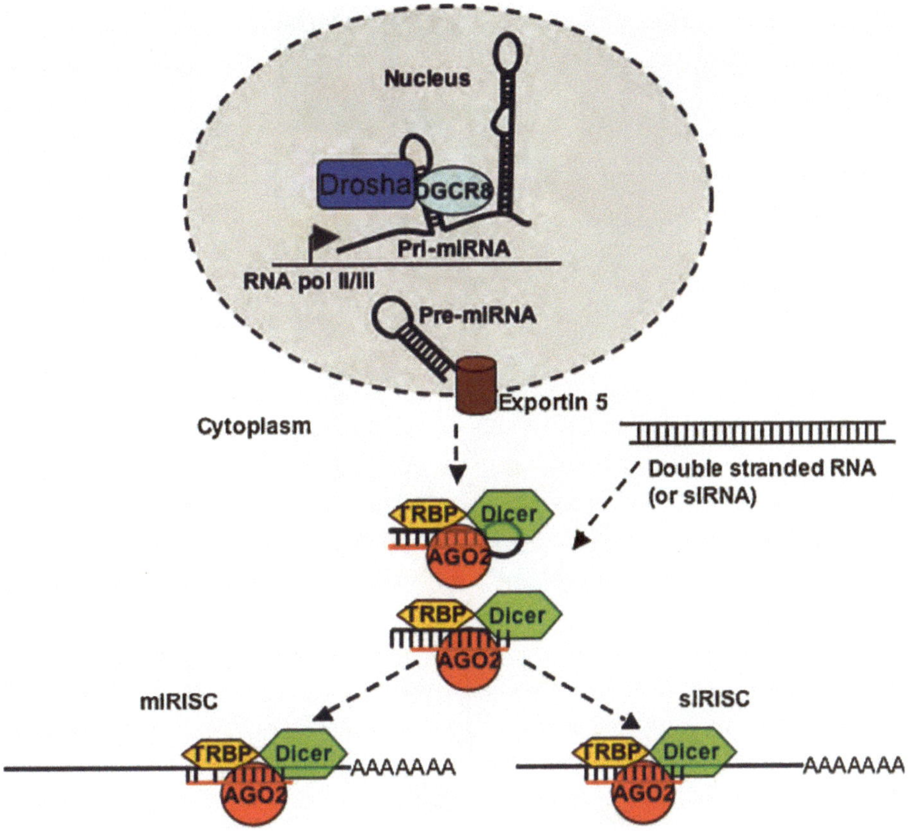

FIG. II.38 The RNAi or miRNA pathway. *Reproduced from Shan, G., 2010. RNA interference as a gene knockdown technique. Int. J. Biochem. Cell Biol. 42, 1243–1251. Elsevier publications.*

II.2.4.2 Methods for gene knockouts and editing

The general method of knocking out or performing other editing functions on a gene in eukaryotes is to create a double-strand break and follow it up with nonhomologous end (NHEJ) repair or homology-directed repair (HDR), respectively. The most critical step in this protocol is to create the double-strand break specifically at the desired site and not on any other sequence in the whole genome. Attaining this level of specificity is a big challenge. Techniques of creating a specific double-strand break fall under three major categories, zinc-finger nucleases (ZFNs), transcription activator-like effector nucleases (TALENs), and CRISPR/Cas9. We will describe these three techniques below.

DNA double-strand break repair

Many processes in the cell generate double-strand breaks. A double-strand break may lead to many untoward consequences and must be repaired. HDR and NHE joining are the two major repair pathways (Fig. II.40). Homologous recombination involves invasion of the broken DNA

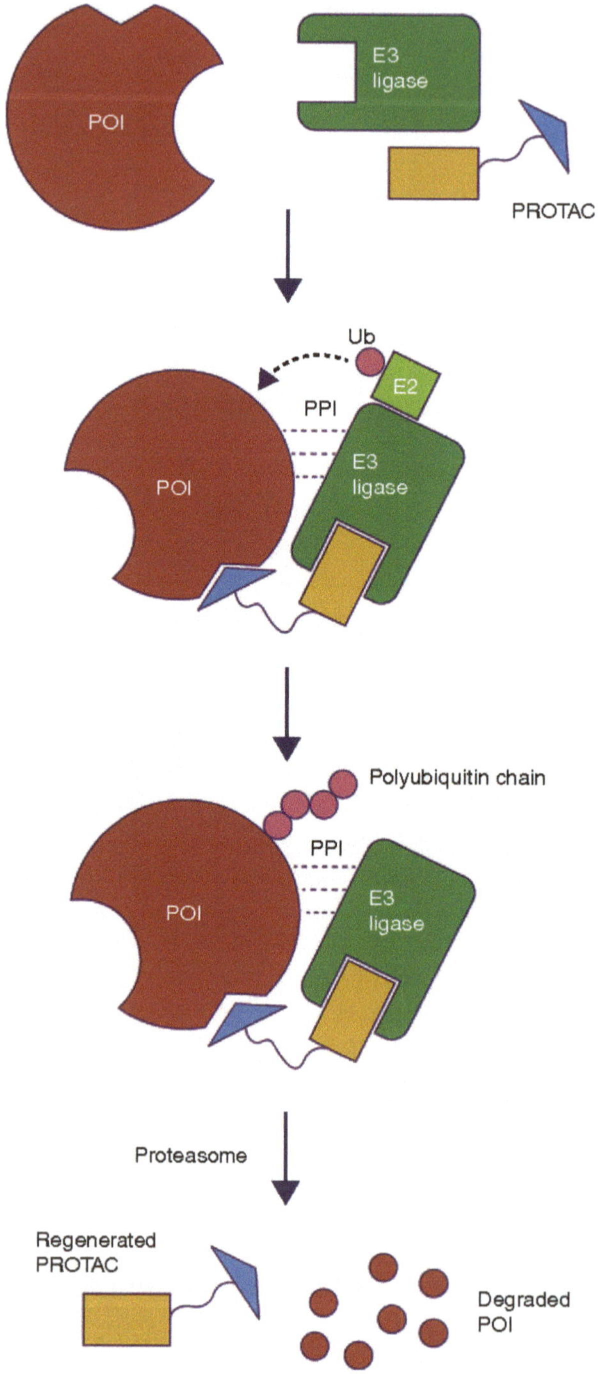

FIG. II.39 Proteolysis-targeting chimera mechanism of action. Proteolysis-targeting chimeras bind a protein of interest (POI) (not necessarily at an active site) and enforce proximity with an E3 ligase. The ternary complex, thus, formed is favored by PPIs (protein-protein interaction) between the POI and the E3. Ubiquitin transfer follows, enabling the formation of the polyubiquitin signal on the POI. Degradation occurs through the proteasome, and the unharmed proteolysis-targeting chimera is regenerated to carry out successive rounds of ubiquitination. *Reproduced from Raina, K., Crews, C.M., 2017. Targeted protein knockdown using small molecule degraders. Curr. Opin. Chem. Biol. 39, 46–53. Elsevier.*

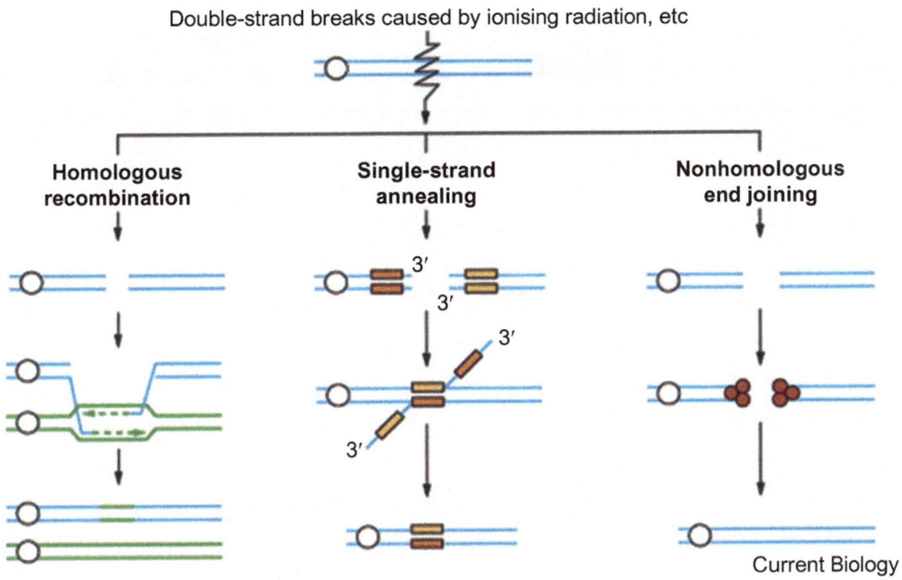

FIG. II.40 Alternative pathways of DNA double-stranded break repair. *Reproduced from Featherstone et al., 1999. Curr. Biol. 9 (20), R759–R761. Cell Press, Elsevier.*

strands into a homologous DNA duplex molecule. Binding of the MRN complex at the DNA ends initiates this process, followed by the recruitment of ATM, and ultimately locating a homologous sequence with which strand invasion and repair can take place. In the NHE-joining pathway, Ku70-Ku80 heterodimer first recognizes the double-strand break, followed by the recruitment of other protein complexes. These proteins or complexes include polymerase, nucleases, and ligases. Resection and polymerization are followed by ligation, usually leading to deletions.

We will discuss the zinc finger-containing DNA-binding proteins in detail in Chapter IV. However, it is pertinent to summarize that the zinc finger is a zinc-stabilized small domain that presents one solvent-exposed α-helix for interaction with the DNA. The α-helix is inserted into the major groove, and amino acids present on one face of the helix make hydrogen bonds and other interactions with base atoms present in the major groove. In a zinc finger containing transcription factors, multiple such fingers are present to augment the interacting interface and enhance specificity and affinity. The zinc-finger domain is a self-contained unit and, in combination with other zinc-fingers, is used to develop artificial transcription factors. When two such artificial transcription factors, specific for two nine-base pair sequences separated by five to seven base pairs, are linked to a subunit of a restriction endonuclease, such as *Fok*I, it creates an artificial ZFN that cleaves the intervening five to seven base pairs to create a double-stranded break (Fig. II.41). TALENs are similar to ZFNs, where the zinc fingers are replaced by a transcription activator-like effector (TALE) domains. TALE domains were originally discovered in Xanthomonas bacteria that infect plants. The domains contain

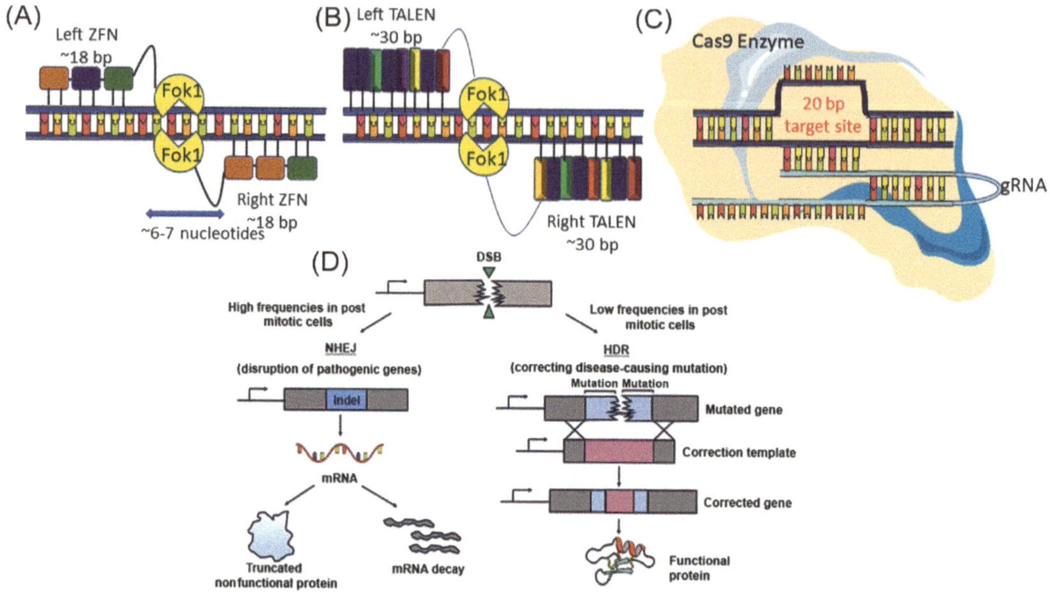

FIG. II.41 Genome editing strategies. Genome editing technologies can be employed using nonenzymatic and enzymatic methods. (A) Zinc-finger modules are linked together in tandem using conserved linker sequences to recognize target sequences that are composed of DNA triplets. Fok1 nuclease is fused to the zinc-finger array. (B) Transcription activator-like effector nucleases are similar to zinc-finger nucleases but confer more flexibility by having each transcription activator-like effector module recognizing only one nucleotide base. Transcription activator-like effector modules can be joined together to form tandem arrays that bind the target site. (C) CRISPR or Cas9 system utilizes a custom-made guide RNA to transport Cas9 nuclease to the target gene. (D) Zinc-finger nucleases, transcription activator-like effector nucleases, and CRISPR or Cas9 methods result in a nuclease-induced DNA double-stranded break at the target site that stimulates the DNA repair mechanism: nonhomologous end joining or homology-directed repair. nonhomologous end joining induces error-prone repair by bluntly joining the cleaved DNA strands together. Hence, it is useful in disrupting pathogenic mutations through indel disruption of the target gene. Homology-directed repair confers high-fidelity correction of the pathogenic mutation using a repair DNA template to drive homologous recombination. It is important to note that nonhomologous end joining is active in both proliferative and nonproliferative cells, but homology-directed repair is restricted to proliferative cells only. (A) A zinc-finger nuclease triplet module is depicted using colored blocks. The varying colors differentiate zinc-finger nuclease triplets of different base combinations. (B) Different DNA bases are depicted as *yellow* blocks. *Reproduced and redrawn from Ho, B.X., Loh, S.J.H., Chan, W.K., Soh, B.S., 2018. In vivo genome editing as a therapeutic approach. Int. J. Mol. Sci. 19, 2721. This is an open access article distributed under the Creative Commons Attribution License.*

approximately 34 amino acids and present 2 hypervariable amino acids at positions 12 and 13 that interact with 1 base pair of the double-stranded B-DNA. There is a simple relationship between the identity of amino acids 12 and 13 and the recognized base. This makes it possible to combine TALE domains to create a highly specific artificial DNA-binding protein.

In the CRISPR or Cas system, the DNA-binding specificity originates from a piece of RNA. In the engineered version of the widely used CRISPR or Cas9, a single-guide RNA (sgRNA) is used, a fusion of the two required RNAs in the wild-type system. Once the sgRNA is hybridized to its target DNA, the Cas9 nuclease can home in, bind, and produce a double-strand break. In the absence of an external DNA having homology to the target region, the NHEJ process will repair the double-strand break to create deletion. If a piece of DNA having homology to both the left and right regions of the double-strand break is present, the HDR process will take over and insert a piece of DNA.

Chapter summary

Methods and techniques play an essential role in explicating complex processes that occur inside the cell. Methods for studying genomes fall under two categories, destructive and nondestructive. Techniques such as Chromosome Conformation Capture belongs to the former class, whereas some types of fluorescence microscopy belong to the latter. We have discussed the principles of techniques that belong to both categories. A significant part of this chapter also deals with methods that elucidate the genome function. Techniques such as chemical genetics, transcriptomics, gene knockdown, and genome editing fall under this category. Chemical principles are the bedrock on which these techniques stand. Our primary focus was on the physical and chemical principles of these methods, and readers may find a more detailed description elsewhere.

- Fluorescence is a type of emission spectroscopy. Its utility for studying the genome is many folds. Fluorescence emission mostly occurs in the visible region of the electromagnetic spectrum, making it particularly suitable for cellular studies.
- Fluorescence-derived techniques such as FRET, fluorescence polarization, have found significant applications in measuring molecular proximity and genome dynamics.
- Fluorescence-based microscopic techniques have revolutionized microscopy and have become essential tools for observing genomes in intact cells. Super-resolution fluorescence microscopy now has overcome the diffraction limit of light and routinely reaches 10 nm, close to the diameter of individual proteins.
- Even higher resolution map of the genome is obtained from chromatin immunoprecipitation and Chromosome Conformation Capture techniques, such as 3C and Hi-C. These are based on chemical cross-linking of proximal intrachromosomal and interchromosomal segments, followed by sequence analysis of cross-linked segments.
- Three powerful methods of decoding gene functions, chemical genetics, transcriptomics, and genome editing, have been described.
- In the chemical genetics approach, the cell is interrogated with a small molecule targeted to a specific protein, followed by observation and analysis of the phenotype.
- Transcriptomics counts the number of different types of mRNA present in the cell and reflects the activities of a significant section of the genomic repertoire.
- The bedrock of genome editing techniques is the ability to create a specific double-stranded break in the genome. Artificial nucleases create such breaks based on the specificity of zinc-fingers, TALE, and in the case of CRISPR-Cas9, gRNAs for particular DNA sequences.

The Further Reading section describes some advanced materials to interested readers.

Further reading

Beck, S., Rakyan, V.K., 2008. The methylome: approaches for global DNA methylation profiling. Trends Genet. 24, 231–237.

Belton, J.-M., McCord, R.P., Gibcus, J.H., Naumova, N., Zhan, Y., Dekker, J., 2012. Hi–C: a comprehensive technique to capture the conformation of genomes. Methods 58, 268–276.

Berg, E.L., 2017. Phenotypic screening. Drug Discovery 27.

Bustin, S., Benes, V., Nolan, T., Pfaffl, M., 2005. Quantitative real-time RT-PCR – a perspective. J. Mol. Endocrinol. 34, 597–601.

Carey, M.F., Peterson, C.L., Smale, S.T., 2009. Chromatin immunoprecipitation (chip). Cold Spring Harb Protoc. 2009, pdb.prot5279. https://doi.org/10.1101/pdb.prot5279.

Chakraborty, C., Sharma, A.R., Sharma, G., Doss, C.G.P., Lee, S.-S., 2017. Therapeutic miRNA and siRNA: moving from bench to clinic as next generation medicine. Mol. Ther. Nucleic Acids 8, 132–143.

De Los Santos, C., Chang, C.W., Mycek, M.A., Cardullo, R.A., 2015. FRAP, FLIM, and FRET: detection and analysis of cellular dynamics on a molecular scale using fluorescence microscopy. Mol. Reprod. Dev. 82, 587–604.

Gaj, T., Sirk, S.J., Shui, S.-l., Liu, J., 2016. Genome-editing technologies: principles and applications. Cold Spring Harb. Perspect. Biol. 8, a023754.

Heller, M.J., 2002. DNA microarray technology: devices, systems, and applications. Annu. Rev. Biomed. Eng. 4, 129–153.

Huang, B., Bates, M., Zhuang, X., 2009. Super-resolution fluorescence microscopy. Annu. Rev. Biochem. 78, 993–1016.

Jerome, W.G., Price, R.L., 2018. Basic Confocal Microscopy. Springer.

Jones, L.H., Bunnage, M.E., 2017. Applications of chemogenomic library screening in drug discovery. Nat. Rev. Drug Discov. 16, 285.

Jradi, F.M., Lavis, L.D., 2019. Chemistry of photosensitive fluorophores for single-molecule localization microscopy. ACS Chem. Biol. 14, 1077–1090.

Karczewski, K.J., Snyder, M.P., 2018. Integrative omics for health and disease. Nat. Rev. Genet. 19, 299.

Knight, Z.A., Shokat, K.M., 2007. Chemical genetics: where genetics and pharmacology meet. Cell 128, 425–430.

Lakowicz, J.R., 2013. Principles of Fluorescence Spectroscopy. Springer Science & Business Media.

Li, T., Liu, B., Spalding, M.H., Weeks, D.P., Yang, B., 2012. High-efficiency TALEN-based gene editing produces disease-resistant rice. Nat. Biotechnol. 30, 390.

Lichtman, J.W., Conchello, J.-A., 2005. Fluorescence microscopy. Nat. Methods 2, 910–919.

Mifsud, B., Tavares-Cadete, F., Young, A.N., Sugar, R., Schoenfelder, S., Ferreira, L., Wingett, S.W., Andrews, S., Grey, W., Ewels, P.A., 2015. Mapping long-range promoter contacts in human cells with high-resolution capture Hi-C. Nat. Genet. 47, 598.

Orlando, V., 2000. Mapping chromosomal proteins in vivo by formaldehyde-crosslinked-chromatin immunoprecipitation. Trends Biochem. Sci. 25, 99–104.

Paiva, S.-L., Crews, C.M., 2019. Targeted protein degradation: elements of PROTAC design. Curr. Opin. Chem. Biol. 50, 111–119.

Pakhomov, A.A., Martynov, V.I., 2008. GFP family: structural insights into spectral tuning. Chem. Biol. 15, 755–764.

Roh, T.-y., Ngau, W.C., Cui, K., Landsman, D., Zhao, K., 2004. High-resolution genome-wide mapping of histone modifications. Nat. Biotechnol. 22, 1013–1016.

Stockwell, B.R., 2000. Chemical genetics: ligand-based discovery of gene function. Nat. Rev. Genet. 1, 116–125.

Suseela, Y., Narayanaswamy, N., Pratihar, S., Govindaraju, T., 2018. Far-red fluorescent probes for canonical and non-canonical nucleic acid structures: current progress and future implications. Chem. Soc. Rev. 47, 1098–1131.

Uchida, T., Rodriquez, M., Schreiber, S.L., 2009. Skeletally diverse small molecules using a build/couple/pair strategy. Org. Lett. 11, 1559–1562.

Usaj, M.M., Styles, E.B., Verster, A.J., Friesen, H., Boone, C., Andrews, B.J., 2016. High-content screening for quantitative cell biology. Trends Cell Biol. 26, 598–611.

Walsh, D.P., Chang, Y.-T., 2006. Chemical genetics. Chem. Rev. 106, 2476–2530.

Wang, Z., Gerstein, M., Snyder, M., 2009. RNA-Seq: a revolutionary tool for transcriptomics. Nat. Rev. Genet. 10, 57–63.

III

Genes, genomes, and genome dynamics

Genes load the gun; environment pulls the trigger.

Scott Kahan

A gene is defined as a physical and functional unit of DNA that is generally the basis for heredity in most organisms except certain viruses that have RNA as their genetic material. Accordingly, genes are made up of a string of deoxyribonucleotides in all these organisms, whereas ribonucleotides replace deoxyribonucleotides in the RNA viruses. Genes themselves are not functional, but the information coded in the genes are transcribed into RNAs followed by their translation into proteins that are the functional entities of a cell, in most cases, in the actual sense. The genome is technically defined as the sum of all the DNA present in a cell, generally within the nucleus (as the DNA present in other organelles such as mitochondria and chloroplasts also contains genes coding for few specific proteins). All the DNA sequences in the genome, however, do not comprise protein-coding genes. Some code for RNAs that do not translate into proteins, whereas other stretches of DNA are not known to code for anything, so far. It is now believed that such noncoding regions of the DNA in the genome also serve as regulatory elements of gene expression in different ways that would be discussed in the subsequent sections. It took almost 100 years since the discovery of the modern concept of genetic inheritance, to establish DNA as the physical basis for genes and the genetic material. Unlike today, that period was not an "era of information." Nevertheless, several independent studies led to the unraveling of the fundamental mystery of life (Fig. III.1).

III.1 Genes: The fundamental unit of DNA

In 1865, Gregor Johann Mendel proposed the laws of inheritance, based on his brilliant experiments using garden pea plants. He postulated that discrete hereditary factors pass down from parents to offspring according to three basic rules or laws, namely the law of dominance, the law of segregation, and the law of independent assortment.

Chemical Biology of the Genome. https://doi.org/10.1016/B978-0-12-817644-3.00003-9

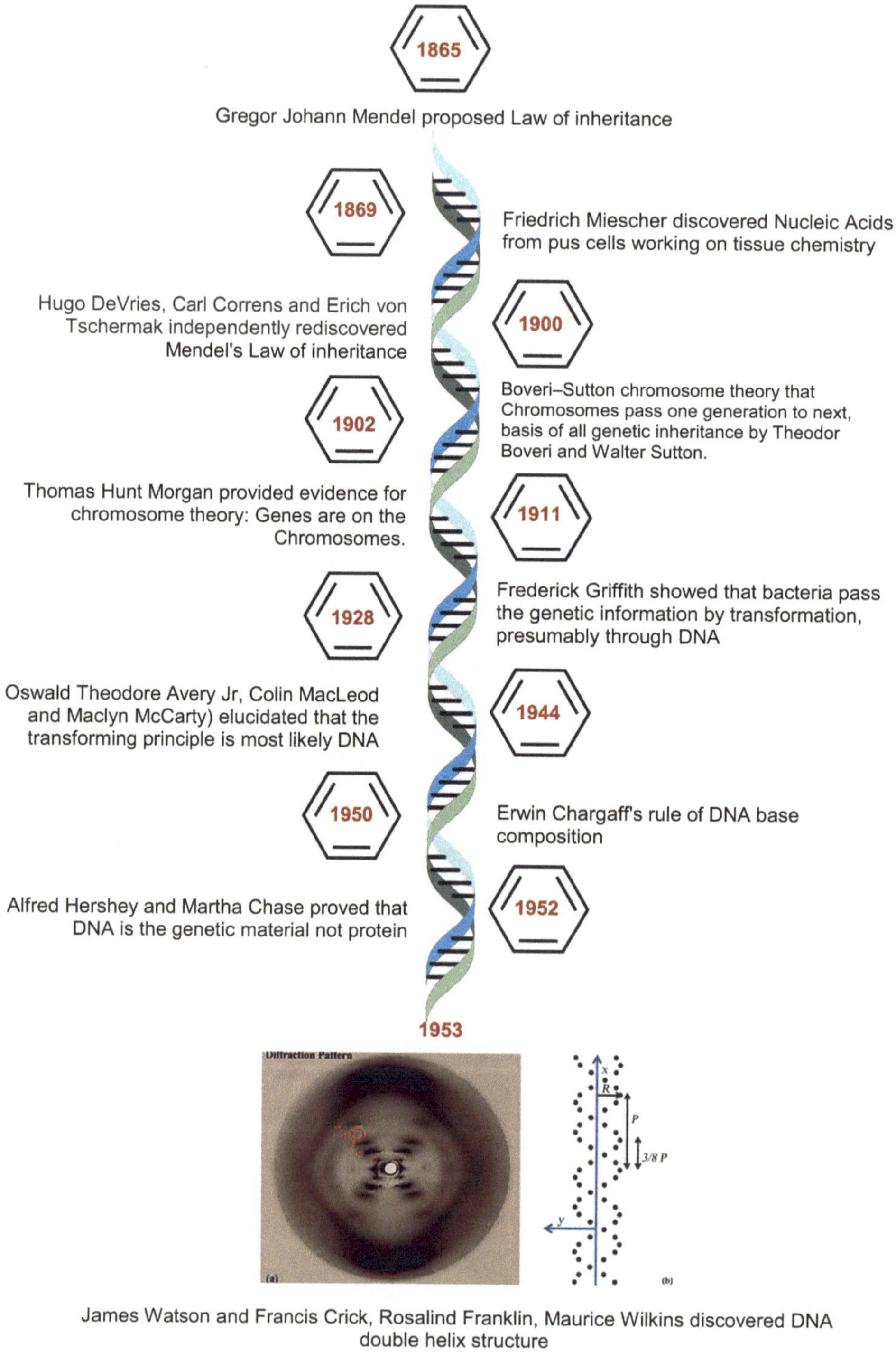

FIG. III.1 Series of great discoveries finally unravel the mystery of the molecular nature of the gene.

Mendel's laws of inheritance

Law of dominance: In a heterozygous state, one of the alleles of a factor (gene), known as the dominant allele, expresses (shows the phenotype), whereas the other allele, known as the recessive allele, remains latent but does not get lost while being transmitted to the next generation. This phenotype for the recessive allele (known as the recessive phenotype) is only expressed under a condition when both the alleles present are the same recessive one.

Law of segregation: During the gamete formation in a diploid organism (with chromosome number 2n), each gamete receives one allele of the gene that passes to the offspring randomly.

Law of independent assortment: An allele of each gene passes to the next generation independently, where one gene does not influence the segregation of the other gene, and every allele of a gene possesses an equal probability to combine with other gene alleles.

This elegant discovery, however, did not get any attention until it was independently rediscovered by three different groups. Even then, the molecular mechanisms of inheritance were largely unknown at that time. In 1902, Walter Sutton and Theodor Boveri proposed the "Chromosome Theory," also known today as the Boveri-Sutton chromosome theory that showed that chromosomes behave like hereditary factors. In 1911, Thomas Hunt Morgan, through a series of experiments using the fruit fly, showed that the physical basis for this heritable factor could be the chromosomes. By that time, it was known that chromosomes are made up of DNA and proteins. So, the natural question was, what carries the heritable characters, protein, or DNA? Frederick Griffith's famous experiments of bacterial transformation using nonvirulent (rough strain) versus virulent (smooth) strains, which employed different chemical techniques for separating different biomolecules from bacteria, suggested that the transforming principle is DNA.

Over the next two decades, Oswald Avery, Colin MacLeod, and Maclyn McCarty, through painstaking experiments, purified Griffith's transforming principle and further confirmed that DNA, and not protein as was popularly believed in that era, was the actual hereditary material. However, final conclusive evidence in this regard was only provided when Alfred Hershey and Martha Chase showed, through differential radiolabeling of protein and DNA in their genetic transformation experiments using bacteriophage T2, that it is the DNA, and not protein, that enters the bacterial cells and produces the active virus.

DNA as a biomolecule was isolated by the Swiss physician and biologist Friedrich Miescher in 1869. His primary interest of study was the chemistry of the cell. In a pus cell nuclei, he found a new type of material, which he named, nuclein that later came to be known as the DNA. By the end of the 19th century, Albert Karl, Ludwing Martin, and Leonard Kossel elucidated the basic chemical components of nuclein that are pentose sugar deoxyribose, phosphates, and four nitrogenous bases, adenine (A), cytosine (C), guanine (G), and thymine (T). Around 1935, Theodore Levene, based on DNA hydrolysis analysis, put forward the basic structure of nucleotides (unit of the DNA macromolecule) and their possible linkage, the phosphodiester bond. However, until 1953, the three-dimensional structure of DNA remained a big hotspot of debate and speculation. Based on all the above-mentioned elegant discoveries, work of Erwin Chargaff and colleagues at Columbia University (known as Chargaff rules) and X-ray crystallographic data of DNA by Rosalind Franklin at Kings College in London, James Watson and Francis Crick uncovered the three-dimensional structure of DNA, the fundamental molecule of heredity.

Franklin's X-ray studies showed that DNA is helical with two strands made of the nitrogenous bases facing inwards. She also predicted that the two strands ran in opposite directions. In 1957, Franklin fell ill and died of ovarian cancer in 1958 at the age of 37. As the rules of the Nobel Committee prohibit posthumous awards, she was not awarded the Nobel Prize for the discovery of DNA structure.

III.2 Genomes: Complexity and diversity

As mentioned previously, all the DNA sequences in the living cells, especially in eukaryotes, do not code for proteins through mRNA. Some of the DNA sequences may code for RNA that are not translated into protein (hence known as noncoding RNAs), whereas some genomic DNA may not code for anything at all. The positions of the coding and non-coding DNA within the genome are not random but were selected through evolution.

In three-dimensional space, the huge length of the DNA is systematically packaged around specific protein molecules with the help of multiple cellular factors, to contain them within the limited confines of the nucleus (Fig. III.2). This packaging is not

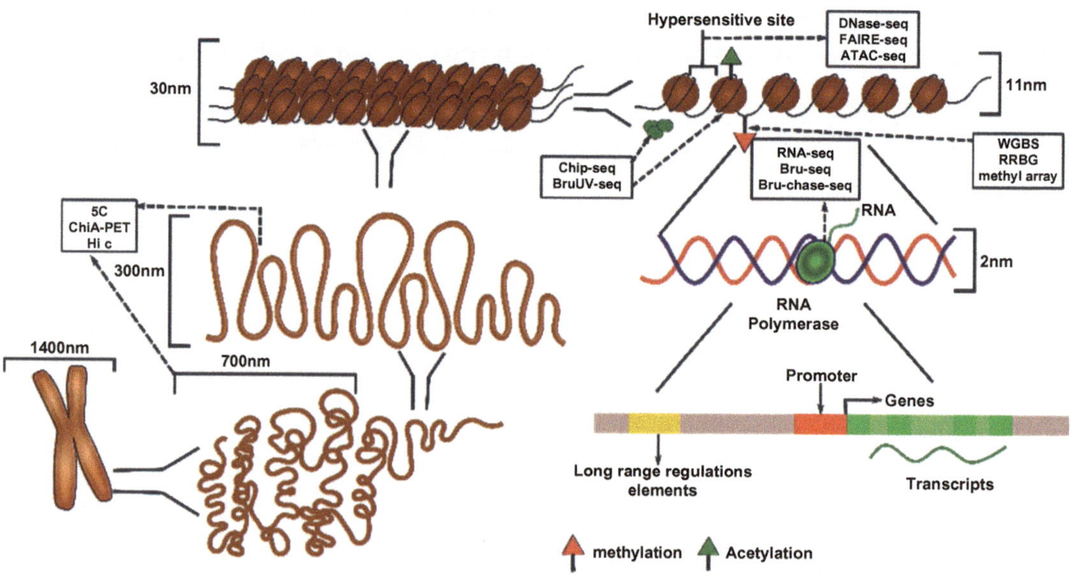

FIG. III.2 The functional gene with the regulatory elements: Packing of DNA into nucleosomes leads to beads on a string structure that is then folded into the chromosome. Different research tools and methodologies are used to understand the gene structure and function as indicated in the boxes. The *dotted arrows* indicate the events.

uniform throughout as some regions are tightly packed, whereas others are loosely packed to different extents. It has been observed that many of the coding genes are present in the relatively loosely packed regions of the genome, whereas much of the noncoding stretches of the DNA tend to be present in the tightly packed regions. The study of the genome organization that is regulated by multiple intrinsic and extrinsic factors is a vast field of research at present that will be discussed further in the subsequent sections.

Chargaff's rules of base composition

In DNA, chemically

1. The nitrogenous bases, A, T, C, and G, are not in equal amount.
2. The amount of bases is the same in a given species but different across the species.
3. The quantity of A is invariably equal to T, so is G to C.

III.2.1 C-value paradox

The size of the genome of each organism (species) is distinct. It is quantified in terms of "C-value," that is, the amount of DNA present in the haploid genome of an organism. Contrary to a logical notion, it has been found that a more complex organism need not always have a larger genome. The C-value of the human genome is about 3.03 gigabases, whereas the C-values of some lungfish genomes are greater than 25 gigabases. This apparent discrepancy in the genome size and complexity of an organism is termed the C-value paradox (Fig. III.3). Humans are physiologically, anatomically, and intelligence-wise more complex than any organism on earth. However, the less complex fish has about 10 times more DNA than humans! It is difficult to provide general reasoning for this observation at this stage. However, the discovery of certain elements known as the transposable elements (TEs) seems to explain this natural fact to some extent.

Despite the differences in genome size and complexity, genomic DNAs from different organisms perform the same conserved function that is to store and regulate the flow of information. Yet, much is unknown about the precise functions of the noncoding regions of the genomic DNA mentioned previously. This dearth of understanding of the functions of such DNA sequences had led to it being referred to as "junk DNA." However, data from the Encyclopedia of DNA Elements (known as the ENCODE project) now suggest that there is little or no "junk DNA" present in the literal sense. The so-called junk DNA may be involved in gene regulatory and finer epigenomic dynamics, the specifics of which constitute a huge field of active research presently. These DNA sequences may also serve as a mutation shock absorber. Nevertheless, these speculations do not provide enough arguments to support the smaller genome size of humans than some fishes. Presumably, there has been a better genome optimization in humans, with more fine-tuned and specific regulation of gene expression (Fig. III.4).

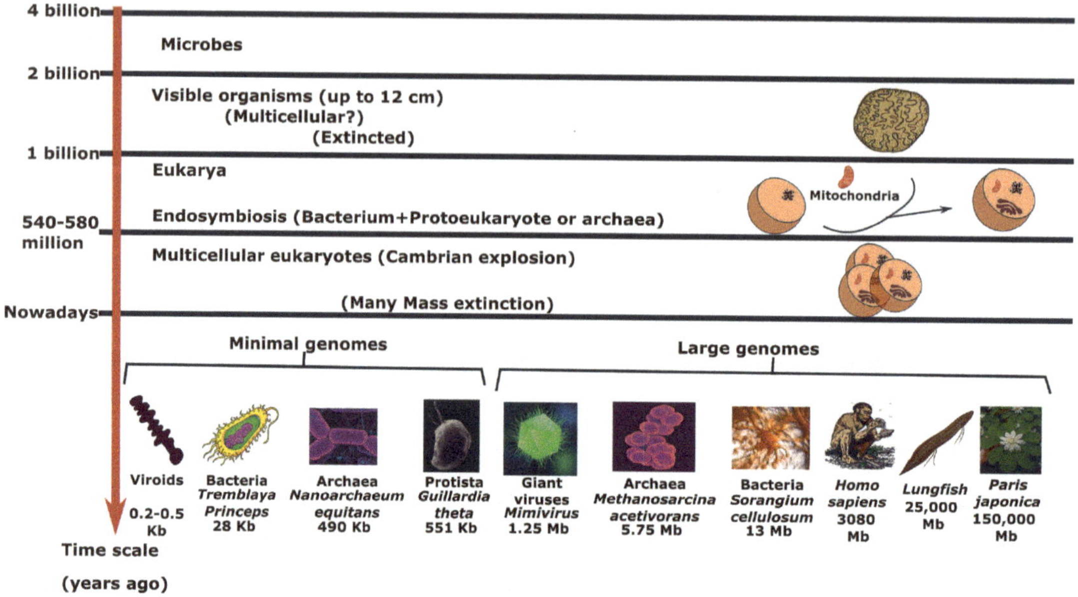

FIG. III.3 The genome size need not be correlated to the complexity of the living organism. The highly evolved human genome size is smaller than that of the lungfish and some of the flowering plants. *Modified from Merhej, V., et al., 2012. Front Cell Infect. Microbiol. 2, 113. Frontiers Media SA.*

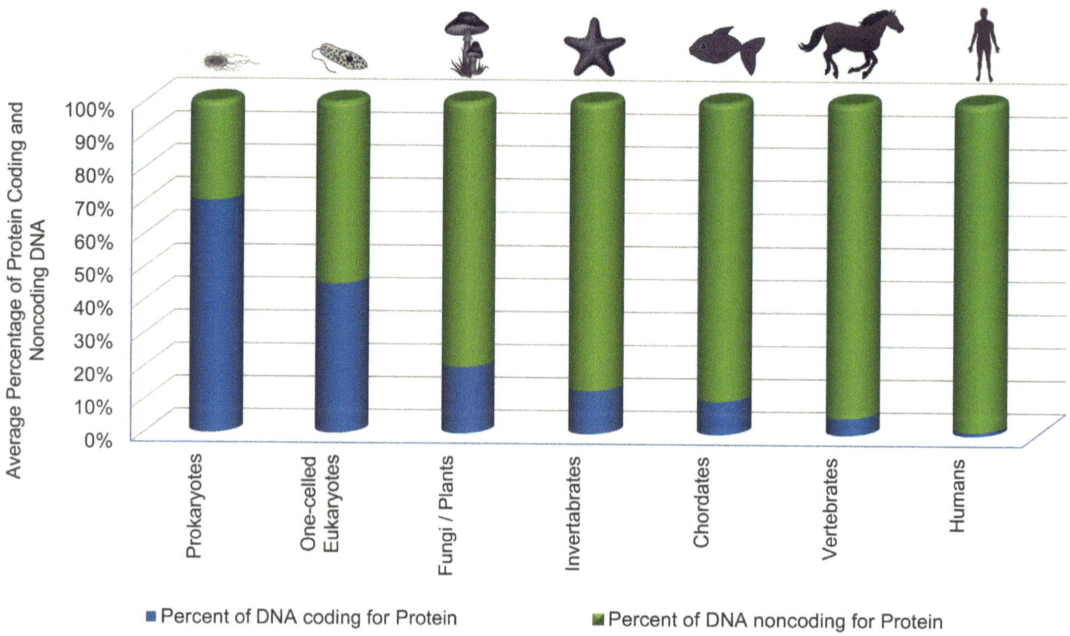

FIG. III.4 During the progression of evolution, the higher organisms accumulated more noncoding DNA in their genomes.

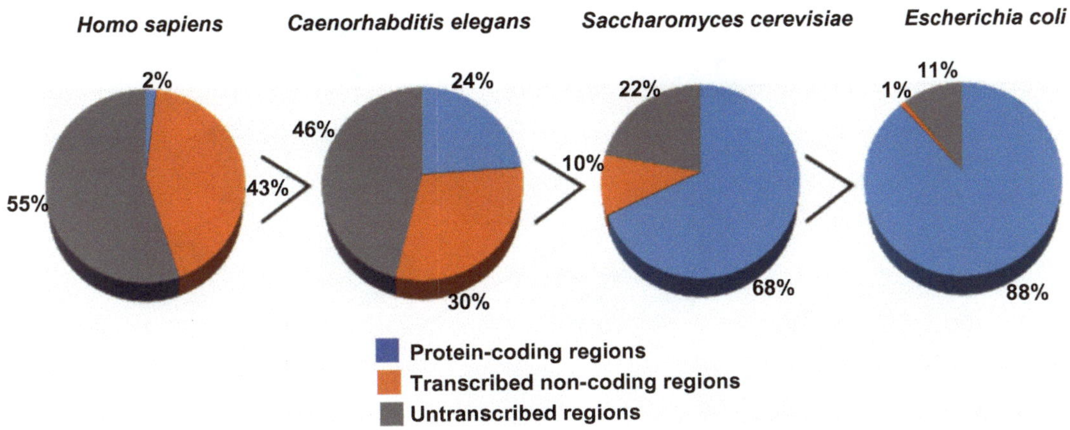

FIG. III.5 Humans devote the lowest percentage of their genomes for protein-coding DNA.

III.2.2 G-value paradox

In the field of genetics, G-value refers to the number of genes in the genome of an organism or species. Besides the C-value paradox, genome science faced another interesting fact upon the genome sequencing of different model organisms. It was found that humans possess about 19,000 protein-coding genes, whereas these numbers for the nematode *C. elegans*, fruit fly *D. melanogaster*, and zebrafish *D. rerio* are 19,700, 14,000, and 26,000, respectively (Fig. III.5). This lack of correlation between the number of protein-coding genes among the eukaryotes and their relative biological complexity is referred to as the G-value paradox. The phenomenon of the G-value paradox may be explained by the fact that in higher eukaryotes, gene products are functionally better optimized due to (1) alternative splicing of the gene transcripts, (2) the presence of multifunctional proteins, (3) regulation of protein function through their posttranslational modifications, (4) diverse interactions between proteins and of course, (5) complex and more fine-tuned transcription regulation. As a result of the evolution of such regulatory mechanisms, the more complex living organisms are capable of producing multiple outcomes from the same genetic material.

III.3 Repetitive DNA and "jumping genes"

III.3.1 Repetitive DNA: A major component of the eukaryotic genome

Genome sequence data from the ENCODE database revealed that a significant proportion of the genomic DNA in higher organisms is present in repetitive forms, an observation that offers a possible explanation to the C-value paradox. These repetitive DNA elements are present in multiple copies in a given eukaryotic genome. The biological significance of these DNA elements is being revealed now. There are two types of repetitive DNA, namely

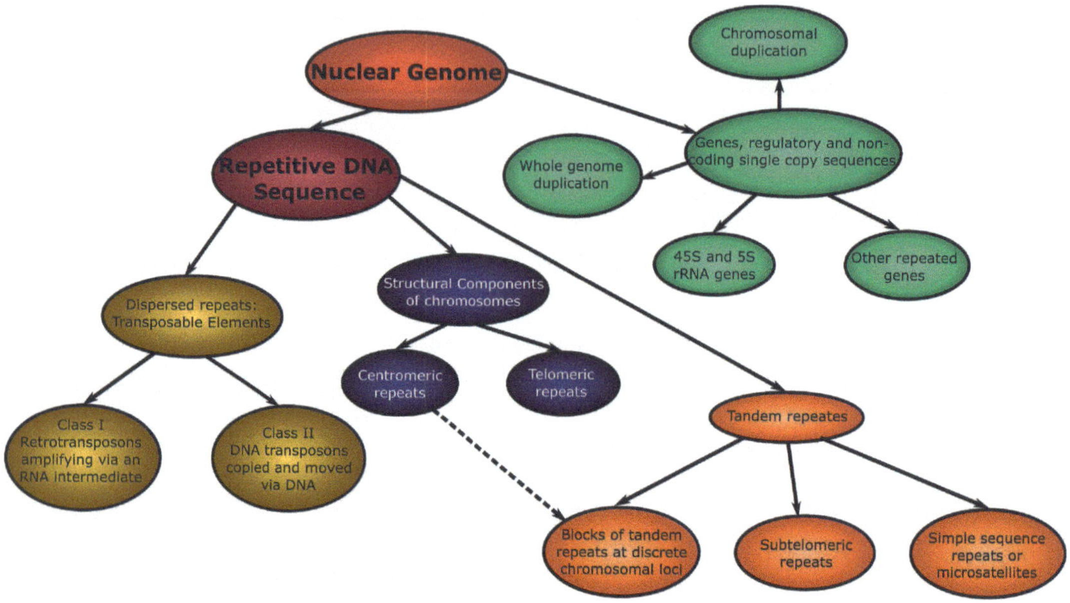

FIG. III.6 DNA elements of the higher eukaryotic genomes. *Modified from Biscotti, et al., 2015. Chromosome Res. 23 (3), 415–420. Springer Nature.*

tandem repeats and dispersed repeats (Fig. III.6). The tandem repeats are the major constituents of heterochromatin and centromeric DNA and are organized in a head-to-tail array fashion. The dispersed repeat DNA constitutes about 45% of the human genome, and several of that belongs to the category of "jumping genes" or the TEs.

III.3.2 Jumping genes (transposons)

One of the most profound discoveries of the 20th century in the field of genetics is the discovery of the "jumping genes" by Barbara McClintock. McClintock, through a series of genetic and cytological experiments using the maize plant (*Zea mays*), concluded that the color variegation in the maize kernel is due to the "jumping"of genes from one site in the chromosome to another. The idea that genes are not stationary and can change their chromosomal location was a revolutionary concept at that time. Later on, it was discovered that the "jumping genes," now known as TEs, are present in all living organisms, both prokaryotic and eukaryotic. After more than four decades of this fundamental discovery, McClintock was awarded the Nobel prize in Physiology or Medicine in 1983.

III.3.3 Transposons in evolution and disease

What could be the connection between TEs and disease? Over the last few decades, several pieces of evidence have established a fundamental role of TEs in not only health and disease but also evolution. There are two types of TEs, classified based on the mechanisms of transposition: DNA transposons and retrotransposons. The DNA transposons, which are rarely found in higher eukaryotes, move through a cut-and-paste mechanism. The retrotransposons, on the other hand, translocate through RNA as an intermediate in the process. There are different kinds of retrotransposons, such as long interspersed nuclear elements (LINEs), short interspersed nuclear elements, Alu (*Arthrobacter luteus*), and SVA (SINE-VNTR-Alus) elements. Among these, LINEs are the only active transposons, which are also autonomous, as they code for the proteins required for their own transposition. The mechanisms and causes of transposition are not fully understood.

The LINEs belong to the non-LTR class of retrotransposons. The transposition or mobilization occurs through an RNA intermediate. All the functional retrotransposons encode two genes, namely Gag and Pol. The Pol gene is polycistronic as it produces three proteins that are a protease, a reverse transcriptase with RNase H activity, and an integrase. The mRNA from the retrotransposon is synthesized by the RNA polymerase II enzyme of the host. The mRNA is then transported to the cytoplasm, where it is translated into Gag and Pol proteins using the host machinery.

The Gag polyproteins form virus-like particles in the cytoplasm, where the reverse transcription of the mRNA, generating the corresponding cDNA, occurs. The cDNA is then transported back into the nucleus, where it integrates at a new location in the genome (Fig. III.7).

The retrotransposons occur in both germ cells and somatic cells in eukaryotes. Under normal conditions, germline retrotransposition is largely silenced in all eukaryotic organisms, whose regulation is not yet fully understood. Occasionally, in normal germ cells or germ cells with deficient silencing mechanisms, transposons in the developing sperm or egg cells' genome may jump from one position to another. Depending on the new locations of the TEs posttransposition, unusual gene products could arise having harmful effects on the cells and eventually on the organism in the form of diseases. Examples of such pathological conditions include hemophilia, cancer, and neurological disorders. Recently it has been observed that the retrotransposons use nurse cells (that nurture developing eggs) to produce the mature TEs for insertion into the egg genomic DNA.

Mammalian cells produce a noncoding RNA called piRNA that suppresses transposition. However, it is possible to overcome the piRNA-mediated control of transposition. Using disruption of piRNA-mediated suppression in the fruit fly (*Drosophila melanogaster*), a method has been developed to track the movements of the "jumping genes." As a result of the disruption of piRNA-mediated suppression, the frequency of retrotransposition was increased, which was then monitored during the egg development process. Several lines of evidence suggest that during human evolution, retrotransposition has played a critical role. Apparently, during chimpanzee, bonobo, and human speciation, the rate of LINE-1 transposition was the lowest in the human lineage. Nonetheless, the variability in the human genome is largely caused by germline retrotransposition whose functional consequences are largely unknown yet.

Although the phenomenon of transposition of genes in somatic cells is quite common in all eukaryotes, it was not established in humans until recently. Examples of such TEs

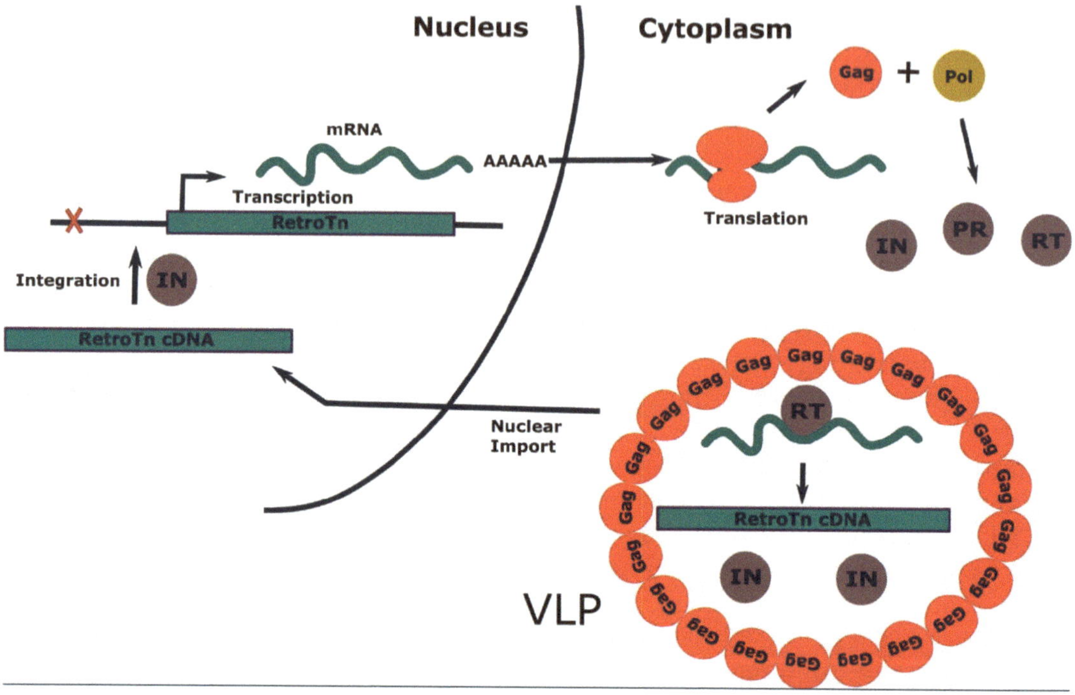

FIG. III.7 Mechanism of retrotransposition. The transcript of a retrotransposon is transported to the cytoplasm from the nucleus. In the cytoplasm, it is translated to Gag and Pol polyproteins. The protease activity of Pol cleaves the polyprotein into integrase and reverse transcriptase. The reverse transcriptase, retrotransposon, and integrase are then assembled into a virus-like particle, where reverse transcriptase synthesizes the cDNA of the retrotransposon, and the complex is transported back into the nucleus for integration at a new location. *Modified from McCullers, T.J., et al., 2017. Mob. Genet. Elements 7 (3), 1–18. Taylor & Francis.*

include Tc1, a DNA transposon in the nematode *Caenorhabditis elegans* and R2, a nonLTR retrotransposon in *Drosophila*. R2 has many common features of mammalian L1 elements. In recent years, our knowledge about somatic transposition in the mammalian system has increased substantially due to several seminal discoveries. Several important methods have been employed to study retrotransposition, such as retrotransposition reporter assays and next-generation genome sequencing approach. Retrotranspositions have been quantified by these methods both in *Drosophila* and mammals. Interestingly, it was found that somatic retrotransposition is higher in neurons than in other cell types. In an individual, somatic retrotransposition can alter synaptic activity and person-to-person variation in cognitive brain functions (even between two genetically identical individuals), depending upon the location and extent of transposition. It can also cause neurodegeneration, aging, and neurodevelopmental disorders. Rett syndrome, an X-linked autism spectrum disorder, was the first neurodevelopmental disorder identified having a high rate of somatic retrotransposon insertions. Increased levels of retrotransposition also occur in some schizophrenia patients.

Somatic cell transpositions are prevalent in cancers. Insertion of LINE-1 causes driver mutations, particularly in gastrointestinal tract tumors. It was first discovered in a colorectal cancer patient in 1992. A 750-bp LINE insertion in this patient disrupted the function of tumor suppressor gene adenomatous polyposis coli. Besides colorectal cancer, head and neck cancers, acute myeloid leukemias, and high-grade central nervous system gliomas frequently show somatically acquired LINE-1 transposition.

Because transposition may cause genomic instability with severe consequences, cells have evolved the ability to suppress the mobility of transposons. The first level of suppression occurs through the inhibition of RNA synthesis (transcription) from the retrotransposons. In the canonical LINE-1 promoter of mice, rat, and human, there are binding sites for SOX2, YY1, RUNX3, TCF (T-cell factor/lymphoid enhancer factor) , and LEF (lymphoid enhancer-binding factor) -like transcription factors (TFs) that are known to be involved in neurogenesis. Evidence suggests that the activity of these TFs along with dynamic DNA methylation control transposition in the context of neural stem cell differentiation. On the other hand, DNA demethylation promotes transposition. Besides TFs and DNA methylation, another layer of epigenetic regulation could be mediated by small noncoding RNAs, affecting both the expression and RNA stability of mobile elements.

"Jumping" of a gene is a common and spontaneous phenomenon in both germline and somatic cells. It is not clearly understood which factors are responsible for the induction of transposition. Growing pieces of evidence suggest that environmental stress induces the mobility of transposons. The extreme stress may lead to pathological consequences upon transposition. For example, chronic alcoholism in humans was found to upregulate transcription of LINE-1 elements, with decreased epigenetic silencing and consequent enhancement of transposition. Interestingly, stress-enhanced fear learning in rats showed a very similar effect.

III.4 Nucleosomal organization of the genome

III.4.1 Evolution of genome organization

In a living cell, the genome is not present as a naked DNA but as a compact nucleoprotein structure. Even in a simple bacterial cell, DNA is compacted up to 400 times. Typically, the circular bacterial genome, which measures 1.6 mm when fully stretched, needs to be fitted into a 2-μm long and 1 μm wide *E. coli* cell. The nucleoprotein complex resides in the central region of the bacterial cell called the nucleoid, and the proteins present in the nucleoid are called nucleoid-associated proteins (NAPs). The composition of NAPs changes during the life cycle of the bacteria, making the genomic DNA available for all the DNA-based molecular processes, namely replication, transcription, DNA repair, and transposition. For example, the concentration of the Fis protein is found to be highest in the early exponential phase of the bacterial life cycle, whereas that of the HU protein increases at a later stage. The binding of these proteins induces structural changes, such as bending, in the DNA that facilitates the recruitment of the transcription machinery. In the stationary phase, the most dominant NAP is the IHF (integration host factor) protein, whereas in the late stationary phase it is the Dps (DNA-binding protein from starved cells). These proteins increase the compactness of the bacterial genome, thereby shutting down the expression of most of the genes. Interestingly, the variable genome composition in accordance with the functional state is well conserved across all living organisms during evolution (Fig. III.8). In parasites such as *Plasmodium* sp., where the first highly ordered nucleosomal organization of the genome was detected, the life cycle is completed in two different hosts, and the composition of nucleosomal histones changes from one stage of the life cycle to another.

This is equally true in the higher eukaryotic system, where differentiation from pluripotent to the differentiated stage of a cell, nucleosomes composition (incorporation of histone variants), and state (upon posttranslational modifications) are dramatically altered. It also alters under stress and aging.

As the living world progressed along the evolutionary path, the genome organization has become increasingly more complex. If we look into the genome of some forms of

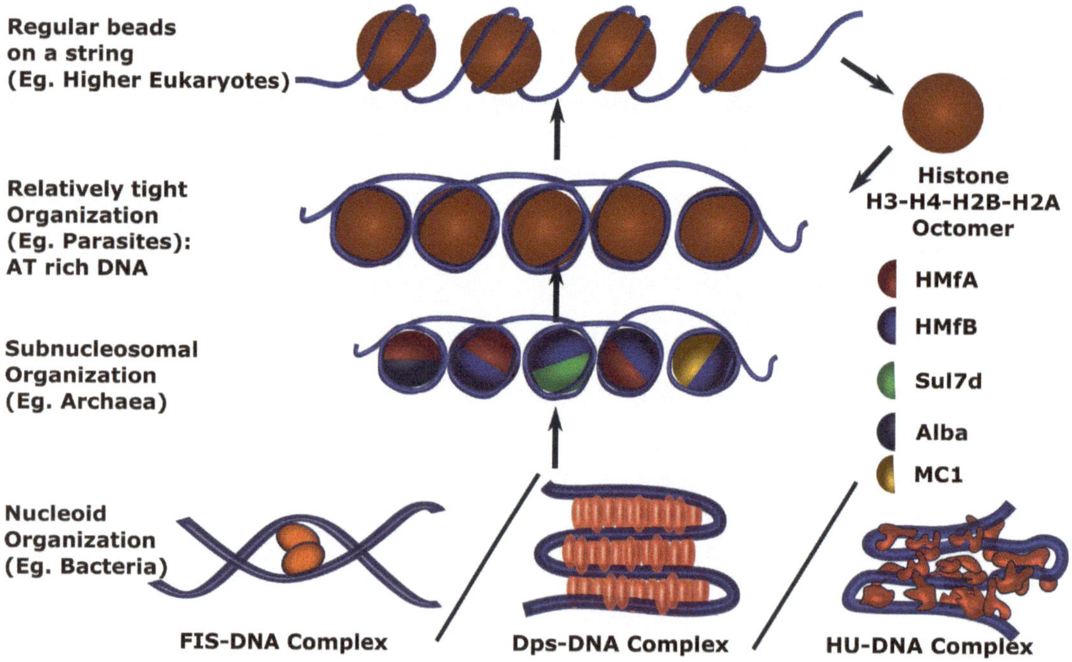

FIG. III.8 Evolution of genome organization: During evolution, the organization of the genome has become more ordered with defined, highly conserved octameric histone core. *Modified from Kundu, T.K., Sikder, S., 2018. J. Biosci. 43 (2), 239–242. Springer Nature.*

Archaea, primitive histone-like molecules could be observed. In eukaryotes, there are four highly conserved core histones, namely H3, H2B, H2A, and H4. In Archaea, only homologs of histone H3 and H4 are found that lack extended N and C terminals that are characteristics of eukaryotic histones. The primitive histones of Archaea, namely HMfA and HMfB, form subnucleosomal particles among themselves or different other proteins.

III.4.2 Nucleosomes and chromatosomes: The fundamental units of genome organization

The genome is organized as highly conserved nucleosome units that are observed in all species ranging from parasites to humans. The nucleosome core particle consists of the core histone octamer wrapped by ~146 bp of DNA in roughly two left-handed superhelical turns. The histone octamer comprises two copies each of the "canonical" core histones H2A, H2B, H3, and H4. The canonical histone-coding genes are clustered in the chromosomes and are present in multiple copies. These proteins are highly conserved across eukaryotes and are abundant in their distribution. Based on the functional demand, the core histones in the nucleosomes may be substituted by respective histone variants that confer specific structural and functional properties to the nucleosomes and consequently, the chromatin. The histone variant-encoding genes are usually single-copy genes and show a substantial degree of variation among species.

All core histones consist of predominantly alpha helices at the C-terminal domain that is essential to form the column-like structure around which the DNA wraps. Although there are hardly any sequence similarities, all the four core histones possess a similar kind of histone fold domain in the C-terminal part. The histone fold domain is responsible for the handshake-like interactions in the nucleosomal core during heterodimerization of H2A with H2B and H3 with H4. The histones H3 and H4 form the tetrameric structure through the H3:H3 interface in physiological conditions. Each of the H2A–H2B dimers binds with either end of H4 or H3:H3 or H4 tetramer. The helical ramp of the tetramer contacts the DNA in almost every helical turn. The high-resolution crystal structure shows that two paired beta loops and a set of alpha-helices contact the DNA backbone, as depicted in Fig. III.9. The *N*-terminal tails of each core histone project outside the core that serves as a site for epigenetic modifications. Inside the nucleosomal core, the DNA touches the core in a regular 10 bp interval that may change due to some posttranslational modifications or replacement of canonical histones with histone variants. Within the genome, nucleosomes are deposited or positioned in a highly regulated manner that is dynamic and functionally significant.

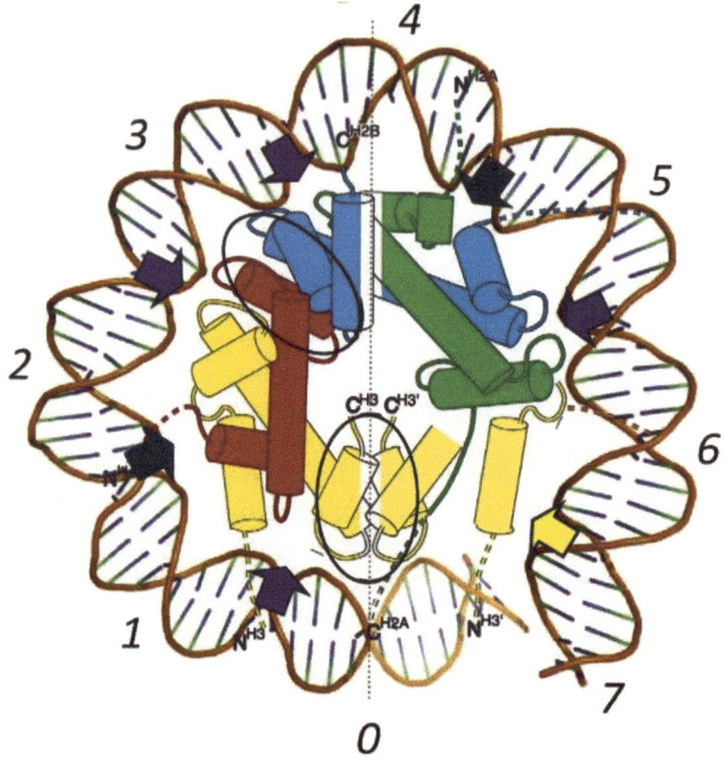

FIG. III.9 Arrangement of the core histone octamer in the nucleosome. The view is down the DNA superhelix axis. The numbers indicate the helix turns. Four alpha helix bundles of different core histones, H3:H3 and H4:H2B, are indicated by ovals. The *arrows* show the paired loop *(blue)* and paired-end-of-helix *(black)* DNA interaction sites and interaction site on the N-helix in H3 *(yellow)*. Different color codes of core histones: H2A, *green*, H2B, *blue*, H3, *yellow*, H4, *red*. *Source: Cutter, A.R., et al., 2015. FEBS Lett. 589 (20) 2914–2922. John Wiley and Sons.*

Besides the core histones, another histone that binds outside the nucleosomal core structure is the linker histone (such as histones H1 and H5). A linker histone-bound nucleosome is termed a chromatosome. Unlike the core histones, the linker histone H1 is not conserved across eukaryotic species. It plays a vital role in the compactness of the core. For a long time, the position of linker histone in the chromatosome and its role in higher-order chromatin organization were a matter of intense debate. Recently, the crystal structure of the nucleosome with chicken linker histone H5 and cryogenic electron microscopy (cryo-EM) structure of nucleosome with vertebrate linker histone H1 have been solved (Fig. III.10).

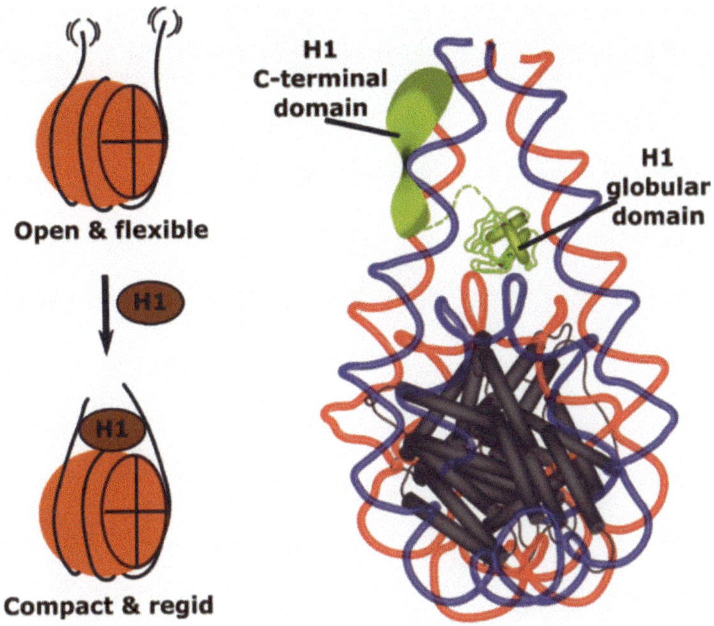

FIG. III.10 Mechanism of nucleosome compaction by linker histone H1. The cryogenic electron microscopy and crystal structures of a 197 bp nucleosome bound to the linker histone H1 have shown that the globular domain of H1 binds on the dyad of the nucleosomal DNA, and the C-terminal domain confines to a single DNA linker that bestows polarity to the nucleosome. This pattern of binding explains the linker histone-mediated compaction of the nucleosome structure. *Modified from Bednar, J., et al., 2017. Mol Cell. 66 (3), 384–397. Elsevier.*

Nucleosome positioning

The positioning of the nucleosome structure on a given DNA sequence plays a pivotal role in all DNA-based molecular processes, including transcription. Extensive research using DNA sequencing has established that nucleosome positioning is not a random phenomenon. In these experiments, cross-linking of the chromatin fiber is followed by micrococcal nuclease (that cuts only the linker DNA) digestion. The digested chromatin is then subjected to tiling microarray or high-throughput sequencing analysis to determine the position of nucleosomes. There are two types of factors: cis (DNA sequence) and trans (chromatin-remodeling complexes, epigenetic modifications, and TFs) that regulate nucleosome positioning. The nucleosome formation demands substantial bending of the DNA around the histone octamer. The affinity of histones for the DNA also differs from sequence to sequence.

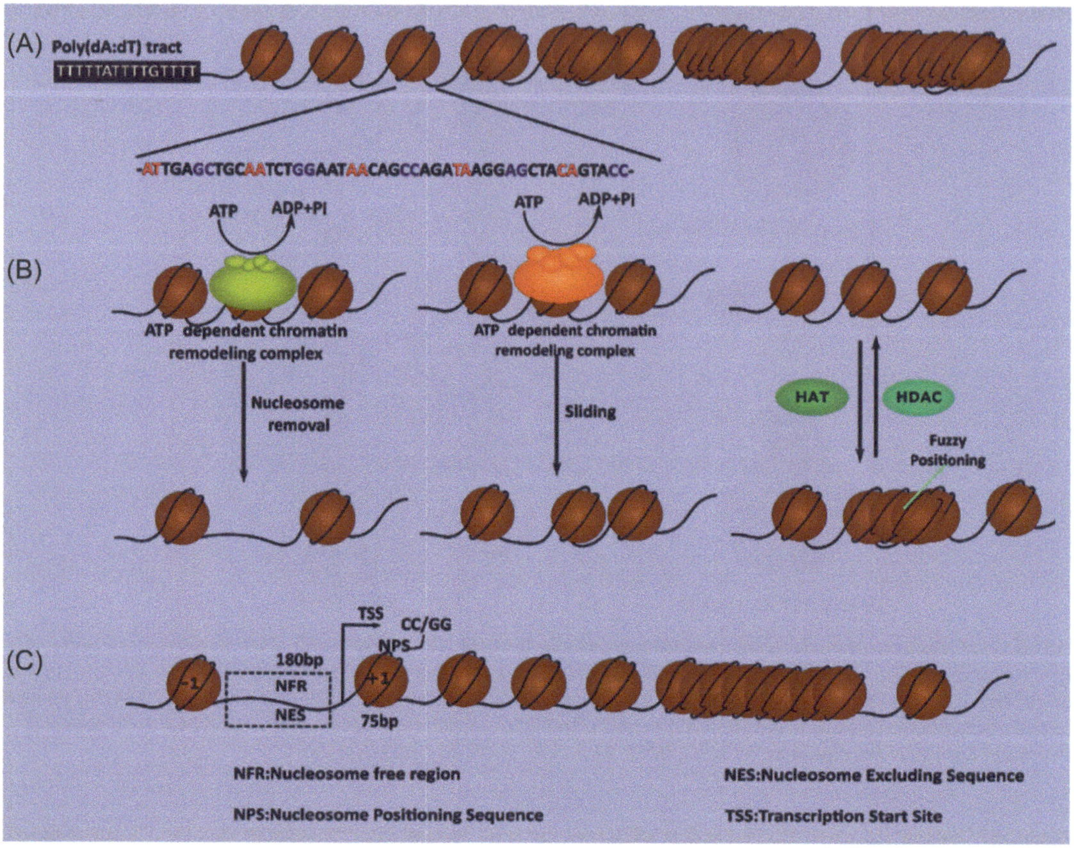

BOX III.1 Nucleosome positioning. The earlier figure shows that the (A) poly (dA/dT) sequence promotes nucleosome exclusion, whereas A or T dinucleotide every 10 bp, G or C dinucleotide every 10 bp, in antiphase with A or T dinucleotides favor nucleosome deposition. The fuzzy nucleosomes are loosely associated due to histone modifications and the replacement of canonical histones by the variants. (B) ATP-dependent chromatin-remodeling complexes are critical for nucleosome removal or sliding for biological functions. DNA-protein interactions in nucleosomes are altered by reversible histone modifications such as acetylation. (C) Dynamics of the nucleosome positioning is essential in transcription activation, where nucleosome-free regions are created for the recruitment of transcription machinery.

These data reveal that the linker histone alters the conformational landscape of the nucleosomes by pulling the linker DNA together at both sides, which reduces the flexibility. The globular domain of the linker histone interacts with the core DNA on the dyad axis and with both the DNA linkers, whereas its C-terminal domain interacts mainly with a single DNA linker. These structural data were validated by the site-directed protein cross-linking and hydroxyl radical footprinting experiments. These arrangements of linker histone also explain its role in higher-order chromatin organization.

Data available from several elegant experiments collectively suggest that the presence of poly(dA or dT) sequence could be responsible for the exclusion of nucleosomes at gene promoters and transcription termination sites (Box III.1). These data also concluded that an intrinsically bendable sequence with 10 bp AA or AT or TA periodicity might favor the nucleosome deposition. However, in mammalian cells, GC or CG or CC or GG sequence in 10 bp periodicity is often observed with nucleosomes. Among the trans factors that influence nucleosome positioning, the most significant role is played by the ATP-dependent chromatin-remodeling systems that are generally recruited at chromatin loci through TFs. The modifications of nucleosomal histones such as acetylation and specific methylations may facilitate the chromatin remodeling-mediated nucleosome removal or sliding.

III.4.3 Higher-order chromatin organization

The genomic DNA is packaged into nucleosomes forming a "beads on a string" structure that is termed the chromatin. The chromatin is about 11 nm in diameter and is further packaged into higher-order structures of a diameter of 30 nm. These 30-nm fibers form chromatin loops with a diameter of about 300 nm, which compress further and form the 250-nm fiber. This condensed chromatin has a diameter of about 700 nm, finally resulting in the metaphase chromosome of a diameter of about 1400 nm. The precise mechanisms of higher-order chromatin compaction are not known. Increasing pieces of evidence suggest that besides linker histone H1, a few other nonhistone proteins play a critical role in chromatin compaction and possibly site-specific heterochromatinization. These nonhistone chromatin-associated proteins include PC4, heterochromatin protein 1 (HP1), and PARP1. However, until recently, the higher-order chromatin organization was almost completely obscure. With the advent of cryo-EM and higher-end X-ray crystallography, the mystery of higher-order genome organization is partially solved.

Guohong Li and group from China reconstituted a 12-nucleosome array in the presence of linker histone, H1 and solved the cryo-EM structure of it. Excitingly, they found that four nucleosomes are arranged back and forth in a zig-zag manner as tetra nucleosomal units through the linker DNA (the DNA between two nucleosomes). When linker histone, H1 was incorporated in these units, it generated a left-handed twist, resulting in a DNA double helix-like chromatin fiber organization. This could be the first experimental proof of the zig-zag configuration of the 30-nm chromatin fiber. However, the existence of other forms of chromatin structure cannot be ruled out under different conditions, especially in the presence of nonhistone chromatin proteins and different ionic concentrations. More recent studies reveal that ionic condition is indeed critical to determine the conformation of the chromatin.

A six nucleosome, linker histone H1-bound crystal structure has been solved at 9.7-Å resolution. Furthermore, by employing cryo-EM and biophysical and biochemical techniques, it has been shown that this array adopts a relatively less condensed ladder-like structure at a particular ionic strength (Fig. III.11). Although these in vitro-reconstituted chromatin fiber structures have immensely contributed to our understanding and visualization of genome structure, in vivo cellular chromatin is much more complex and dynamic in nature. The chromosome conformation capture, high-resolution light, and electron microscopy are now making it possible to understand better the four-dimensional (considering space and time) folding of the genome and its function in an unprecedented high resolution. The upcoming results suggest that genome folding is much more heterogeneous than previously believed based on the structural information from the purified

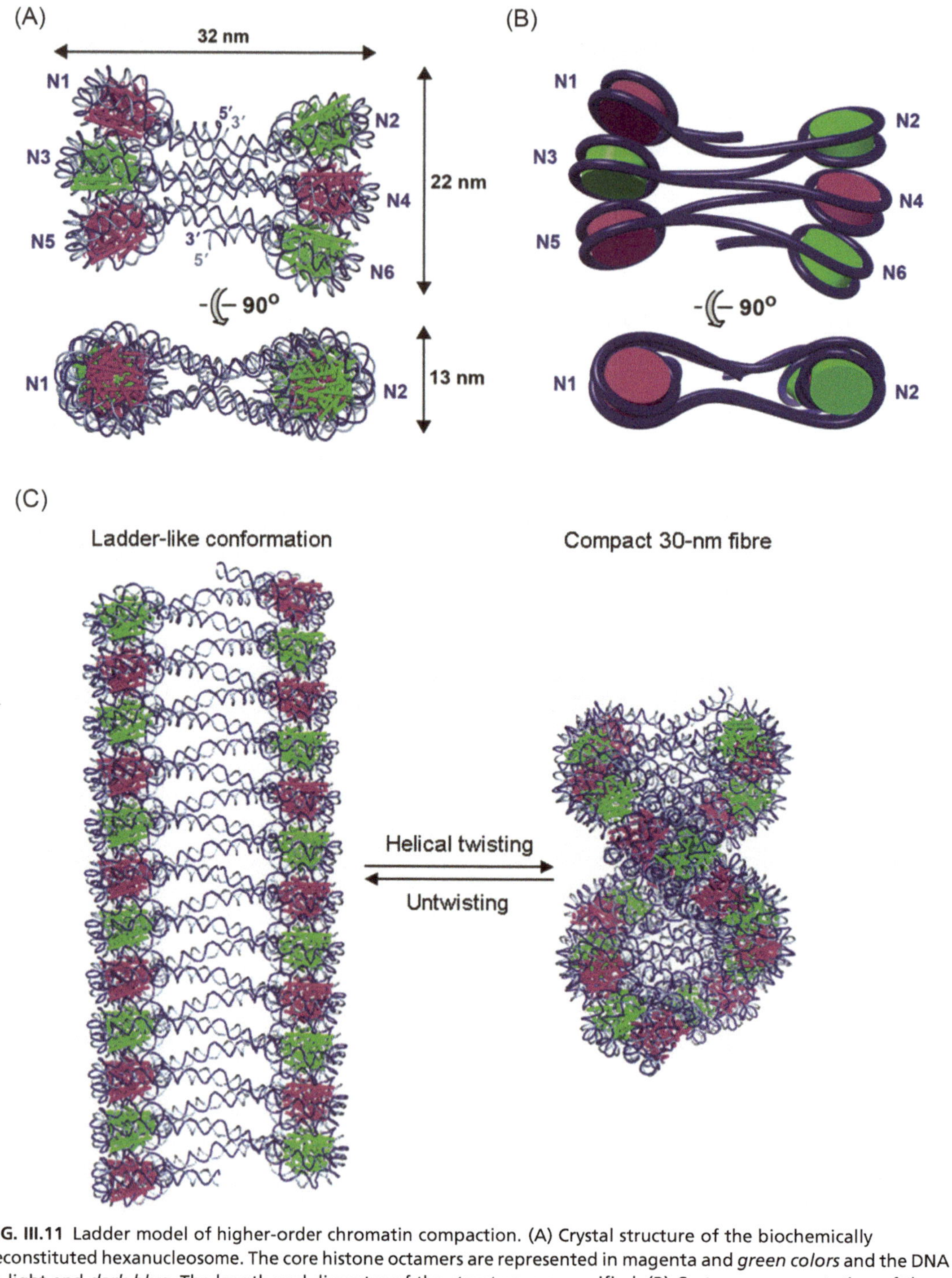

FIG. III.11 Ladder model of higher-order chromatin compaction. (A) Crystal structure of the biochemically reconstituted hexanucleosome. The core histone octamers are represented in magenta and *green colors* and the DNA in light and *dark blue*. The length and diameter of the structure are specified. (B) Cartoon representation of the hexanucleosome structure as displayed in (A) highlighting the connections between the nucleosomes. (A and B) The individual nucleosomes are numbered, and the structures have been visualized from two different angles. (C) Model showing the reversible transformation of the chromatin between the open ladder-like and the compact 30-nm fiber conformation that occurs with a shift in the environmental ionic conditions. *From Garcia-Saez, I., et al., 2018. Mol Cell. 72 (5), 902–915. Elsevier.*

nucleosomal array. Epigenetic modifications of DNA and histones, different histone variants, nonhistone chromatin-associated proteins, and long noncoding RNAs could significantly contribute to the dynamic and variable functional chromatin organization.

In the metaphase stage of cell division, the chromatin is in the most compact state. However, in the interphase, each homologous chromosome is present as distinct as a chromatin territory (CT). The space between two CTs is known as an interchromatin territory. In CTs, chromatin fibers are present in highly folded chromatin domains. The relatively open, euchromatin regions are generally spaced toward the interior of the nuclear space, whereas the highly compact heterochromatin regions are dispersed toward the nuclear periphery as these are part of a rather silent genome. The chromatin territories are organized in distinct chromatin domains known as topologically associating domains (TADs). In each TAD, there may be a 100 kilobases to 1–2 megabases of DNA packed as nucleosomes. Functionally, the TADs may interact within the same territory or across the territories. These interactions are highly essential for the regulation of gene expression. Presently by employing different elegant techniques such as chromosome conformation capture (3C) and Hi-C, the functional dynamics of TADs are being studied.

III.5 Epigenome

Chemical modifications of DNA and the associated histones fine-tune gene expression affecting different processes of development, differentiation, and gene transposition. The global modification state of the genome could be referred to as the epigenome. These modifications include methylation and hydroxy-methylation of DNA and a variety of histone modifications predominantly on the lysine (K), arginine (R), and Serine (S) residues. The epigenome is highly dynamic based on environmental signals. The cross-talks and sequence of these modifications are a matter of active research at present.

III.5.1 The dynamic epigenome

Inside the living cell of an organism, the chromatin needs to be highly dynamic to attend to the different DNA-templated processes such as replication, repair, and transcription. This dynamicity is ensured and brought about by the action of various factors. These include the replacement of canonical histones by their respective variants, covalent modifications of the core and linker histones, methylation or hydroxyl methylation of DNA, and also the actions of the ATP-dependent chromatin-remodeling proteins.

Chromatic remodeling

ATP-dependent chromatin-remodeling complexes are multiprotein complexes that assist in the restructuring of nucleosomes by sliding or removing. This remodeling is essential for transcriptional regulation (both activation and repression) and also DNA repair. Each remodeling complex possesses an ATPase subunit. There are five different complexes identified so far in eukaryotes: SWI or SNF (SWItch/sucrose non-fermentable), ISWI (imitation switch), NuRD (nucleosome-remodeling deacetylase complex) or Mi-2 (auto-antigen for dermatomyositis) or CHD (chromodomain helicase DNA-binding), INO80 (inositol auxotroph 80), and SWRI (SWI2/SNF2-related 1 chromatin remodeling complex). Although all the complexes fundamentally perform the same function, their application differs based on the cellular signals and their recruitment by the specific factor(s) at specific sites (Box III.2). In almost all cancers, one or more mutations have been identified in the subunits of these complexes. These complexes are also involved in senescence and aging. Several antineoplastic drugs have been developed, targeting the functions of ATP-dependent remodeling complexes.

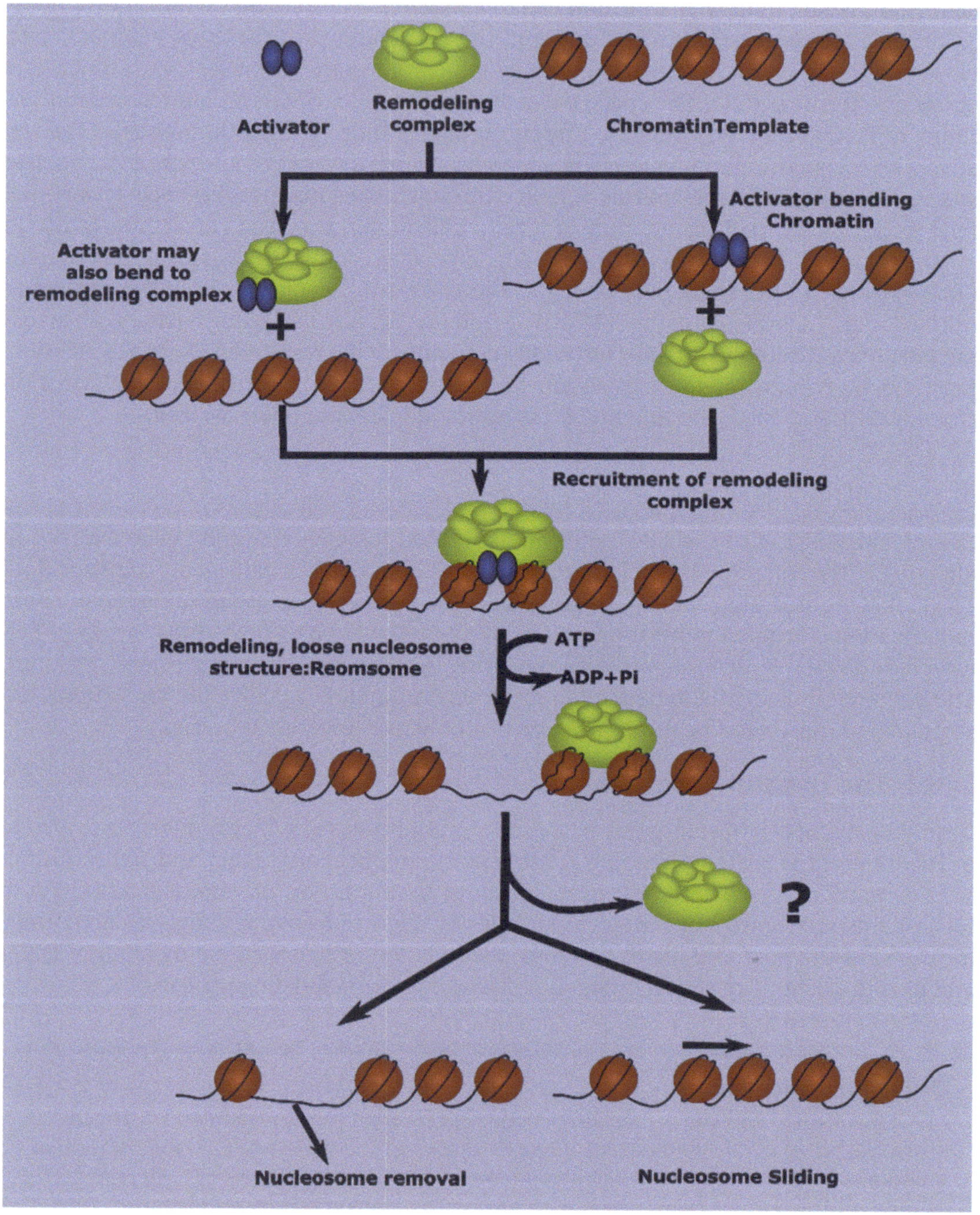

BOX III.2 Role of transcriptional activator in the recruitment of ATP-dependent chromatin-remodeling complexes. The activator may bind to its cognate sites on the nucleosomal DNA or interact with the complex itself to guide it for recruitment at the desired chromatin sites.

Table III.1 Cofactors essential for epigenetic modifications.

Cofactor	Structure	Metabolic pathways	Chromatin modifications	Reader domains
Acetyl-CoA	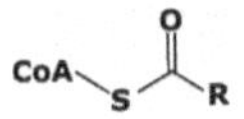	Glycolysis; TCA cycle; FA oxidation; amino acid catabolism; acetate metabolism	Histone acetylation	Bromo; YEATS; DPF
-Acyl-CoA	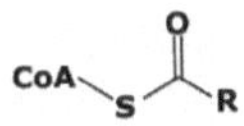	FA oxidation; ketogenesis; amino acid catabolism	Histone acylation	YEATs; DPF
SAM	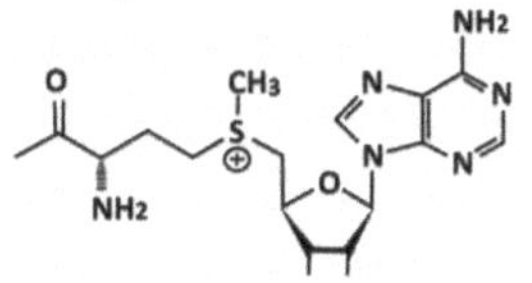	1C metabolism; (methionine and folate cycles); phospholipid synthesis	Histone methylation; DNA methylation	Chromo; PHD; Tudor, PWWP; TTD; MBD
NAD$^+$	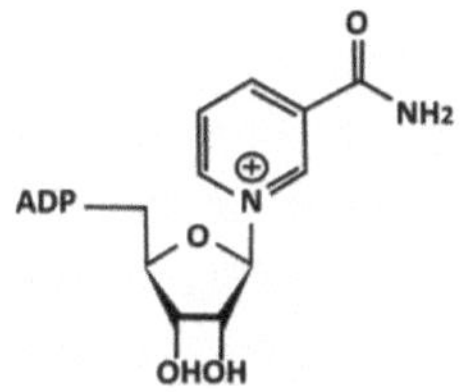	Amino acid catabolism; Preiss-Handler pathway; NAM salvage	Histone deacetylation; histone deacylation; histone ADP-ribosylation	Bromo; YEATS; DPF; PBZ; WWE; Macro
α-KG	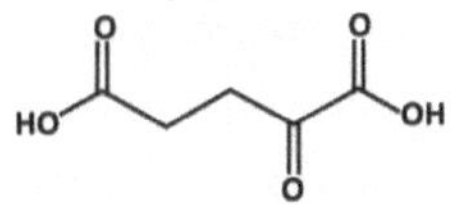	TCA cycle; amino acid metabolism	Histone demethylation; DNA demethylation	Chromo; PHD; Tudor, PWWP; TTD; MBD
FAD	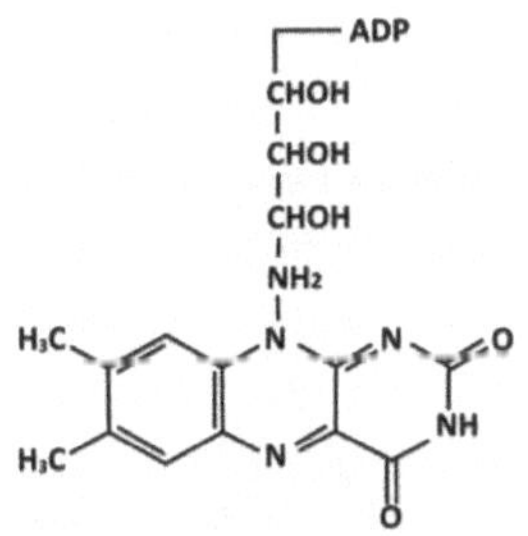	TCA cycle; oxidative phosphorylation	Histone demethylation	Chromo; PHD

Notes: *DPF*, double PHD finger; *αKG*, α-ketoglutarate; *SAM*, S-adenosyl methionine; *PHD*, plant homeo domain; *TCA*, tricarboxylic acid; *NAM*, nicotinamide; *PAR*, poly(ADP-ribose); *PBZ*, PAR-binding zinc finger; *MBD*, methyl CpG binding domain; *TTD*, tandem tudor domain; *PWWP*, proline-tryptophan-tryptophan-proline domain; *WWE*, tryptophan-tryptophan-glutamate domain; *FA*, Fanconi anemia; *Yaf9, ENL, AF9, Taf14, Sas5*, (YEATS) domain; *ADP*, adenosine diphosphate.

Modified from Haws, S.A., et al., 2020. Trends Biochem. Sci. S0968–0004. Elsevier, ahead of print.

Covalent modifications of the chromatin (both DNA and histones) are a major part of the epigenome. These modifications are altered dynamically based on the metabolic conditions of the cell, as the acetyl states, acyl states, and methyl states of the epigenome are dependent on different cofactors that are the products of various metabolic pathways (Table III.1).

III.5.1.1 Reversible acetylation of histones

Although acetylation of histones in the context of transcriptional activation was discovered by Vincent Allfrey in 1964, it took almost three decades to discover the first histone acetyltransferase (HAT, presently termed as KAT) by Charles David Allis in 1995. The entire field of epigenetics received a resurrection with the discovery of the first HAT enzyme, GCN5. In principle, acetylation of a lysine residue in histone neutralizes the electrostatic interactions essential for higher-order chromatin organization. Furthermore, the acetylated lysine residue is recognized by "reader" proteins with bromodomain, YEATS domain, and double PHD finger domain that then recruit the transcriptional activation machinery, creating a transcriptionally active open chromatin region. There are a large number of HATs, mostly grouped into three major families, namely GNAT, p300 or CBP, and MYST.

In the bi-substrate acetylation reaction, acetyl-CoA donates the acetyl group. The acetyl-CoA is synthesized in the mitochondria by pyruvate dehydrogenase complex through different metabolic processes (Fig. III.12). The mitochondrial acetyl-CoA is then

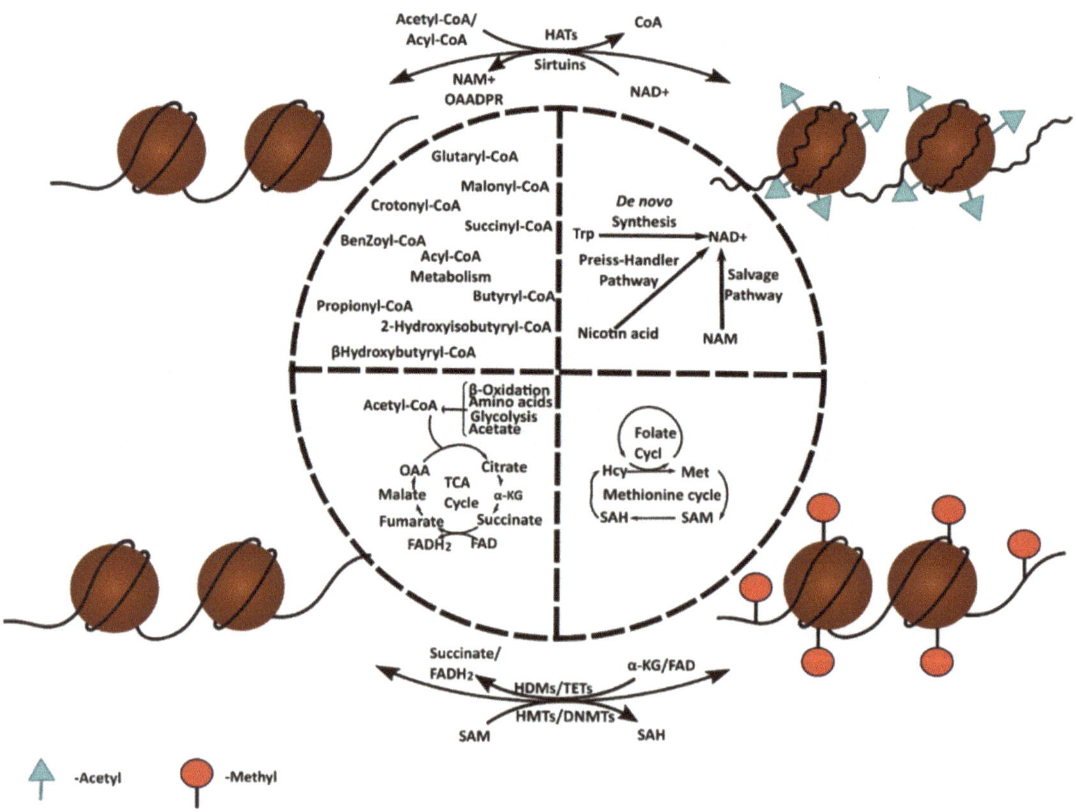

FIG. III.12 Cofactors for epigenetic modifications generated from metabolic pathways. Acetylated and methylated dinucleosomes represent the active and repressed state of chromatin, respectively. *α-KG*, α-ketoglutarate; *CoA*, coenzyme A; *Dnmt*, DNA methyltransferase; *HAT*, histone acetyltransferase; *Hcy*, homocysteine; *HDM*, histone demethylase; *HMT*, histone methyltransferase; *NAM*, nicotinamide; *OAA*, oxaloacetic acid; *SAH*, S-adenosyl homocysteine; *SAM*, S-adenosyl methionine; *TCA*, tricarboxylic acid; *TET*, ten eleven translocation DNA demethylase. *Modified from Haws, S.A., et al., 2020. Trends Biochem. Sci. S0968–0004. Elsevier, ahead of print.*

converted to citrate upon reacting with oxaloacetate and transported to either the cytoplasm or the nucleus, where different enzymes convert citrate to acetyl-CoA and supply it to various acetylation reactions in a context-dependent manner. Acetylated histones are the signature of transcriptionally active open chromatin. The reverse reaction catalyzed by histone deacetylases (HDACs) induces chromatin compaction (Fig. III.13).

There are four classes of deacetylases in mammals, namely class I, II, III, and IV. Class III HDACs, also known as sirtuins, need NAD+ as a cofactor for the catalysis reaction. There are seven sirtuins in mammalian cells, SIRT1 to SIRT7. SIRT1, SIRT3, and SIRT5 deacetylate proteins in the mitochondria. Histones in the nucleus are deacetylated by SIRT2, SIRT6, and SIRT7. In a cell, although a minor amount of NAD+ is synthesized de novo, generated through the Preiss-Handler pathway, the major amount of NAD+ is produced by the NAM salvage pathway. The functional activity of sirtuins is directly correlated with the concentration of NAD+.

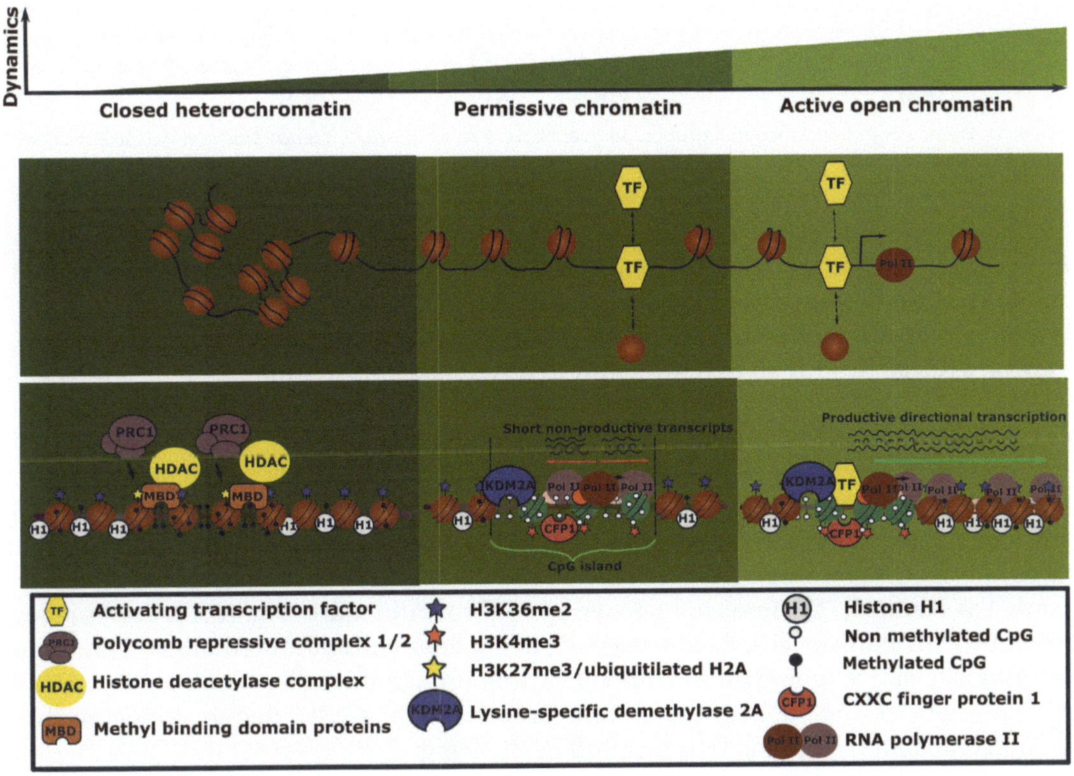

FIG. III.13 Transcriptional states of the chromatin. In highly dynamic epigenome, genes are present in different states—closed or repressed, permissive (where RNA polymerase is stalled), and active or open (where transcription is initiated). In repressed chromatin, DNA is highly methylated, histones are deacetylated, and H3K36me2 modification mark is enriched. In the active state, DNA is demethylated, histones are acetylated at multiple sites, and methylated at the H3K4 site (H3K4me3). The closed state is more stable and less accessible. *Modified from Klemm, S.L., et al., 2019. Nat Rev. Genet. 20 (4), 207–220. Springer Nature and Blackledge, N.P., et al., 2011. Epigenetics 6 (2), 147–152. by Taylor & Francis.*

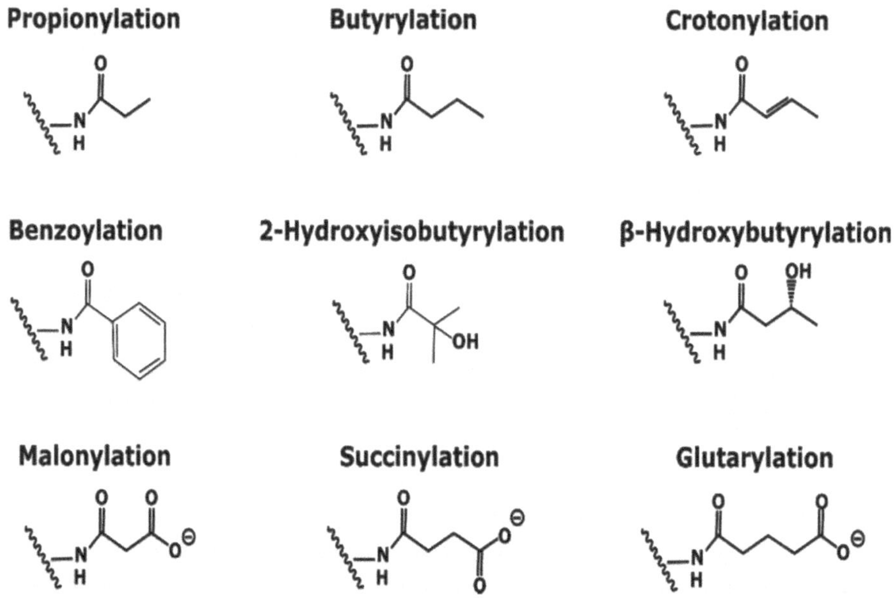

FIG. III.14 Histone acylation functional groups. *Source: Haws, S.A., et al., 2020. Trends Biochem. Sci. S0968–0004. Elsevier (ahead of print.*

Besides acetylation, histones are also subjected to different types of acylations by the KATs (Fig. III.14). So far, nine different types of acylation have been identified. The metabolic sources of these acyl groups are currently being investigated. Although their confirmed physiological roles are not known, indirect evidence suggests that acylation marks may be associated with active chromatin in a gene-specific manner.

III.5.1.2 Reversible methylation of histones and DNA

Histones can be methylated at both lysine and arginine residues. Lysine (K) methylation of histones contributes the most to the chemical diversity of the epigenome. The mono-, di-, and trimethylation of K in histones can create a huge permutation of diversity. Depending upon the specific residue(s), K methylation may function as an activation or repression mark. For example, H3K4me3 is an active open chromatin mark (Fig. III.13) that is recognized by PHD domain TFs, facilitating ATP-dependent chromatin remodeling and transcriptionally active preinitiation complex formation. On the other hand, H3K9me2/3 is recognized by chromo domain-containing proteins such as HP1, thereby inducing chromatin compaction resulting in transcription silencing.

DNA is methylated at the fifth carbon of cytosine (5mC) in the CpG dinucleotide. Methylation of DNA is an important factor for chromatin compaction, heterochromatinization, and, consequently, gene silencing. The single-molecule fluorescence resonance energy transfer studies of mononucleosomes showed that DNA methylation induces the nucleosome to be more compact and rigid in nature, which explains its presence in heterochromatin or tightly packed regions of the genome.

CpG island and genome organization

CpG islands are stretches of the DNA sequence of 300 bp–3000 bp long with CG: GC ratio of more than 0.6. Therefore this is the region of the genome where there is a high frequency of CpG dinucleotide sites. In the human genome, there are around 30,000 CpG islands, predominantly found in the promoter regions of genes, particularly housekeeping and some tissue-specific genes. In a heavily methylated mammalian genome, CpG islands are generally kept methylation free through mechanism(s) yet to be fully understood. CpG islands play a significant role in chromatin organization and gene regulation. There are several TFs such as SP1, CREB, and CTCF, whose canonical DNA-binding sites contain CpG islands. Furthermore, CpG islands may nucleate heterochromatinization and gene silencing through methylation, followed by binding of methyl CpG-binding protein MeCP2 and deacetylation of CpG island chromatin.

Genome-wide analysis of chromatin modifications suggests that methylated heterochromatin is also highly enriched with histone H3K27 trimethylation (H3K27me3) and K9 trimethylation (H3K9me3). These areas of chromatin are called the constitutive heterochromatin. The euchromatin or the relatively open, transcriptionally active chromatin, on the other hand, predominantly contains acetylated histones, with histone H3K4 trimethylation mark (H3K4me3). Similarly, histone variants such as H3.3 and H2A.Z usually are the signature marks of the active chromatin. Under certain conditions (e.g., tissue-specific signal), the histone variants macroH2A and H2A.Bbd could also be present in the active region of the chromatin.

The linker histone H1 stabilizes the nucleosome and helps in the formation of a stable higher-order chromatin structure. Further site-specific compaction, driven by HP1, is a critical epigenetic mechanism for heterochromatinization and gene silencing. HP1 recognizes the repressive histone modification mark, lysine 9 di or trimethylated histone H3 (H3K9me2/3).

It was elegantly shown that HP1 directly interacts with DNA methyltransferases (Dnmts) and recruits them on the site of action to create a repressive epigenetic landscape (Fig. III.15).

The Dnmts then recruit H3K9 methyltransferases (for example, Suv39h1) required for heterochromatin formation. Interestingly, Dnmts can also recruit G9a or GLP methyltransferases associated with H3K9 methylation in euchromatin regions, indicating the overlapping reversible network of the dynamic epigenome. By specifically recognizing the H3K9 methylation mark, Dnmt3 also makes sure that no DNA methylation occurs near the H3K4me3 mark that is associated with transcriptionally active chromatin. However, epigenetic modifications are highly reversible. We put forward these examples to provide a glimpse of the highly complex networks of epigenomic activity to fine-tune the genome function. For each modification of DNA or histones, there are a set of enzymes or protein complexes that get recruited at the required sites in a signal-dependent manner.

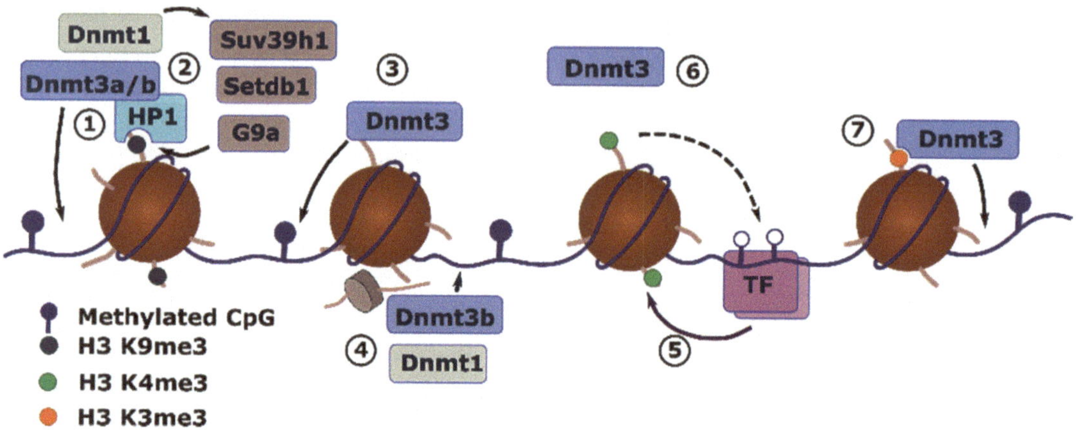

FIG. III.15 DNA and histone methylation function together for dynamic chromatin organization. (1) Heterochromatin protein 1 binds to H3K9me2/3 and recruits DNA methyltransferases to maintain the state of DNA methylation. (2) Heterochromatin protein 1 and DNA methyltransferases recruit histone methyltransferases (Suv39h1, Setdb1). (3) Deacetylated histone H3 tail recruits DNA methyltransferase 3. (4) Linker histone H1 help in the recruitment of DNA methyltransferase 3b. (5 & 6) Methylation of histone H3, H3K4me3, prevents DNA methyltransferases 3 recruitment that allows the binding site of the transcription factor. (7) In the gene bodies, H3K36 methylation is used to recruit DNA methyltransferases 3 that maintains the DNA methylation state of the silent chromatin. *Modified from Soshnev, A.A., et al., 2016. Mol Cell. 62 (5), 681–694. Elsevier.*

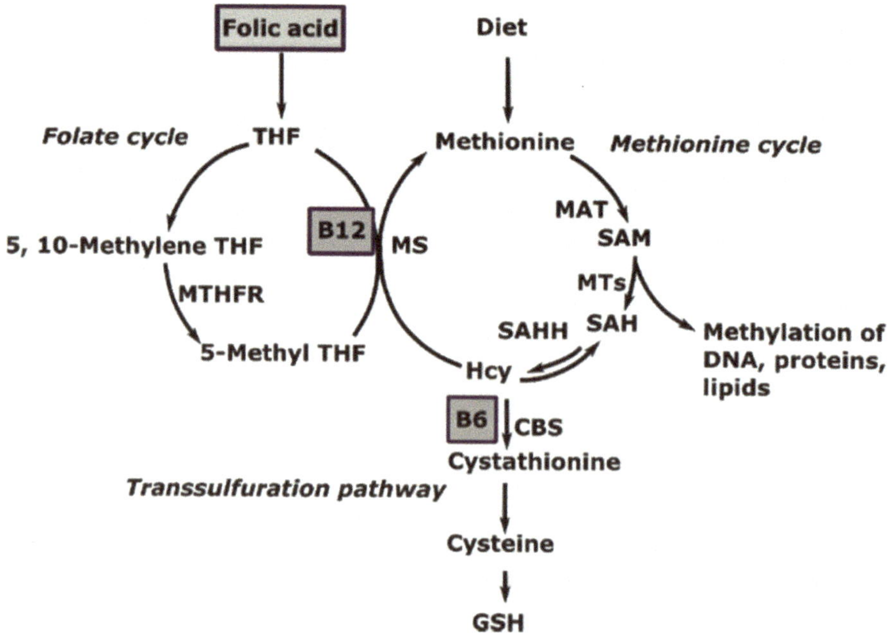

FIG. III.16 One-carbon metabolism of SAM biosynthesis. Abbreviations: *SAM*, S-adenosylmethionine; *SAH*, S-adenosylhomocysteine; *Hcy*, homocysteine; *GSH*, glutathione; *THF*, tetrahydrofolate; *B12*, vitamin B12; *B6*, vitamin B6; *MTHFR*, methylenetetrahydrofolate reductase; *MS*, methionine synthase; *MAT*, methionine adenosyltransferase; *MTs*, methyltransferases; *SAHH*, S-adenosylhomocysteine hydrolase; *CBS*, cystathionine-β-synthase. *Adapted from Cavallaro, R.A., et al., 2017. Antioxidants (Basel) 6 (4), 76. MDPI.*

In all methylation reactions, S-adenosylmethionine (SAM) is the universal methyl donor. In the mammalian system, two methionine adenosyltransferases (MATs), MATIα (in liver only) and MATIIα (in all tissue) synthesize SAM, utilizing ATP and folic acid or dietary methionine through one-carbon metabolism as depicted in Fig. III.16.

The availability of SAM directly regulates histone methylation and, thereby, the epigenome organization and gene expression. Recent pieces of evidence suggest that 5mC also assists in the process of transcriptional activation indirectly by blocking the recruitment of the polycomb repressive complex (PRC2).

Polycomb repressive complex 2 repressive complex

DNA methylation inhibits the recruitment of the repressive complex PRC2. Active RNA polymerase II-generated transcripts bind to the PRC2 complex to prevent its binding to the chromatin. Besides, both H3K4 and H3K36 methylations inhibit the binding of the PRC2 complex. EZH2 methyltransferase, a constituent of the PRC2 complex, methylates H3K27 that also helps its recruitment (Box III.3).

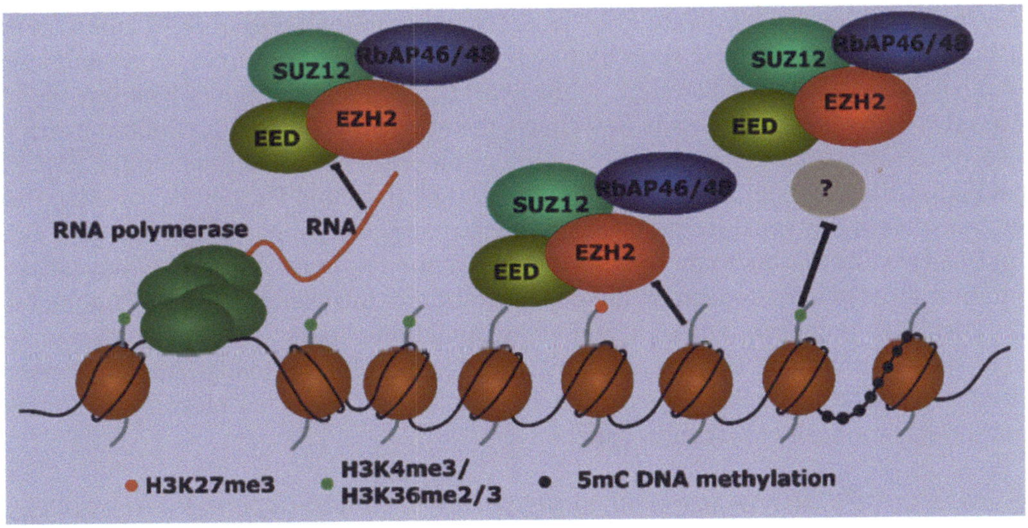

BOX III.3 Role of polycomb repressive complex 2 in chromatin regulation. *Modified from Holoch, D., et al., 2017. Trends Biochem. Sci. 42 (7), 531–542. Elsevier.*

There are two different classes of histone demethylases that directly remove the methyl groups from the histones, namely lysine-specific demethylase (LSD) and Jumonji C (JmjC). LSD needs FAD as a cofactor for the demethylase reaction and catalyzes the removal of mono and dimethylation. JmjC, on the other hand, utilizes Fe(II) and α-ketoglutarate (α-KG) and demethylates both lysine and arginine methylation by dioxygenase mechanism. Both FAD and α-KG are integral parts of the tricarboxylic acid cycle indicating the role of mitochondrial metabolism in epigenetic modifications.

III.5.1.3 Metabolic influence on epigenome dynamics during physiological, pathophysiological, and dietary alterations

From the previous sections, it is certain that chemical modifications of the genome largely depend upon the availability of the specific cofactors. Because the process of metabolism is highly dynamic, varying with age, infection, and even day and night cycle (circadian cycle) of our life, cofactor concentrations that are dependent on metabolic states of the cells also alter accordingly, thereby altering the epigenetic state of the genome. With age, the level of acetyl-CoA increases promoting hyperacetylation of histones at multiple sites, such as H4K12, H3K9, H3K14, H3K23, and H4K16. This hyperacetylation with aging is also promoted due to the decrease in cofactor NAD+ and associated deacetylase enzyme, SIRT1. The hyperacetylation at multiple sites induces the expression of several age-related genes. Calorie restriction is generally helpful to reduce acetylation and methylation and enhance the longevity of the individual. Nonvegetarian foods (such as meat, egg, and fish) provide more methionine and, thereby, induce the enhanced synthesis of SAM. Increased levels of SAM could be used for hypermethylation of DNA and histones. Furthermore, foods that are a source of high energy, such as those rich in fat and sugar, also alter the level of both acetyl-CoA and SAM, thereby altering the epigenome.

Besides diet, our daily life cycle also affects the epigenome due to the fluctuation of the rate syntheses of cofactors, such as NAD+ (nicotinamide adenine dinucleotide), FAD (flavin adenine dinucleotide), and SAM in the day and night hours. The cellular or organismal metabolism is highly sensitive to different disease conditions, including cancer and infection. In a cancer cell, due to the Warburg effect, a phenomenon of increased uptake and fermentation of glucose to lactate, the concentration of intracellular acetyl-CoA, elevates dramatically. As a consequence, global hyperacetylation occurs that is a signature in several types of neoplasms, such as solid tumors and hematological malignancies. Moreover, due to the mutation(s), the demethylase activity of JmjC and ten eleven translocation could be metabolically blocked in cancer.

III.5.1.4 The order and sequence of epigenetic modification-dependent functional genome organization

The dynamic epigenome, mediated by the coordination and establishment of different events, such as reversible DNA methylation, reversible histone modifications, ATP-dependent chromatin remodeling, and replacement of histone variants, among others, is essential to achieve different patterns of gene expression from the same genome. These events are highly regulated and often processive. The local modifications of DNA and histones finally lead to altered higher-order folding of the chromatin. For any given signal, the epigenetic processes are highly interconnected. Functionally, the epigenetic landscape or epigenetic language could be distinct for the class of target genes or chromatin domains. DNA methylation may be the most fundamental chemical modification of the epigenome,

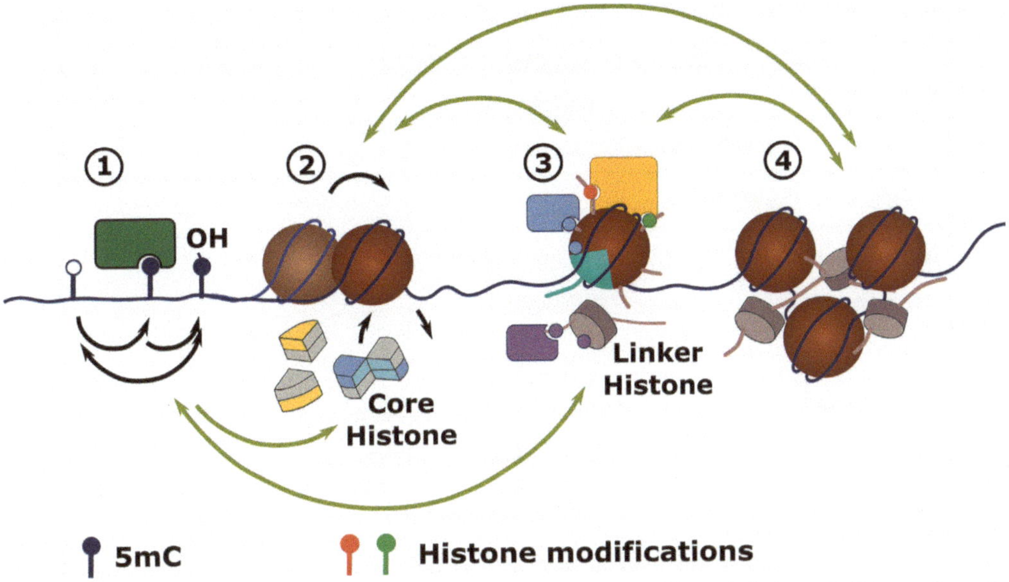

FIG. III.17 Sequence of epigenetic regulation in the dynamic epigenome. (1) Methylation of DNA may favor the binding of methyl CpG-binding protein to displace the transcription factor. Hydroxy-methylation leads to demethylation and reversal of the process. (2) Nucleosome reorganization occurs by the exchange of histone variants in the place of canonical histones with the assistance of histone chaperones (not shown here). Linker histones also may be removed or exchanged in a similar manner. (3) Core histone tails may be subjected to modifications and then recognized by the reader proteins to recruit factors either for further opening and transcription activation or repression depending upon the modification. (4) Upon subjected to required histone modifications and DNA methylation, chromatin undergoes a higher-order organization to a closed silent conformation. The *bold green arrows* indicate cross-talk between the events of dynamic chromatin organization. *Modified from Soshnev, A.A., et al., 2016. Mol Cell. 62 (5), 681–694. Elsevier.*

which also plays a vital role in the dynamics of histone modifications and recruitment of nonhistone chromatin-associated proteins (such as HP1) for genome compaction and repair. Replacement of the canonical histones in the nucleosomes with the histone variants and removal of histones in a context-dependent manner with the help of histone chaperones and chromatin-remodeling complexes, both in transcription and DNA repair, are essential for nucleosomal dynamics. Higher-order chromatin compaction in a region-specific manner is assisted by linker histone H1, where core histone modifications and DNA methylation also play a significant role (Fig. III.17).

Genome-wide analysis of histone modifications in the context of transcription regulation has shown a near definitive signature of euchromatin and heterochromatin modifications. Enrichment of histone variants H2A.Z along with the acetylation of histone H3 and H4 (Fig. III.18) are characteristics of active eukaryotic promoters. The effect of histone acetylation was found to be cumulative.

It loosens the histone-DNA interactions and helps to remove, replace, or slide the nucleosomal histones. Besides, multiple readers that recruit the regulatory proteins

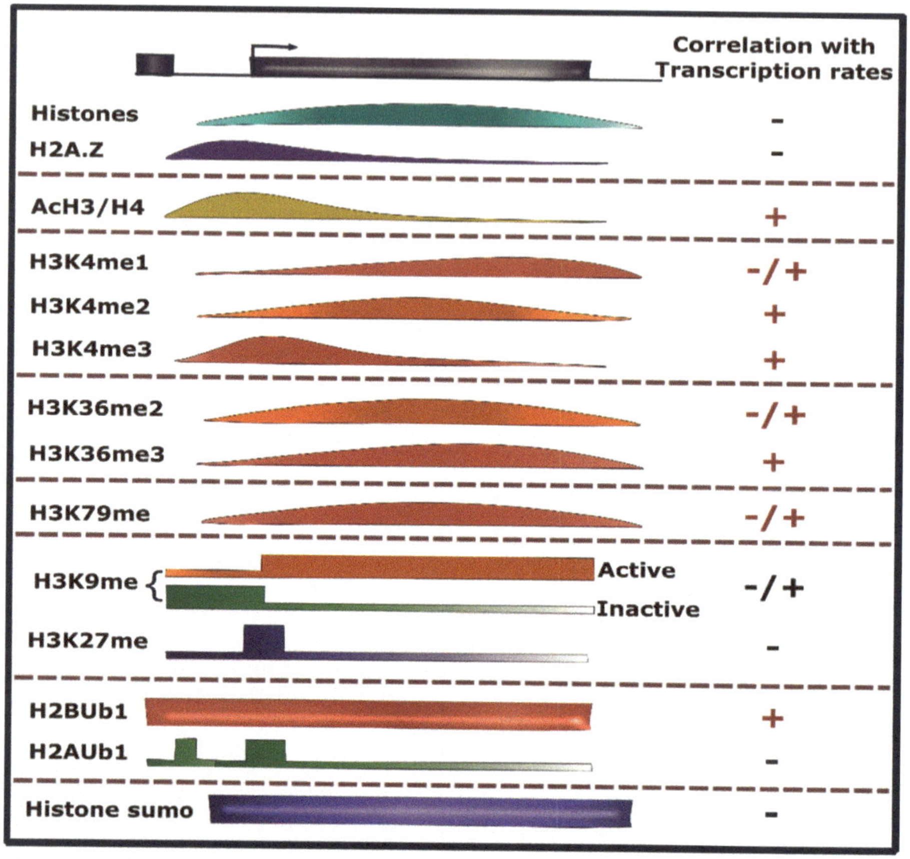

FIG. III.18 Genome-wide distribution of histone variant H2A.Z and various histone modifications as shown, over transcriptionally active and inactive genes. *Adapted from Li, B., et al., 2007. Cell 128 (4), 707–719. Elsevier.*

recognize these acetylation marks. However, histone H4K16 acetylation inhibits the higher-ordered folding of chromatin. Although H3K4 trimethylation (H3K4me3) is predominant in the promoter region, the gene body is enriched with mono- and dimethylation. Histone methyltransferase Set2-mediated methylation of histone H3K36 is enriched in the overactive gene body, but artificial methylation on the promoter region at H3K36 by Set2 inhibits the transcription.

In a bivalent promoter, both the activation mark (H3K4me3) and repressive mark (H3K27me3) are present together (Box III.4). The state of DNA methylation directly regulates the active, poised, and repressed state of the promoter and enhancer, with a signature pattern of histone modifications (Fig. III.19). In an embryonic stem cell, this was

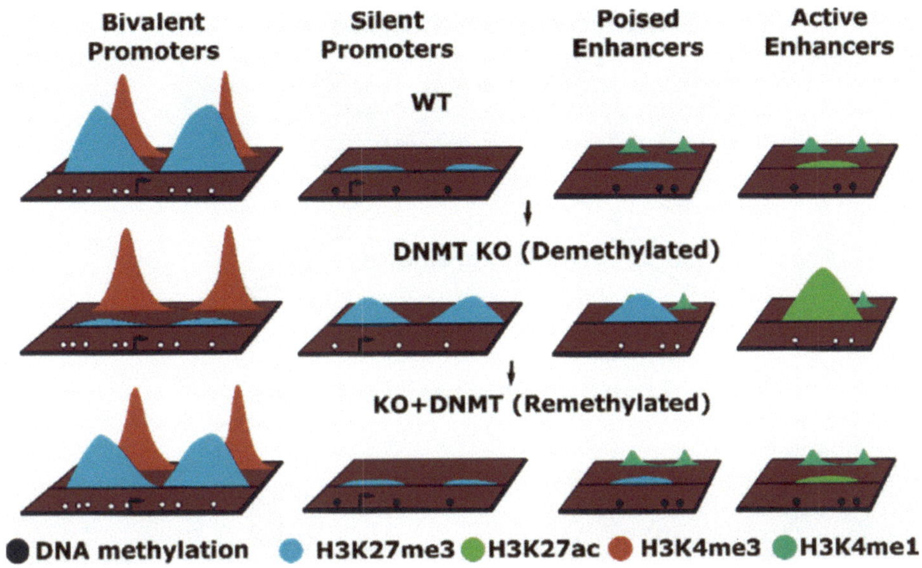

FIG. III.19 DNA methylation guides the establishment of epigenetic landscape in the dynamic epigenome. Bivalent promoters possess both H3K4me3 and H3K27me3 under the hypomethylated DNA background. Demethylation by knocking down of DNA methyltransferase and remethylation by ectopic expression of DNA methyltransferase, alter histone modifications dramatically both in the promoter and the enhancer region of the chromatin. *Adapted from King, A.D., et al., 2016. Cell Rep. 17 (1), 289–302. Elsevier.*

experimentally shown by knocking out the DNA methyltransferase and reexpressing it to rescue the phenotype.

The field of epigenomic dynamics is an expanding area of active research at present. Here, we have presented a glimpse of it from the perspective of chemical modifications and metabolic regulations. The cross-talk of these processes and their functional implications have been briefly described with a focus on transcription regulation and its physiological effects.

Bivalent chromatin

The bivalent chromatin domains possess both the repressive and activating epigenetic signatures. Depending upon the signal and context, they may shift either toward activation or repression of the gene. Bivalent chromatin is generally located in the pluripotent embryonic stem cells and imprinted genes. These domains are enriched with histone H3K4me3 and H3K27me3. However, these modifications could be present in different histone H3 molecules in the same octamer. The H3K27me3 represses the gene expression to stop the differentiation and helps to maintain the pluripotency. Depending upon the developmental signal, a regulatory factor may use the H3K4me3 signature to activate the transcription and proceed toward differentiation. The ATP-dependent chromatin remodelers assist in the eviction of the polycomb group of proteins to push the switch toward differentiation (Box III.4).

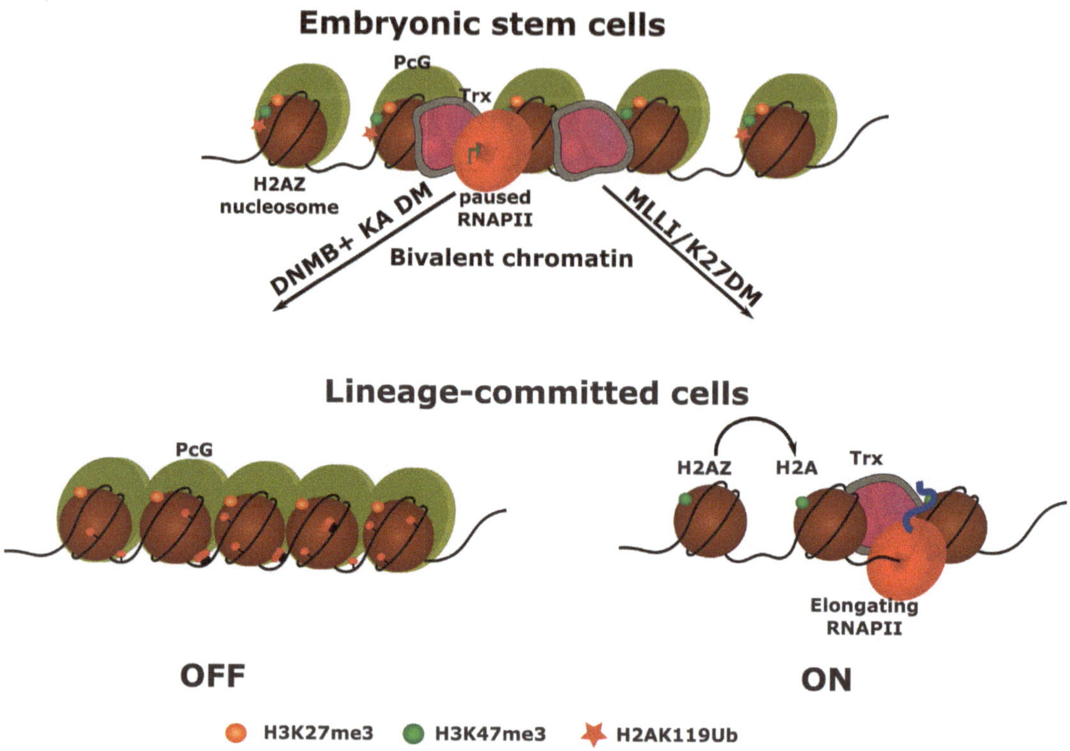

BOX III.4 Nature of bivalent chromatin state in embryonic stem cells. The transcriptional state of the bivalent chromatin may shift in either direction, activation or repression, depending upon necessary epigenetic modifications such as deacetylation or demethylation. *Modified from Sha, K., Boyer, L.A., StemBook, 2009.*

All the other DNA-templated molecular processes, namely DNA replication, recombination, and repair, are subjected to epigenetic regulation as well. To keep the chapter more conceptual than descriptive, we have not discussed several other important histone modifications such as phosphorylation, SUMOylation, and ubiquitination. The role of noncoding RNAs in epigenome dynamics, being out of the scope of the present chapter, has also not been discussed. The fundamentals and chemistry behind some of these processes will be described in detail in subsequent chapters.

III.6 Genome editing

Genome editing is a gradually improving technology (Fig. III.20) to insert, delete, modify, or replace genes (DNA) in the genome of a living cell or organism. In genome editing, precise targeting and insertion of genes are possible. The ability of site-specific insertion of a functional gene to replace a defective one has opened the golden opportunity to more

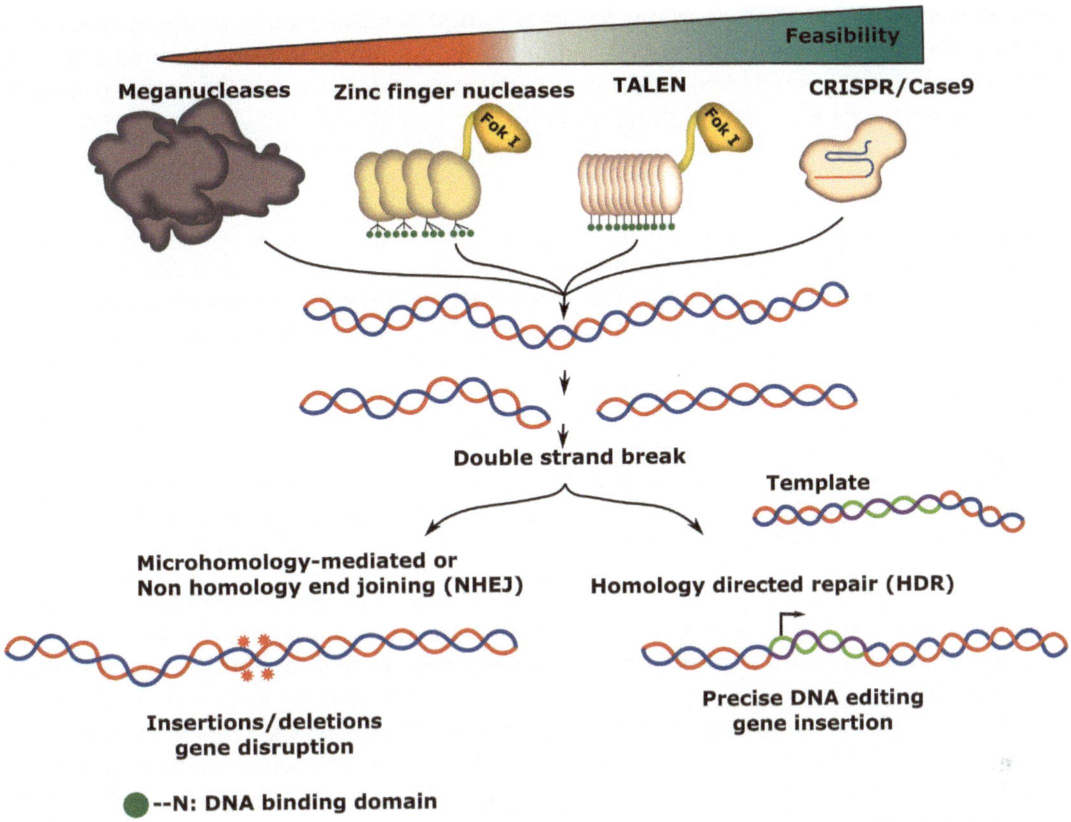

FIG. III.20 The evolving genome editing tools. *Adapted from Adli, M., 2018. Nat Commun. 9 (1), 1911. Springer Nature.*

precise gene therapy. Besides, there are several other uses of genome editing, such as the introduction of specific mutations and deletions to understand the gene function.

Mechanistically, in genome editing, creation of site-specific double-stranded breaks at a desired site in the genome are essential. Discovering a nuclease that can perform this job faithfully was the real challenge. Over the past three decades, dramatic improvement has been made in this direction. We shall briefly discuss four different types of nucleases in this context (more detailed descriptions are given in Chapter V). These are (i) meganucleases, (ii) zinc finger nucleases (ZFNs), (iii) transcription activator-like effector nucleases (TALEN), and (iv) clustered regularly interspaced short palindromic repeats (CRISPR or Cas9) system.

Meganuclease is a microbial enzyme that can recognize more than 12 bp of DNA, thus making it very specific. Unlike meganucleases, in ZFNs and TALEN technology, the DNA-cutting catalytic domain has no DNA sequence recognition capability. The nuclease catalytic domain is linked to specific DNA sequence-recognizing peptides, such as zinc fingers and transcription activator-like effectors.

CRISPRs consist of short sequences that are originally from a bacteriophage genome. Bacteria use this system as their acquired immune system against bacteriophages. The CRISPR-associated protein Cas cuts the viral DNA matching to CRISPRs. In an in vitro construct, plasmids containing the Cas genes and specifically constructed CRISPRs are introduced into eukaryotic cells to make gene editing possible, as depicted in Fig. III.20.

Chapter summary

In this chapter, we have described the fundamental basis for genome organization from an evolutionary perspective. The molecular organization of the genome and genome dynamics, focusing on epigenetic mechanisms, have been described. The coordination of these simple chemical modifications and their implications in life and health have been put forward.

- We have briefly presented the historical perspective of the multidisciplinary efforts of scientists around the world, over a century to discover the molecular nature of the gene.
- The fundamental nature of genome evolution has been presented in the context of the C-value and G-value paradoxes, where the size of the genome and the number of functional genes in an organism in the context of evolutionary complexity have been described.
- Over time, genomes have accumulated a huge amount of DNA, whereby several repetitive DNA elements have been accommodated and utilized for multiple purposes. We have described the jumping gene or transposon and its significance in life and health in this context.
- The molecular organization of the genome has been discussed from an evolutionary perspective. The ordered nucleoprotein organization of the genome, from the fundamental particle that is the nucleosome to the higher-order chromatin structures, has been described.
- An account of the unique epigenomic organization has been described. The dynamic nature of the epigenome connected to metabolism and cross-talk between various epigenetic modifications to fine-tune gene expression has been presented.
- Recent technological advances of genome editing to manipulate the gene function have been briefly described.

Further reading

Adli, M., 2018. The CRISPR tool kit for genome editing and beyond. Nat. Commun. 9 (1), 1911. https://doi.org/10.1038/s41467-018-04252-2.

Allfrey, V.G., Faulkner, R., Mirsky, A.E., 1964. Acetylation and methylation of histones and their possible role in the regulation of rna synthesis. Proc. Natl. Acad. Sci. U. S. A. 51, 786–794.

Bednar, J., Garcia-Saez, I., Boopathi, R., Cutter, A.R., Papai, G., Reymer, A., Syed, S.H., Lone, I.N., Tonchev, O., Crucifix, C., Menoni, H., Papin, C., Skoufias, D.A., Kurumizaka, H., Lavery, R., Hamiche, A., Hayes, J.J., Schultz, P., Angelov, D., Petosa, C., Dimitrov, S., 2017. Structure and dynamics of a 197 bp nucleosome in complex with linker histone H1. Mol. Cell 66 (5), 384–397.

Blackledge, N.P., Klose, R., 2011. CpG island chromatin: a platform for gene regulation. Epigenetics 6 (2), 147–152.

Brownell, J.E., Allis, C.D., 1995. An activity gel assay detects a single, catalytically active histone acetyltransferase subunit in Tetrahymena macronuclei. Proc. Natl. Acad. Sci. U. S. A. 92 (14), 6364–6368.

Burns, K.H., 2017. Transposable elements in cancer. Nat. Rev. Cancer 17 (7), 415–424.

Cavallaro, R.A., Nicolia, V., Fiorenza, M.T., Scarpa, S., Fuso, A., 2017. S-Adenosylmethionine and superoxide dismutase 1 synergistically counteract Alzheimer's disease features progression in TgCRND8 mice. Antioxidants (Basel) 6 (4), 1–17.

Cutter, A.R., Hayes, J.J., 2015. A brief review of nucleosome structure. FEBS Lett. 589 (20 Pt A), 2914–2922.

Das, C., Lucia, M.S., Hansen, K.C., Tyler, J.K., 2009. CBP/p300-mediated acetylation of histone H3 on lysine 56. Nature 459 (7243), 113–117.

Garcia-Saez, I., Menoni, H., Boopathi, R., Shukla, M.S., Soueidan, L., Noirclerc-Savoye, M., Le Roy, A., Skoufias, D.A., Bednar, J., Hamiche, A., Angelov, D., Petosa, C., Dimitrov, S., 2018. Structure of an H1-bound 6-nucleosome array reveals an untwisted two-start chromatin fiber conformation. Mol. Cell 72 (5), 902–915 e907.

Haws, S.A., Leech, C.M., Denu, J.M., 2020. Metabolism and the epigenome: a dynamic relationship. Trends Biochem. 45, 731–747.

King, A.D., Huang, K., Rubbi, L., Liu, S., Wang, C.Y., Wang, Y., Pellegrini, M., Fan, G., 2016. Reversible regulation of promoter and enhancer histone landscape by DNA methylation in mouse embryonic stem cells. Cell Rep. 17 (1), 289–302.

Kundu, T.K., Sikder, S., 2018. Evolution of genome organization and epigenetic machineries. J. Biosci. 43 (2), 239–242.

Lakadamyali, M., Cosma, M.P., 2020. Visualizing the genome in high resolution challenges our textbook understanding. Nat. Methods 17, 371–379.

Li, B., Carey, M., Workman, J.L., 2007. The role of chromatin during transcription. Cell 128 (4), 707–719.

Radman-Livaja, M., Rando, O.J., 2010. Nucleosome positioning: how is it established, and why does it matter? Dev. Biol. 339 (2), 258–266.

Reilly, M.T., Faulkner, G.J., Dubnau, J., Ponomarev, I., Gage, F.H., 2013. The role of transposable elements in health and diseases of the central nervous system. J. Neurosci. 33 (45), 17577–17586.

Song, F., Chen, P., Sun, D., Wang, M., Dong, L., Liang, D., Xu, R.M., Zhu, P., Li, G., 2014. Cryo-EM study of the chromatin fiber reveals a double helix twisted by tetranucleosomal units. Science 344 (6182), 376–380.

Soshnev, A.A., Josefowicz, S.Z., Allis, C.D., 2016. Greater than the sum of parts: complexity of the dynamic Epigenome. Mol. Cell 62 (5), 681–694.

IDT, 2010. A Brief History of DNA Tutorial. Integrated DNA Technologies, pp. 1–6.

IV

Chemical modifications of genome

IV.1 Review of chemical and electronic properties of DNA and protein

IV.1.1 Atomic and molecular orbitals

Atomic and molecular orbitals

An electron behaves both like a particle and wave (wave-particle duality). The electron is a quantum object, implying that its position and velocity cannot be determined simultaneously with absolute precision (Heisenberg's uncertainty principle). One can, however, calculate the probability of finding the electron in a given volume. In an atom, without going into a rigorous definition, we may consider the region of space where an electron is likely to be found with a very high probability as an atomic orbital.

When atoms bond to form a molecule, the overlap of atomic orbitals allows the unpaired electrons to be shared for forming covalent bonds. Again, without going into a rigorous definition, we may consider that analogous to the atomic orbital, the regions of space, where shared electrons can be found in a molecule with high probability as molecular orbitals.

Carbon is the most abundant element in biological molecules; its electronic properties play a pivotal role in building up biological macromolecules. Apart from carbon, elements such as nitrogen, hydrogen, and oxygen also play crucial roles. The phenomenon of hybridization of atomic orbitals holds a key to understanding some important structural, chemical, and electronic properties of DNA and proteins. We will recapitulate some important features of the hybridization of atomic orbitals first.

Aufbau principle, Hund's rule, and Pauli exclusion principle

Rules stated earlier govern how electrons fill atomic orbitals as an atom is built-up. Filling up of electrons starts from 1 s orbital, followed by 2 s, then 2p, and so on. The Pauli exclusion principle defines that only two electrons of opposite spin can populate a subshell, such as 1 s or $2p_x$. Hund's rule defines the principle of adding electrons when subshells of the same energy (degenerate) are available. It states that electron must fill the degenerate subshells with single electrons of the same spin before adding more electrons to the subshells. An application of Hund's rule is the electronic configuration of the neutral nitrogen atom. Electronic configuration of nitrogen is $1s^2 2s^2 2p^3$. Degenerate p-shells, p_x, p_y, and p_z will have one electron each, giving the nitrogen atom an electronic ground state of $1s^2 2s^2 2p_x^1 p_y^1 p_z^1$.

Chemical Biology of the Genome. https://doi.org/10.1016/B978-0-12-817644-3.00001-5

A neutral carbon atom has an electronic configuration of $1s^22s^22p_x^1p_y^1p_z^0$. Hybridization of 2s and 2p orbitals occur commonly after promotion of a 2s electron to the vacant p orbital creating an electronic state of $1s^22s^12p_x^1p_y^1p_z^1$. The promotion is followed by the mixing of 2s and 2p orbitals. In the sp^3 hybridized state, the result is the creation of four equivalent orbitals (Fig. IV.1). In the sp^2 hybridized state (Fig. IV.2), one 2s and two 2p orbitals mix, leaving a p orbital in the original state. In an sp^2 hybridized state, the three hybrid orbitals are planar, with the unhybridized p orbital lying perpendicular to that plane. Two sp^2 hybridized carbon atoms could come together to form a π-bond (in addition to the C—C σ-bond) that has electron clouds above and below the plane containing sp^2 hybridized orbitals. Ethene is an example of such a bonding mode (Fig. IV.3).

The overlap of p-orbitals plays an essential role in generating the aromatic character. To properly understand the properties of aromatic side chains of proteins, we recapitulate

FIG. IV.1 Orbitals of a carbon atom in the sp^3 hybridized state.

FIG. IV.2 Orbitals of a carbon atom in the sp^2 hybridized state. The hybridized orbitals are in gray lying in the same plane and the remaining unhybridized p-orbital is lying above and below the plane of hybridized sp^2 orbitals.

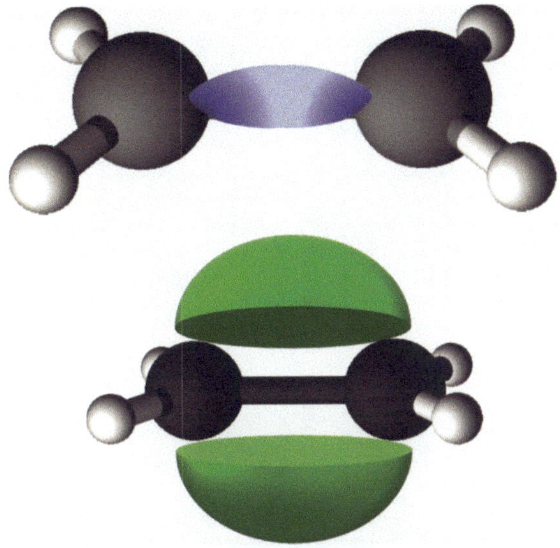

FIG. IV.3 *Top*: Formation of a σ-bond between two sp^2 hybridized carbon atoms. *Bottom*: Orbital structure of ethene molecule. The *green* orbitals are the π-orbital formed by the overlap of two unhybridized p-orbitals of two carbon atoms.

here the basic rules. Three essential criteria for aromaticity are a planar ring, a continuous conjugated closed system of π electrons, and the number of electrons in the π-electron system satisfying the Huckel's 4n+2 rule, where n is either zero or an integer. These conditions are met by the benzene ring (Fig. IV.4).

IV.1.2 Electronic properties of constituents of protein

Many naturally occurring amino acids possess side chain atoms having one or more lone pairs of electrons or can lose a proton under physiological conditions to generate one. These lone pairs of electrons are nucleophiles that may participate in nucleophilic substitution and addition reactions. The most potent of these nucleophiles is the ionized sulfur atom of the cysteine residue (Fig. IV.5). A solvent-exposed cysteine residue on a protein surface has a pK_a of around 8 and, hence, is significantly ionized at the physiological pH making it a powerful nucleophile.

Another potent nucleophile present in proteins is the ε-amino group of the lysine residue. It has a lone pair of electrons in the unprotonated state and is capable of participating in nucleophilic addition and substitution reactions. A solvent-exposed lysine has a pK_a around 10 and remains mostly protonated at physiological conditions (Fig. IV.6). However, in many enzymatic reactions, the proton is first removed to generate a lone electron pair, making its participation in nucleophilic reactions possible. Many chemical modifications, such as acetylation and methylation, occur on a lysine residue because of its nucleophilic character.

FIG. IV.4 Orbital arrangement in benzene molecule. Unhybridized p-orbitals *(top)* overlap to produce π-molecular orbitals above and below the plane of the ring *(below)*.

The imidazole ring of histidine contains two nitrogen atoms in its ring system. To understand how nitrogen atoms in the imidazole confer aromaticity and influence the electron density in the ring, and other properties of imidazole, we have to understand the electronic configuration of the ring nitrogen atoms. A nitrogen atom in its lowest energy state has an electronic configuration of $1s^2 2s^2 2p_x^1 2p_y^1 2p_z^1$. In an sp^3 hybridization state, the nitrogen atom will have a tetrahedral geometry with one orbital occupied by the lone electron pair (see Fig. IV.6). This hybridization state occurs in ammonia and amines.

Aromatic rings must be planar to create effective orbital overlaps above and below the ring plane, which is necessary for the delocalization of electrons of the unhybridized p-orbitals. A nitrogen atom must be in an sp^2 hybridized state to become a member of a planar aromatic ring system. The sp^2 hybridized orbitals of nitrogen atoms point toward

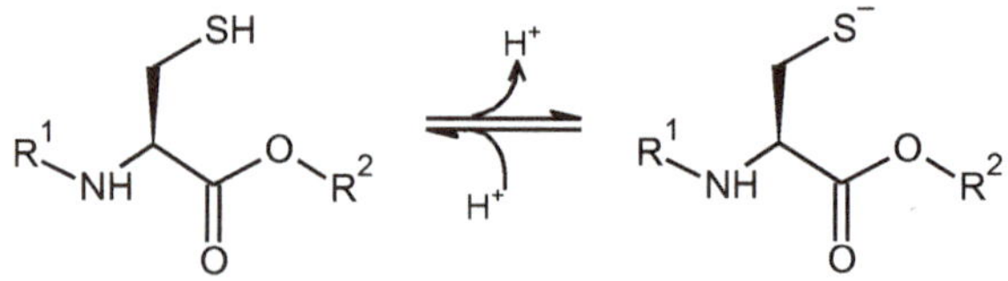

FIG. IV.5 Chemical structures of cysteine *(left)* and cysteine anion *(right)*.

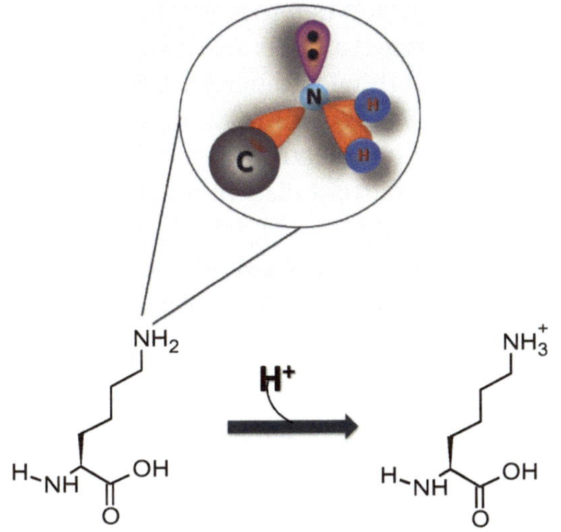

FIG. IV.6 Protonation of ε-amino group of the lysine. The lone pair is shown in purple color.

vertices of an equilateral triangle in a plane. The remaining p orbital is perpendicular to this plane. Pyridine is a six-membered heterocycle containing one sp^2 nitrogen atom in its ring system (Fig. IV.7). Its nitrogen lone pair occupies an sp^2 hybridized orbital in the ring plane, and the unhybridized p orbital, containing an unpaired electron, is perpendicular to the ring plane. This out-of-plane p-orbital participates in the formation of the aromatic π-electron system, and the other two sp^2 planar orbitals, having an unpaired electron each, participate in σ-bonds with two adjacent carbon atoms in the ring. Thus the lone pair is coplanar with the aromatic ring and free to accept a proton. Thus pyridine is a moderately

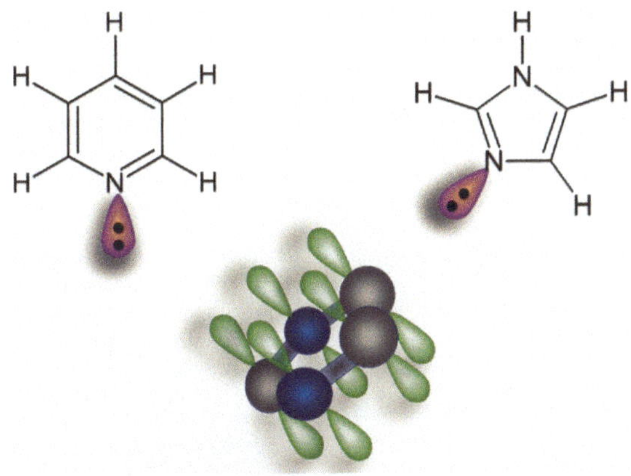

FIG. IV.7 Structure of pyridine and imidazole showing the lone pair of electrons *(top)*. Unhybridized p orbitals (green) of the imidazole ring tilted for sake of clarity, above and below the plane *(bottom)*.

good base. Pyrrole is a five-membered heterocycle containing one nitrogen atom in its ring system. Like pyridine, the nitrogen atom is also sp^2 hybridized; however, here the lone pair resides in the out-of-plane p orbital and becomes a part of the ring π-electrons. The participation of the lone pair is necessary for fulfilling the Huckel's $4n+2$ rule (four electrons from four carbon atoms and two electrons from the nitrogen lone pair). The participation of the lone pair in the ring current distributes six electrons over five atoms in the ring, making the ring system electron rich. Thus the pyrrole undergoes facile attacks from electrophiles and undergoes electrophilic substitution reactions.

The five-membered imidazole ring has two nitrogen atoms in the ring system (Fig. IV.8). To satisfy the $4n+2$ rule, one of the nitrogen atoms must donate two electrons to the π-electron system and the other a single one. Thus the nitrogen donating a single electron assumes a pyridine nitrogen-like character and possesses a free lone pair, whereas the other nitrogen has a pyrrole nitrogen-like character (Fig. IV.7). The nitrogen atom having the free lone pair is a moderately good nucleophile. This lone pair is readily available for protonation, making histidine side chain pK_as around 6.0. At physiological conditions, the imidazole ring exists in a mixture of protonated and unprotonated states (Fig. IV.8), giving it the ability to act as either a general acid or a general base.

Tyrosine is another frequently modified amino acid in proteins. In vivo, the exocyclic oxygen is the primary site of modifications. The exocyclic oxygen of tyrosine donates electrons to the aromatic ring, making the ring electron rich and the oxygen electron deficient (Fig. IV.9). Electron sharing makes the exocyclic hydroxyl group weakly acidic, remaining mostly protonated at physiological conditions. Deprotonation occurs during enzymatic modifications. Deprotonated tyrosine is a moderately powerful nucleophile and participates in nucleophilic substitutions, such as phosphoryl transfer reactions.

Side chains of serine and threonine residues contain hydroxyl groups attached to aliphatic carbon atoms. The absence of a conjugated aromatic ring and electron withdrawal makes them weaker acids than tyrosine, thus, making corresponding conjugate bases very powerful nucleophiles. Serine and threonine residues are targets of phosphorylations by serine or threonine protein kinases.

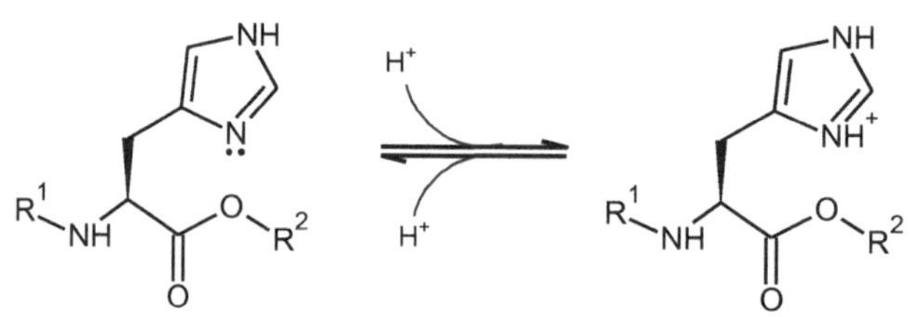

FIG. IV.8 Protonated and deprotonated states of imidazole side chain of histidine.

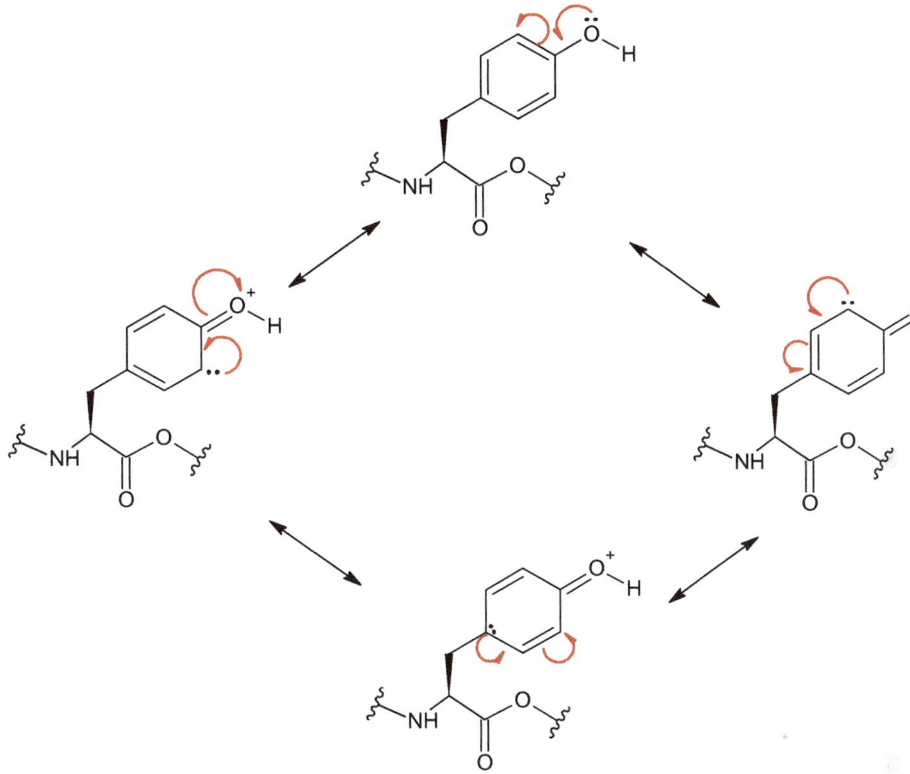

FIG. IV.9 Resonance structure of tyrosine.

The guanidinium group of arginine has a distinctly different character from other amino acid side chains. At physiological conditions, the guanidinium group is always protonated, leading to the essential equivalence of all the nitrogen atoms (Fig. IV.10). A "Y" shaped delocalized π orbital forms in the guanidinium cation, giving it high stability and high pK_a. This stability makes it difficult to deprotonate the guanidinium cation, even

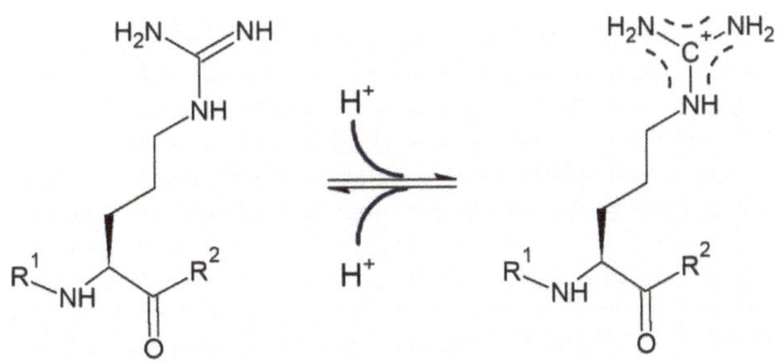

FIG. IV.10 Charge delocalization in the protonated form of guanidinium group of arginine.

FIG. IV.11 Atomic orbitals of the guanidinium cation. The *green* orbitals are p orbitals containing the nitrogen lone pair, and the *gray* orbital over the carbon atom indicates a vacant p orbital. All the C and N atoms are sp^2 hybridized with one p orbital perpendicular to the plane that contains all the atoms. The electrons in the p orbitals (three lone pairs, each from one nitrogen atom) delocalize over one carbon and three nitrogen atoms.

in hydrophobic protein interiors. Fig. IV.11 shows the atomic orbitals (schematic representation) of the protonated guanidinium group.

IV.1.3 Electronic properties of DNA bases and the phosphodiester backbone

DNA consists of two kinds of bases, the purines and the pyrimidines. In DNA, the most abundant purines are adenine and guanine, whereas the most abundant pyrimidines are thymine and cytosine (Figs. IV.12–IV.15).

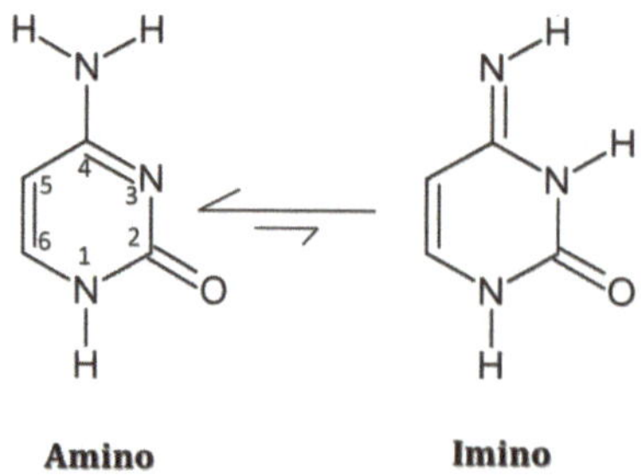

FIG. IV.12 Tautomeric forms of cytosine. The amino form predominates in most situations.

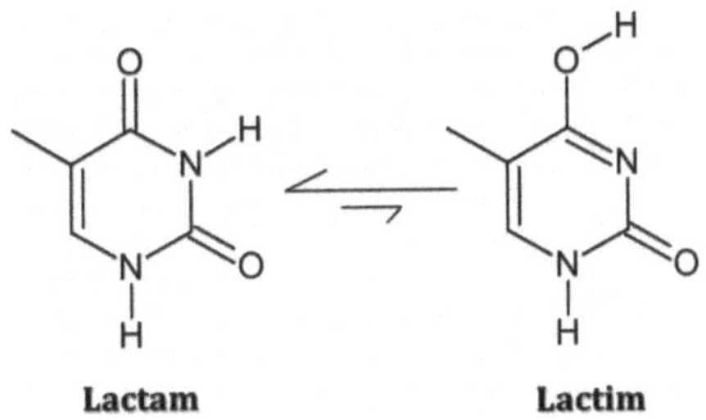

FIG. IV.13 Tautomeric forms of thymine. The lactam form predominates in most situations.

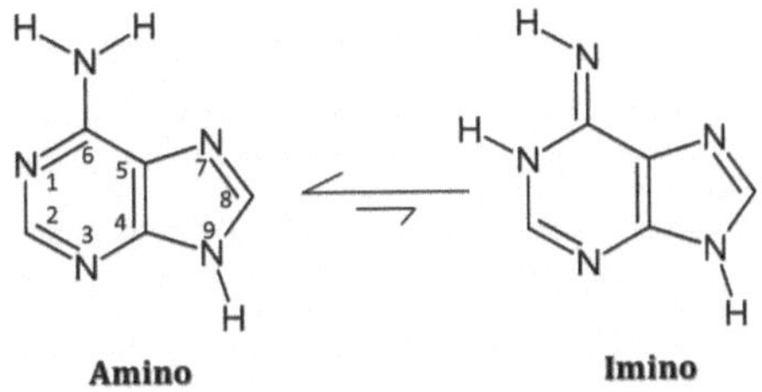

FIG. IV.14 Tautomeric forms of adenine. The amino form predominates in most situations.

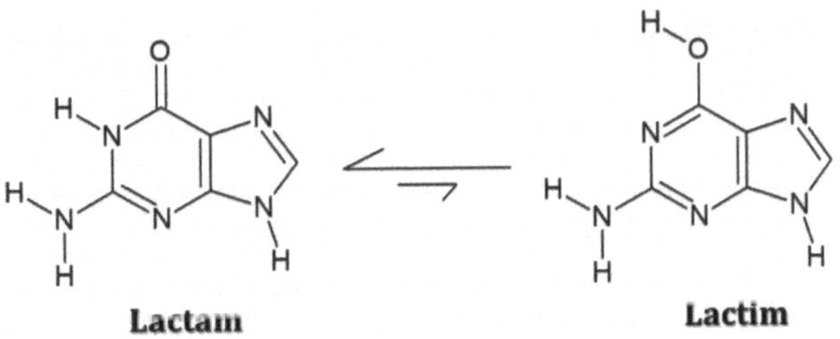

FIG. IV.15 Tautomeric forms of guanine. The lactam form predominates in most situations.

Electronic structures of DNA bases

All the DNA bases are aromatic ring systems and contain π-electron clouds similar to benzene. Purine and pyrimidine rings are heterocycles containing multiple nitrogen atoms in their ring systems. The presence of atoms of different electronegativities in the ring causes a nonuniform distribution of the π-electrons. Electron density is the highest at C5 of the pyrimidine ring making it most susceptible to electrophilic attack. In the purine ring, C8 has sufficient electron density to make it susceptible to electrophilic substitution. N7 lone pair is also the target of electrophilic reagents. Exocyclic nitrogen and oxygen atoms are modestly susceptible to electrophilic attack due to the presence of lone pairs of electrons that are conjugated to the ring π-electron system.

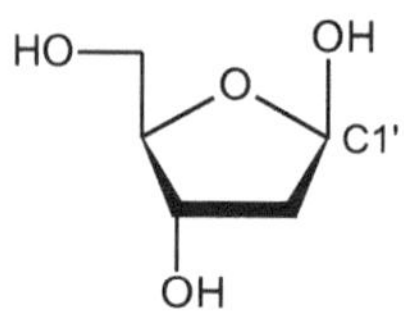

FIG. IV.16 Structure of 2′-deoxyribose sugar.

The pyrimidine ring is a six-member heterocyclic ring containing two nitrogen atoms. Both the nitrogen atoms donate one electron each to the ring π-electron system, thus, making the nitrogen atoms pyridine like. Thymine has two exocyclic hydroxyl groups, at C2 and C4, and a methyl group at C5. Cytosine has an exocyclic amino group at C4 and a hydroxyl group at C2. Both thymine and cytosine undergo tautomerism. Tautomerism is a rapid equilibrium between two isomers, where a proton shifts from one atom to the other. The best-known example is the keto-enol tautomerism. The lactam and the amino form predominate in thymine and cytosine, respectively (Figs. IV.12 and IV.13). In thymine and cytosine, the N1 bears a hydrogen atom. This hydrogen is replaced by C1′ of the deoxyribose sugar to produce the corresponding nucleosides, thymidine, and cytidine.

The purine ring is a fusion of an imidazole ring and a pyrimidine ring. In the five-member imidazole portion, N9 has a pyrrole-like character and N7 is pyridine like. N9 is covalently linked to the 2′-deoxyribose in nucleosides, leaving the N7 lone pair free. The N1 position is hydrogen bonded in a duplex DNA (Watson-Crick base pairing) and resistant to chemical reactions, whereas the N3 position does not participate in hydrogen bonding and is susceptible to attack by electrophiles. Exocyclic polar atoms possessing lone electron pairs also participate in nucleophilic substitution reactions (Figs. IV.14 and IV.15).

The oxygen atoms in 2′-deoxyribose sugars (Fig. IV.16) in the DNA chain participate in covalent bonds and, hence, cannot ionize and create good nucleophilic groups. Thus, under mild conditions, they do not participate in nucleophilic substitution reactions. However, they are susceptible to free-radical reactions causing strand breaks. We will describe these reactions in the following section.

IV.1.4 Review of principles of chemical equilibrium and reaction rate

Chemical reactions of small molecules have been studied since the dawn of modern chemistry. A chemical reaction can proceed to a significant extent only when the free energy of the products is lower than that of the reactants. The quantitative formulation of these principles is now an essential component of modern equilibrium chemical thermodynamics. Even though the product(s) having lower free energy than that of the reactant(s) is a necessary condition for the progress of a chemical reaction, it is not always sufficient. Not all chemical reactions having favorable free energy will proceed toward completion in a reasonable time frame. An oft-cited example of the insufficiency of the criterion of just having favorable free energy is that of a mixture of hydrogen and

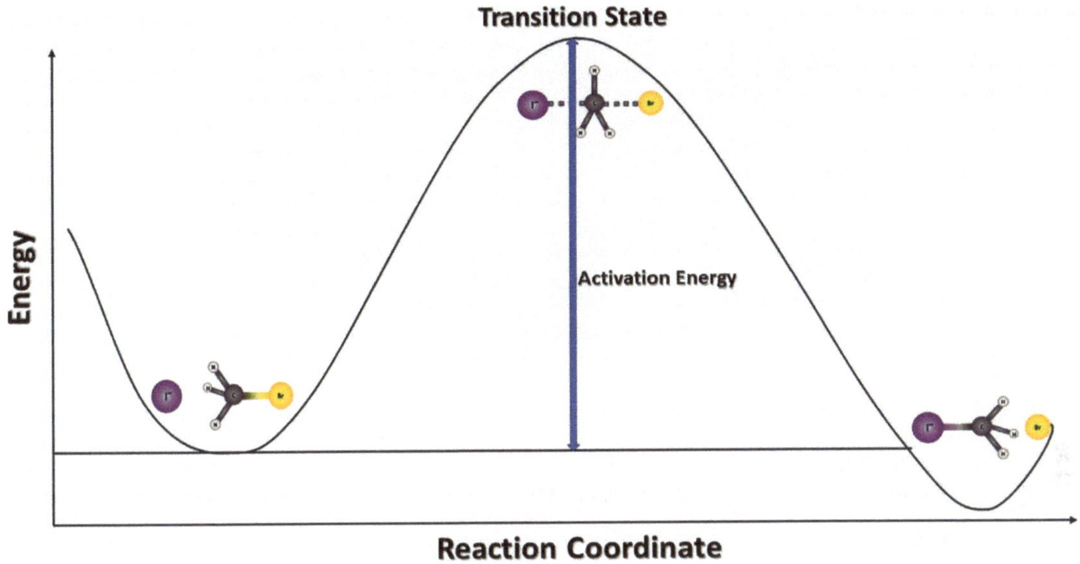

FIG. IV.17 Energy profile of the reaction of $I^- + CH_3Br$ to $Br^- + CH_3I$.

oxygen gases. Although the free energy of the product, water in this instance, has lower free energy than that of a mixture of hydrogen and oxygen, the mixture will not spontaneously react to form water, unless ignited externally. Initiation of the reaction at a rate sufficient enough for completion within a reasonable time frame may require input from outside.

The concept of rate and the equilibrium of chemical reactions can be integrated into a single free-energy diagram (Fig. IV.17). The X-axis of this diagram depicts the progress of a reaction, represented as a reaction coordinate, as some bonds break and new ones form. The reaction proceeds by crossing over a free-energy barrier. The energy difference between the reactants and the highest point of the barrier, called the transition state, is the activation energy of the forward reaction. Clearly, both the conditions mentioned in the previous section are necessary for significant progress of spontaneous chemical reactions.

Although these two conditions are sufficient for predicting the progress of reaction for small molecule reactants, reactions between small and macromolecules face additional complexity. Macromolecules often fold into compact structures, making the interior inaccessible to reactants. Thus the accessibility of reacting atoms in macromolecules becomes an additional criterion for reactions to occur. We cite the case of three cysteines of λ-repressor as an example. All three cysteine residues in the protein are unreactive to the thiol-reactive reagent, DTNB (5,5′-dithiobis-(2-nitrobenzoic acid). Fig. IV.18 shows the crystal structure of the dimeric λ-repressor. Three cysteine residues are partially buried inside the protein, explaining their unreactive nature. However, we must remember that reactant accessibility is not solely determined by static accessibility, as the dynamic nature of macromolecules may transiently open up the interior, allowing access to the interior residues.

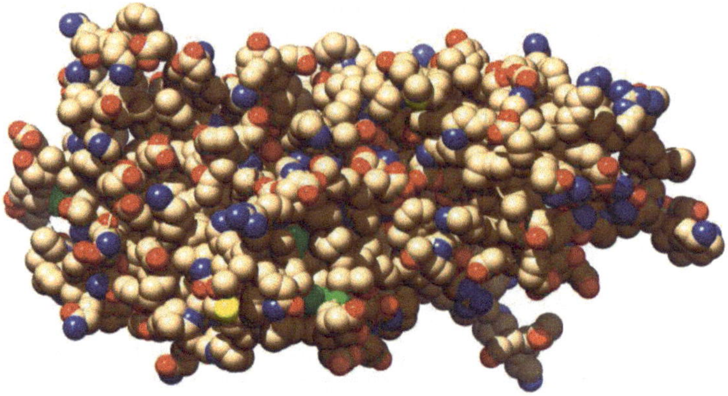

FIG. IV.18 Space-filling model of λ-repressor monomer (PDB3BDN). The *green-colored* residues are three cysteine residues present in the protein.

Steady-state concentrations of modified proteins

Many genomic proteins undergo constant modifications and their removal. These modifications are in a dynamic state. A kinetic explanation of the dynamic state is as follows:

$$A \underset{k_2}{\overset{k_1}{\rightleftarrows}} B,$$

where A is the unmodified protein, and B is the modified protein. We cite the example of a kinase-phosphatase duo. The kinase may be phosphorylating a serine, and the phosphatase is removing the phosphate. This reaction is not at equilibrium but are two separate reactions. After some time, the rate of the forward reaction will reach the rate of the backward reaction. At that point, the reaction will reach the steady state. The following equation gives the rate of change of $[B]$:

$$\frac{d[B]}{dt} = k_1[A] - k_2[B]$$

At steady state, $k_1[A] - k_2[B] = 0$. Because $[A] + [B]$ is constant, let us say, C, the equation can be written as follows:

$$k_1([C] - [B]) - k_2[B] = 0$$

$$\text{or } \frac{k_1}{k_1 + k_2} = \frac{[B]}{[C]}$$

The right-hand side of the equation is the fraction of the total protein having the modification. The rate constants k_1 and k_2 are dependent on enzyme activities. Thus, by controlling the activity of the enzymes, the steady-state concentration of B can be changed. The regulation of the rates of two reactions is an excellent way to regulate the concentration of the modified protein.

IV.1.5 Review of enzyme catalysis

Enzymes catalyze almost all the reactions that occur in the genome. Our understanding of the enzyme catalysis is reasonably well developed but incomplete. There are many ways by which an enzyme catalyzes a reaction. We will primarily focus on the chemical mechanisms, that is, how nucleophiles and electrophiles present at or near the active site help the reaction to proceed smoothly. As pointed out earlier, many of the protein side chains are powerful nucleophiles or general acids and bases. Thus they can form covalent intermediates, protonate a substrate, or deprotonate it. These actions make it possible for the reaction to proceed along a path of favorable energy barrier.

Role of acids and bases in enzyme catalysis

In chemical reactions in water, the transfer of protons often plays a vital role. The proton transfer occurs in water through different routes. H_3O^+ may donate a proton, and OH^- may remove a proton from reacting species or reaction intermediates. We call H_3O^+ a specific acid and OH^- a specific base. Solutes may donate or accept protons as well. Solutes that donate protons are called general acids, and solutes that accept protons are called general bases. Many groups on the enzyme, primarily ionizable side chains of amino acids, can act as general acids or general bases.

Enzymes enhance the rate of reaction enormously and show high specificity toward their substrates. A small change in the substrate structure may drop the reaction rate to insignificant levels. Many factors contribute to this high specificity toward their cognate substrate. We will not discuss these aspects here as they are described in many textbooks. Instead, we will discuss aspects of enzyme action most relevant to the chemistry of genomes.

A key feature of chemical modifications that occur on chromatin constituents is that they are limited to small regions. This locale specificity may be necessary to impart selective regulation of genes. Such locale-specific modifications may be achieved through tethering of modifying enzymes to a specific DNA sequence. Additional layers of regulatory mechanisms may exist. A recent demonstration that binding of p300 to p53 leads to enhanced autoacetylation suggests that p300 may only be active when it finds the right partner.

Enzyme kinetics

Chemical modifications play a significant role in gene regulation and epigenetic inheritance. Understanding the chemistry of protein and DNA modifications is immensely important for developing enzyme inhibitors. The study of enzyme kinetics offers a path toward understanding enzyme mechanisms. Enzymes that catalyze the conversion of a single substrate are generally kinetically simple and follow the well-known Michaelis-Menten kinetics at substrate concentrations considerably higher than the enzyme concentration. Enzymes that act on multiple substrates are kinetically more complex and have several mechanisms. In multisubstrate enzymes, two major classes of kinetic mechanisms are observed, the sequential and ping-pong. In the sequential mechanism, two substrates come in and form a ternary complex. Two substrates can bind either in a random manner (a random mechanism) or in an ordered manner (an ordered mechanism). In the latter case, one substrate must bind first for the second substrate to bind. A special case of the sequential mechanism is the Theorell–Chance kinetics. In Theorell-Chance scheme, the ternary complex is transient, and its existence cannot be detected under normal conditions. In the ping-pong mechanism, the first substrate binds, modifies the enzyme, and leaves as a product; this is followed by binding of the second substrate that receives the modified group from the enzyme and leaves as the second product.

IV.2 Reactions of proteins and DNA

IV.2.1 Major classes of reactions that proteins and DNA participate in vivo

IV.2.1.1 Electrophilic substitution reactions

In electrophilic substitution reactions, the incoming reacting species is electron deficient, and the molecule undergoing substitution is electron rich. The incoming electrophilic species forms a new bond(s) with the electron-rich species undergoing substitution and displaces an atom or a group of atoms, called the leaving group (Fig. IV.19). Electrophilic substitution reactions are divided into two broad classes: Electrophilic aromatic substitutions and electrophilic aliphatic substitutions. Electrophilic substitution reactions are scarce in vivo.

IV.2.1.2 Nucleophilic substitution reactions

Nucleophilic substitution reactions frequently occur in the genome. In nucleophilic substitution reactions, a nucleophile—most nucleophiles possess lone electron pairs—displaces an atom or a group of atoms (the leaving group) attached to an electron-deficient atom. From previous discussions on the electronic properties of amino acids and nucleotides, we know that many of them possess lone electron pairs or ionize under physiological conditions to generate one, making them powerful nucleophiles. A well-known example is that of carboxy-methylation of cysteine that is routinely performed before protein analysis. The ionized thiol group of cysteine attacks the carbon atom of

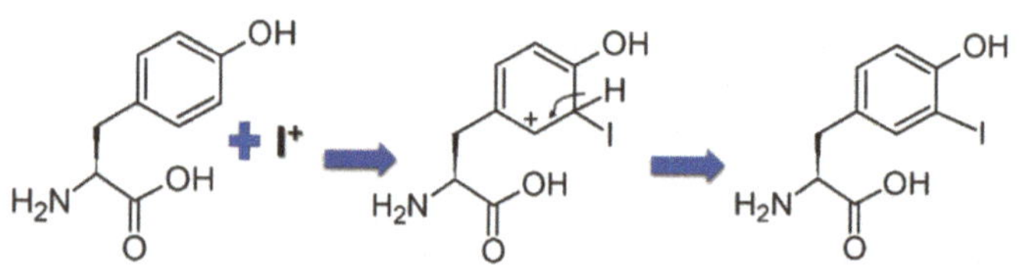

FIG. IV.19 Example of an electrophilic aromatic substitution reaction. I^+, a highly reactive species generated in situ is the incoming electrophile that substitutes a hydrogen atom in the aromatic ring of the tyrosine.

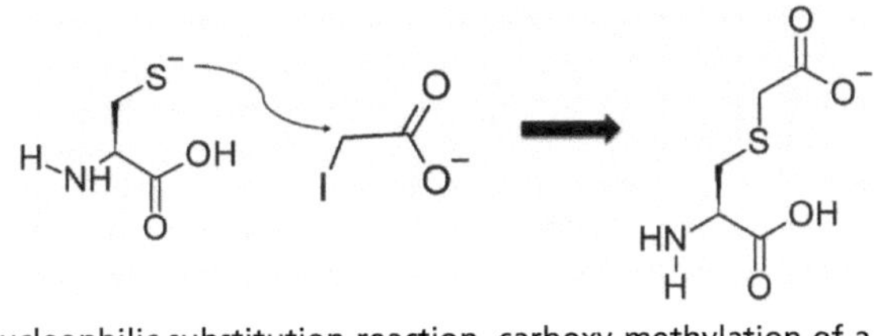

FIG. IV.20 An example of nucleophilic substitution reaction, carboxy-methylation of a cysteine residue.

iodoacetic acid that bears the iodine atom. The thiol displaces the iodine to produce carboxymethylated cysteine (Fig. IV.20).

IV.2.1.3 Elimination reactions

In elimination reactions, two atoms or groups are eliminated from a molecule to generate a double bond. In β-elimination reactions, two leaving groups from adjacent atoms are eliminated to create a double bond. An example of the β-elimination reaction is the generation of alkenes from alkyl halides. Elimination reactions are often part of consecutive reactions that may ultimately result in a substituted product. Lyases are enzymes that catalyze bond cleavage. A common cleavage reaction is the cleavage of a carbon-carbon bond. Mechanistically, cleavage reactions catalyzed by lyases are β-elimination reactions (Fig. IV.21).

IV.2.1.4 Addition reactions

Addition reactions occur when an atom or a group of atoms is combined with another species to generate a larger molecule, the adduct. There are two broad classes of addition reactions: electrophilic and nucleophilic, depending on the nature of the first incoming species. In electrophilic addition reactions, an electrophile first combines with a molecule, creating a cationic molecule that then reacts further to produce the final addition products. A well-studied example of electrophilic addition is the addition of hydrogen halides across an alkene (carbon-carbon double bond) (Fig. IV.22).

In nucleophilic addition reactions, a nucleophile first adds to a molecule; this intermediate further reacts to produce the final addition product. An example of a nucleophilic addition reaction is the addition of cyanide ion to the carbonyl group (Fig. IV.23). Carbonyl groups are prone to nucleophilic addition reactions due to significant charge separation

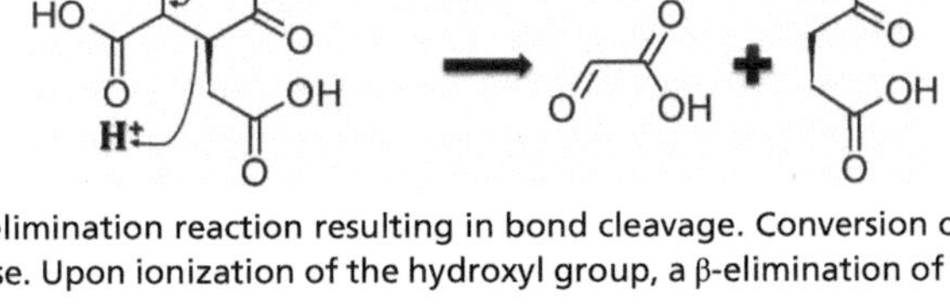

FIG. IV.21 An example of elimination reaction resulting in bond cleavage. Conversion of isocitrate to succinate and glyoxylate by isocitrate lyase. Upon ionization of the hydroxyl group, a β-elimination of the succinate moiety cleaves the carbon-carbon bond.

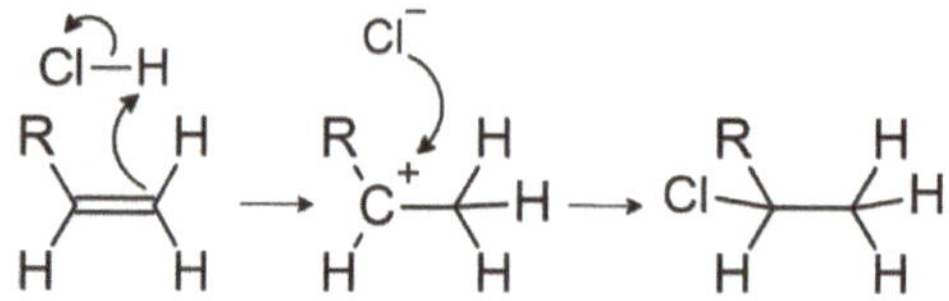

FIG. IV.22 Electrophilic addition of hydrogen chloride across a carbon-carbon double bond.

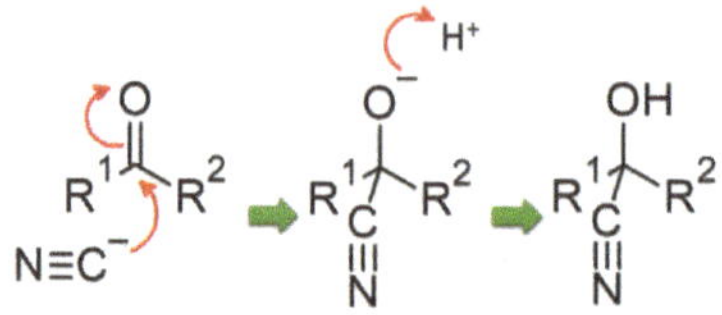

FIG. IV.23 Nucleophilic addition of cyanide to a carbonyl group.

and consequent electron deficiency at the carbon atom. Charge separation occurs because of the strongly electronegative nature of the oxygen atom. The electron-deficient carbon atom is prone to attack by nucleophiles. We will encounter this mode of reaction in subsequent sections.

IV.2.2 Mechanism of elimination reactions, nucleophilic substitution, and nucleophilic addition reactions

Many different types of chemical reactions occur in the genome. The most common types are the elimination, the nucleophilic substitution, and the nucleophilic addition reactions. We will explore these three types of reactions in greater detail. A common type of elimination reaction is the β-elimination, in which the two leaving groups leave from adjacent atoms. β-elimination can take place by E1, E2, or E1cb mechanism. The structure of the reactant and the nature of the leaving groups play a role in determining which mechanistic path will be followed. In the E1 mechanism, elimination takes place sequentially. One of the leaving groups is an excellent leaving group that leaves first, resulting in a carbocation, that then loses a proton forming an unsaturated double bond (Fig. IV.24). In the E2 mechanism, both the leaving groups leave simultaneously (Fig. IV.25). In contrast, the E1cb mechanism proceeds through a carbanion intermediate. In this mechanism, the leaving group is a poor one, and the elimination is base catalyzed. At first, a base removes a proton, generating a carbanion. The leaving group is then eliminated from an adjacent atom, producing an unsaturated bond (Fig. IV.26).

Nucleophilic substitution reactions generally take place on sp^3 hybridized carbons. They also occur on other atoms such as phosphorus. In nucleophilic substitution reactions, two main mechanisms, S_N1 and S_N2, have considerable similarity with the E1 and E2 mechanisms of the elimination reaction, respectively. In the S_N1 mechanism, the leaving group dissociates first, producing a carbonium ion intermediate (a positively

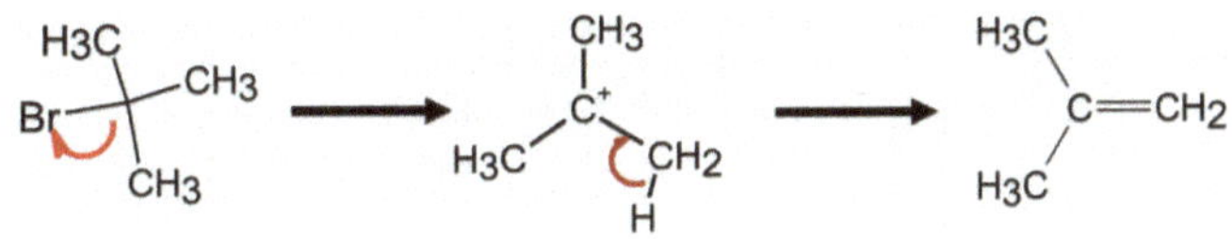

FIG. IV.24 β-elimination reaction by E1 mechanism.

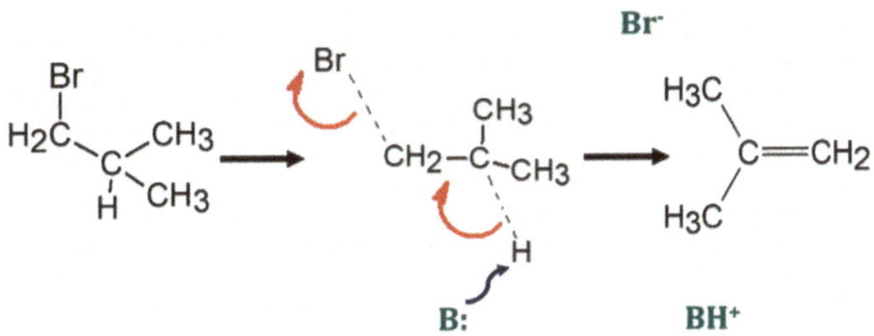

FIG. IV.25 β-elimination reaction by E2 mechanism.

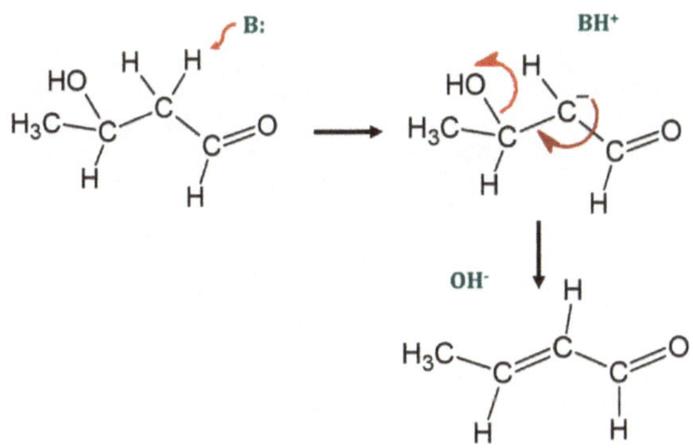

FIG. IV.26 β-elimination reaction by E1cb mechanism. B: stands for an external base.

charged carbon). The intermediate carbonium ion then combines with the incoming nucleophile to produce the product (Fig. IV.27). In the S_N2 mechanism, the incoming nucleophile and the leaving group approach and leave, respectively, in a concerted manner (Fig. IV.28).

Nucleophilic addition reactions are quite common in biology. An example of a nucleophilic addition reaction is the NADH-dependent reduction of unsaturated compounds. It involves the direct attack of a hydride ion from NADH or NADPH; the hydride ion acts as the incoming nucleophile in the nucleophilic addition reaction. As described before, for a strong nucleophile, the nucleophilic attack occurs first like the cyanide addition to the carbonyl group. In some cases, the oxygen atom of the carbonyl group may be protonated first to create a carbocation before the nucleophilic attack occurs.

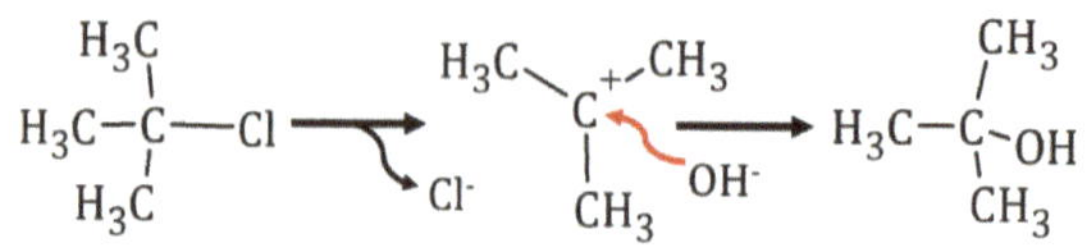

FIG. IV.27 Nucleophilic substitution reaction by S_N1 mechanism, hydrolysis of t-butyl chloride.

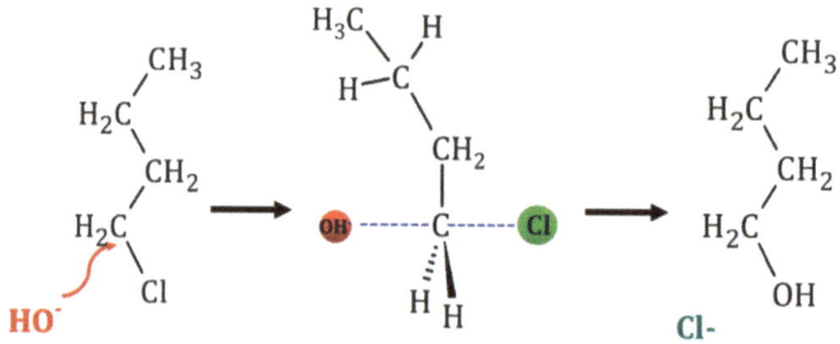

FIG. IV.28 Nucleophilic substitution reaction by S_N2 mechanism, showing the transition state of the hydrolysis of n-butyl chloride.

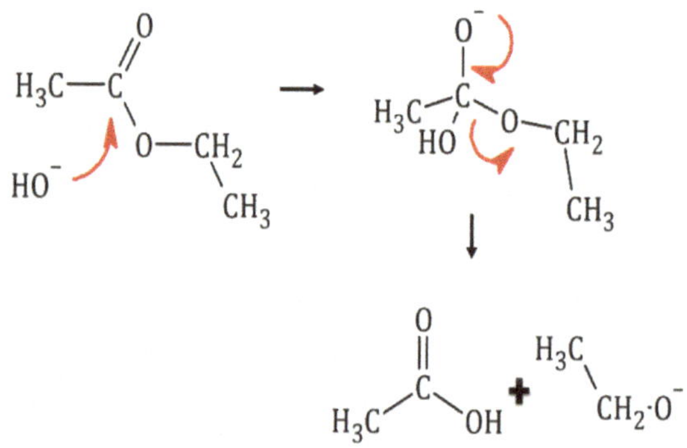

FIG. IV.29 Nucleophilic addition-elimination reaction mechanism; hydrolysis of ethyl acetate.

Another reaction mechanism often encountered in biology is the nucleophilic addition-elimination mechanism. In this mechanism, the incoming nucleophile first attacks the carbonyl carbon (sp^2 hybridization) and creates a tetrahedral intermediate (sp^3 hybridization). This intermediate then eliminates the leaving group to return to the carbonyl carbon (sp^2) form. Hydrolysis of ethyl acetate is an example of such a nucleophilic addition-elimination mechanism (Fig. IV.29).

IV.2.3 Mechanism of free-radical reactions

Free radicals are molecules or atoms that possess one or more unpaired electrons and have the potential to exist independently. In general, they are also extremely reactive. Although free radicals can participate in a variety of reactions, two important reactions from our perspective are the hydrogen atom abstraction from a C—H bond and addition to a double bond.

The hydrogen atom abstraction occurs during sunlight-catalyzed chlorination of alkanes. At first, the sunlight generates a chlorine radical by homolytic fission of the Cl—Cl bond. The reactive chlorine radical then abstracts a hydrogen atom from an alkane molecule to produce a hydrocarbon radical. This highly reactive hydrocarbon radical then reacts with a Cl_2 molecule to produce a chloroalkane molecule and releasing another Cl• radical, thus propagating the chain.

$$Cl-Cl \rightarrow Cl\bullet + Cl\bullet \rightarrow Cl\bullet + CH_4 \rightarrow HCl + CH_3\bullet \rightarrow$$

$$CH_3\bullet + Cl-Cl \rightarrow CH_3Cl + Cl\bullet$$

The rate of reaction depends on the C—H bond strength. The weakest C—H bonds occur in alkanes with tertiary hydrogens, making them the most reactive in free-radical substitution reactions.

Another common reaction of free radicals in biological systems is their addition to a double bond. This reaction is best demonstrated by the addition of hydrogen halides to alkenes. When the halogen atom is generated from a homolytic cleavage of H—X, where X is a halogen atom, the addition results in alkyl halides. However, the products of the free-radical reaction are somewhat different from the products of the corresponding electrophilic addition (see Fig. IV.22). In the free-radical mechanism, the halogen atom attacks first. It prefers a primary carbon for two reasons: the primary carbon is sterically least crowded as all the attached atoms are hydrogen atoms; the resulting carbon radical has its lone electron residing on a secondary carbon atom that is more stable than on a primary carbon (Fig. IV.30). The result is a primary halide, whereas the electrophilic addition results in a secondary or tertiary halide.

IV.3 Chemical modifications of DNA

IV.3.1 Methylation and demethylation of base atoms

Chemical modifications of nucleic acids are widespread. Transfer RNAs are one class of highly modified nucleic acids. Methylation of the DNA bases is ubiquitous in nature.

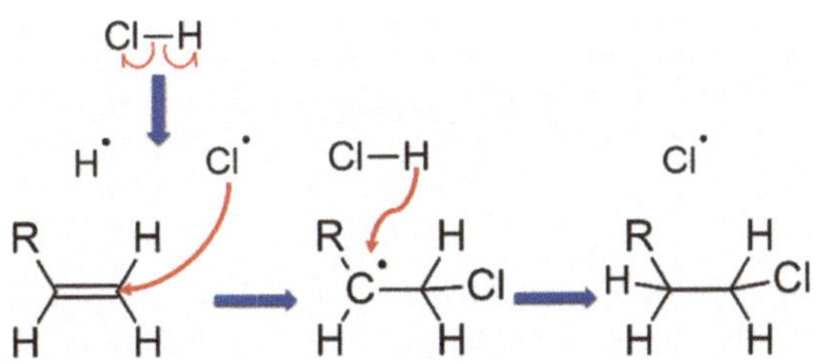

FIG. IV.30 Addition of hydrogen chloride to an alkene by free-radical mechanism.

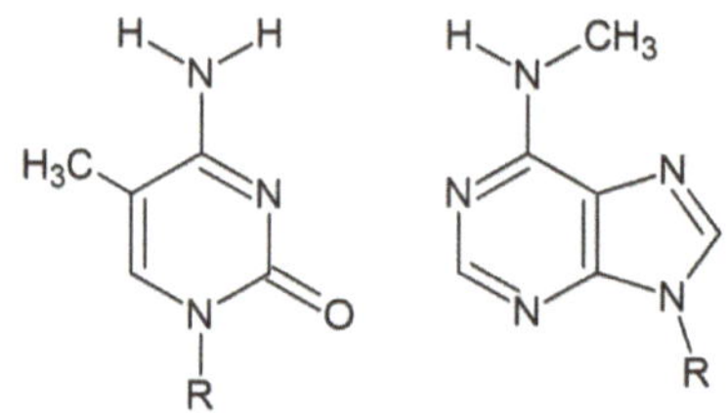

FIG. IV.31 Left panel, the structure of 5-methyl cytosine; right panel, the structure of 6-methyl adenine.

The two most common natural methylations occur on the C5 of cytosine and the exocyclic N6 of adenine (Fig. IV.31). The adenine methylation is widespread in the prokaryotic domain of life, whereas the cytosine methylation is widespread in eukaryotes. Enzymatic methylation at C5 of cytosine in the duplex DNA occurs in multiple steps (Fig. IV.32). First, the enzyme DNA methyltransferase binds the DNA, followed by the targeted cytosine flipping out from the duplex structure, and nucleophilic addition of a cysteine sulfhydryl group from the methyltransferase to the C6 position of the cytosine. This covalent adduct with the enzyme is highly electron rich in the C5 position of cytosine that then attacks the methyl group of S-adenosyl methionine in a nucleophilic displacement reaction. The nucleophilic displacement reaction, likely by an S_N2 mechanism, is facilitated as the

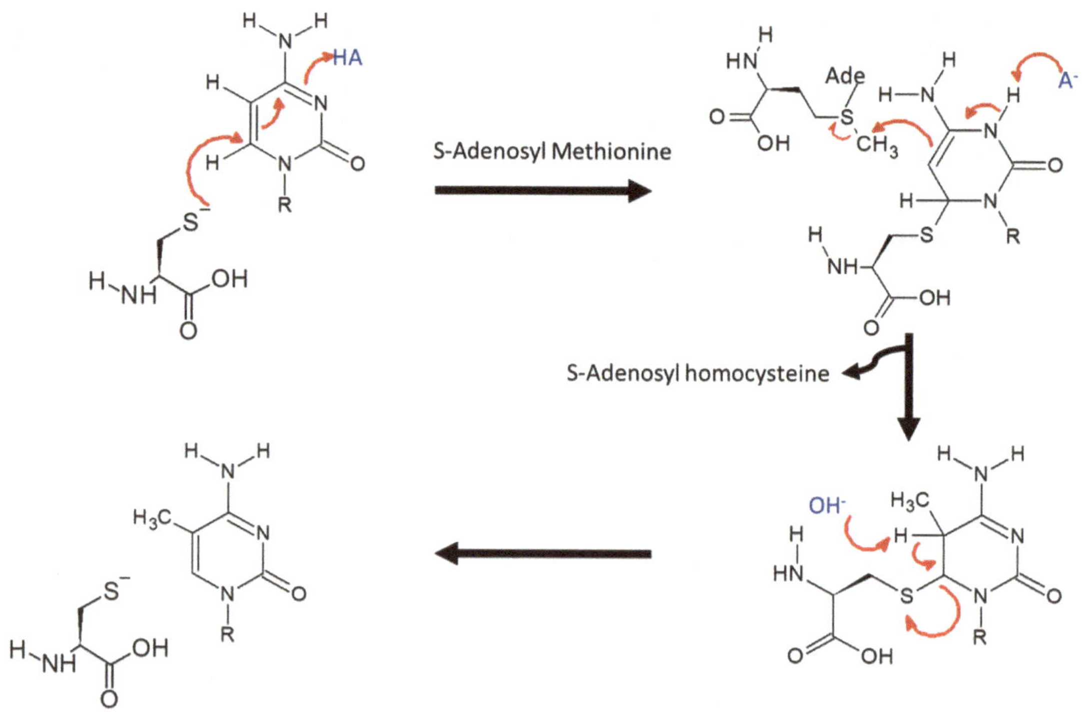

FIG. IV.32 General mechanism of methyl transfer to fifth position of cytosine.

S-adenosyl homocysteine is an excellent leaving group. N6-DNA methyltransferases (DNMTS) work similarly. They also use S-adenosyl methionine as the methyl donor. The lone pair of electrons on the N6 of adenine attacks the methyl group of S-adenosyl methionine directly, displacing S-adenosyl homocysteine in an S_N2 reaction.

In mammals, the majority of DNA methylation occurs in the C5 position of cytosine. Competing processes of methylation and demethylation maintain a steady-state level of methylation that may change depending on the cellular state. A major demethylation pathway is through oxidation of 5mC, followed by base excision repair (Fig. IV.33). The pathway is initiated by oxidation of 5-mC to 5-hydroxymethyl cytosine (5hmC), followed by further oxidation to 5-formyl cytosine (5fC), and 5-carboxyl cytosine (5caC). The base-excision repair (BER) system actively removes 5-formyl cytosine or 5-carboxyl cytosine to generate abasic sites, ultimately producing single-strand breaks. Polβ- or XRCCI- or Ligase

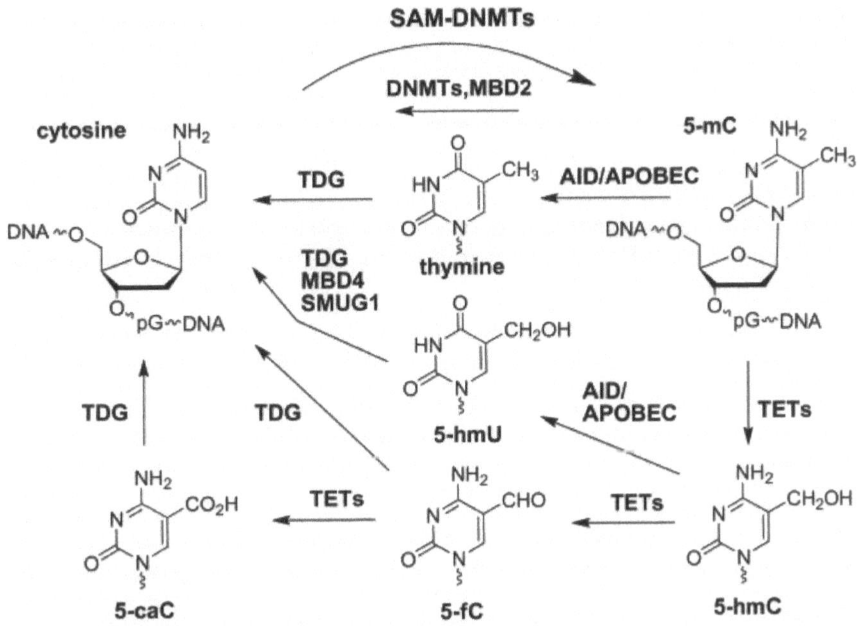

FIG. IV.33 Pathways of cytosine methylation and 5-methylcytosine demethylation. C is methylated to 5mC by DNA methyltransferases using S-adenosyl-L-methionine as a methyl donor. Several pathways have been proposed to explain active demethylation of 5mC (1–4): (1) single-enzyme removal of the methyl group by DNA methyltransferases or DNA methyl-binding domain protein 2; (2) deamination of 5-mC to thymine by activation-induced deaminase or apolipoprotein B mRNA-editing enzyme complex, followed by the removal of thymine by thymine DNA glycosylase; (3) oxidation of 5mC to 5-hydroxymethylcytosine (5hmC) by ten eleven translocation proteins, followed by deamination by activation-induced deaminase or apolipoprotein B mRNA-editing enzyme complex and removal of 5-hydroxymethyluracil by thymine DNA glycosylase, methyl-binding DNA domain protein 4, or single-strand-specific monofunctional uracil DNA glycosylase 1; (4) iterative oxidation of 5mC to further oxidation products, 5fC or 5caC, followed by the removal of the latter modifications by thymine DNA glycosylase. In addition, there may be several as yet not-well characterized proteins in mammalian cells that may catalyze the removal of different intermediates in 5mC demethylation. *Reproduced from Cadet, Wagner, 2014. Mutation Res. 764–765, 18–35. Elsevier.*

III-mediated gap repair system then repairs the single-strand breaks and replaces them with unmodified cytosine. Other pathways of 5mC demethylation may exist but they may be system specific.

Nature of reactions that occur on genomes

The chemical reactions inside the cell are highly specific, and in many cases localized, including those occurring on the genome. DNA sequence recognition is the pivot on which the localization of genome modifications rests. An example of such localized reactions is the reaction that demethylates 5-methyl cytosine, as described earlier. Primary targets of the demethylation machinery are the CpG dinucleotides. Once the substrates of the thymine DNA glycosylase (TDG) are generated (5fC and 5caC), the BER system takes over. However, if the successive reactions of base excision, generation of a single-stranded break, and gap-filling are not processive, it may lead to double-strand breaks. Recent studies suggest that the process of replacing a modified C with C is highly processive, preventing the occurrence of double-strand breaks. The abundance of supramolecular ensembles in the nucleus suggests that many genomic reactions are similarly localized.

IV.3.1.1 DNA methyltransferases

Methylation of C5 position of cytosine is catalyzed by a family of enzymes called DNMTS. There are three major DNMTS, DNMT1, DNMT3a, and DNMT3b in mammals. DNMT1 primarily methylates hemimethylated DNA, generated during replication, whereas DNMT3a and DNMT3b methylate unmodified DNA (de novo DNA methylation). Thus DNMT1 is primarily a maintenance methyltransferase; whereas, the latter two are de novo methyltransferases. The catalytic domains of all three enzymes are very similar. All three enzymes use CpG dinucleotides as substrates. For DNMT1, a strong preference exists for the hemimethylated CpG (mCpG in one strand and GpC on the other strand). Catalytic mechanisms appear to be very similar. In human DNMT1, Cys1226 is the residue that adds to the C6 position of the cytosine, and Glu1266 is the residue that donates the proton to the N3 position of the cytosine (AH in Fig. IV.32). Fig. IV.34 shows the active site architecture of human DNMT1.

IV.3.1.2 DNA demethylases

The oxidation of the 5-methyl group of cytosine is catalyzed by the ten eleven translocation (TET) family of proteins. In mammals, there are three distinct TET proteins, TET1, TET2, and TET3. All the three proteins catalyze the oxidation of 5-mC to 5-hydroxymethylcytosine (5hmC), 5-formyl-cytosine (5fC), and finally to 5-carboxyl-cytosine (5caC). All isoforms contain the CD domain (cysteine rich and Double-stranded beta helix), but the longer isoforms of TET1 and TET3, in addition, contain the CXXC (C for cysteine and X for any amino acid) domain. TET proteins are α-ketoglutarate dependent, nonheme Fe containing dioxygenases. The metal-binding residues reside in the DSBH domain.

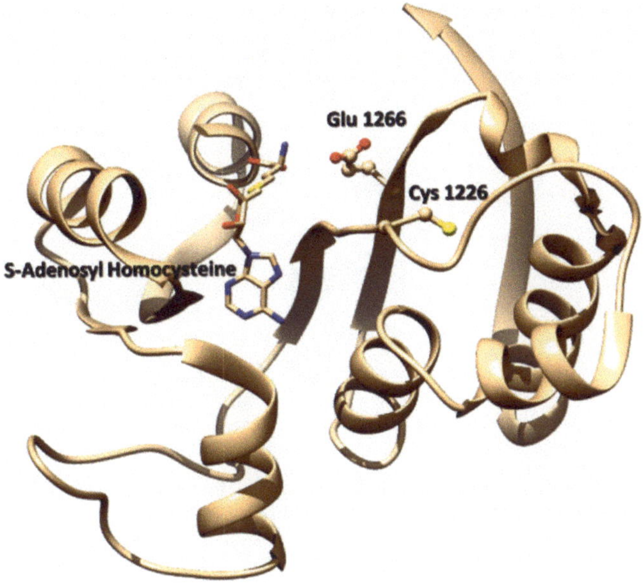

FIG. IV.34 Active site structure of human DNMT1. Only residues from 1150 to 1300 are shown. *Generated from PDB3pta.*

TET proteins use molecular oxygen to oxidize 5mC, 5hmC, or 5fC with concomitant generation of succinate and CO_2. A critical step, probably the rate-determining step, is the hydrogen atom extraction from the methyl group of 5mC to produce the corresponding radical that then combines with the hydroxyl radical to produce the 5hmC (Fig. IV.35). Two histidines and one aspartic acid coordinate the Fe atom in the protein (Fig. IV.36). As described earlier, the generation of 5fC or 5caC makes the base susceptible to removal by TDG. We will describe the base excision repair in the following section.

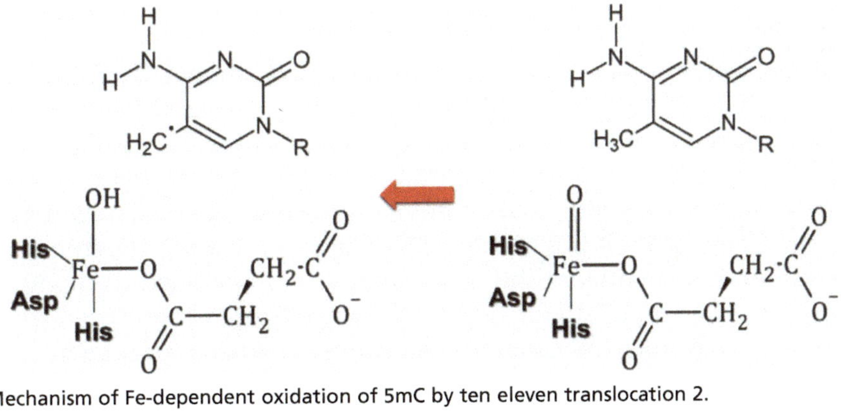

FIG. IV.35 Mechanism of Fe-dependent oxidation of 5mC by ten eleven translocation 2.

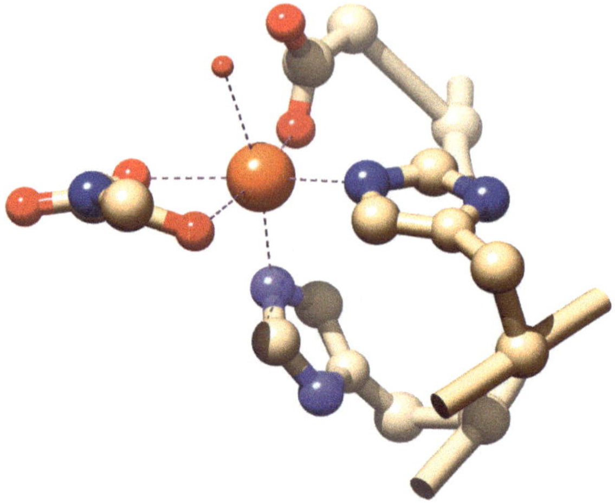

FIG. IV.36 Structure of the active site of ten eleven translocation 2. The Fe atom *(orange ball)* is coordinated by Asp, His, and a α-ketoglutarate analog, oxalyl glycine. *Generated from PDB4NM6.*

IV.3.2 Deamination reactions

Exocyclic amine groups of cytosine, adenine, and guanine are prone to deamination reactions to a varying extent. Cytosine loses the exocyclic amino group most readily among the three bases in the DNA to yield uracil. Deamination of adenine and guanine in DNA yields hypoxanthine and xanthine, respectively. Mechanistically, cytosine deamination, an addition-elimination reaction, proceeds through the nucleophilic addition of a water molecule to the C4 atom of cytosine to produce a tetrahedral intermediate. Protonation of the exocyclic amine group may occur before or after the nucleophilic addition of the water molecule. The protonated amino group then leaves, leaving the hydroxyl group that quickly tautomerizes to produce the most abundant tautomer of uracil (Fig. IV.37).

IV.3.3 Free-radical reactions with DNA bases

Evolution of life in an oxygen-rich environment and establishment of oxidative phosphorylation exposed living forms to many highly reactive oxygen-based free radicals. Some of these radicals are generated during the normal metabolic processes in the cell, and some are generated externally and enter cells through different routes. Because of their high reactivity, they tend to react with many different types of molecules, including the DNA.

The oxygen molecule itself is a di radical having two unpaired electrons of the same spin (triplet state). We can understand this electronic configuration using the molecular orbital theory. The overlaps of atomic orbitals from two oxygen atoms produce molecular orbitals for the O_2 (Fig. IV.38). As the molecular orbitals are filled, starting from the lowest

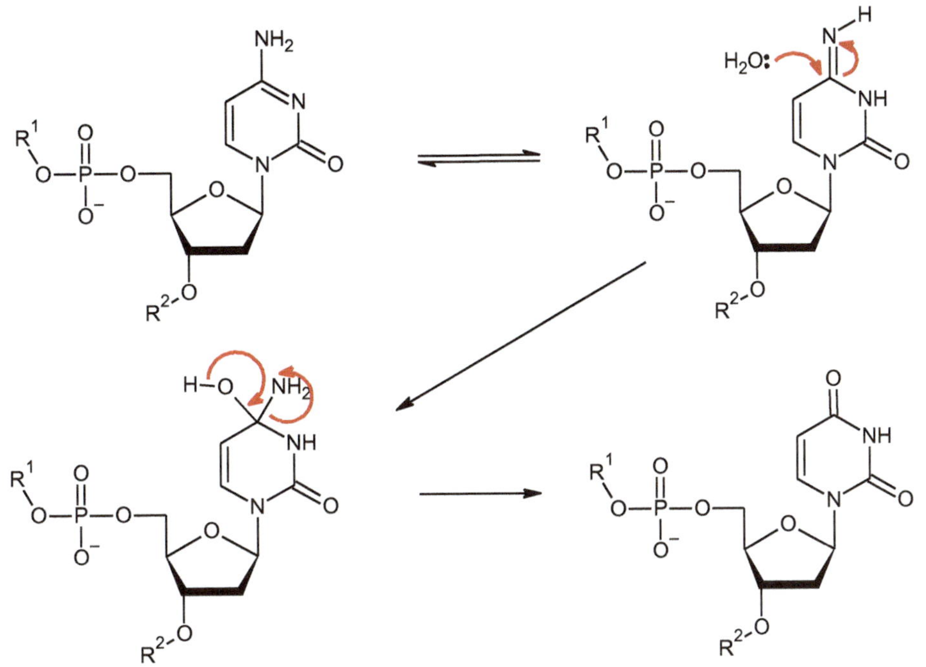

FIG. IV.37 Mechanism of spontaneous deamination of cytosine. R1 and R2 represent substituents in a continuous phosphodiester backbone.

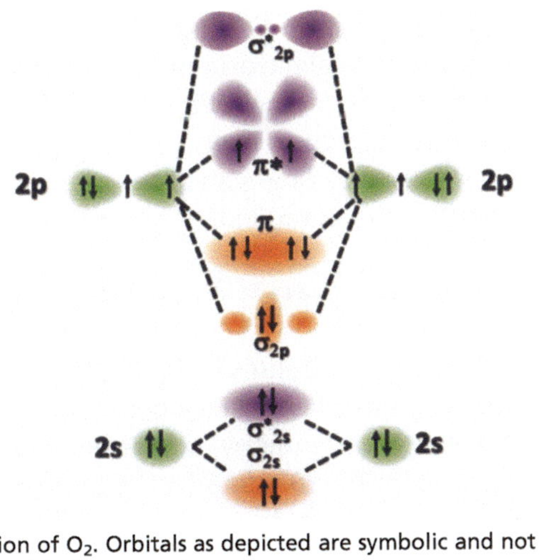

FIG. IV.38 Electronic configuration of O_2. Orbitals as depicted are symbolic and not meant to represent the actual atomic or molecular orbitals. 1s orbitals are not shown.

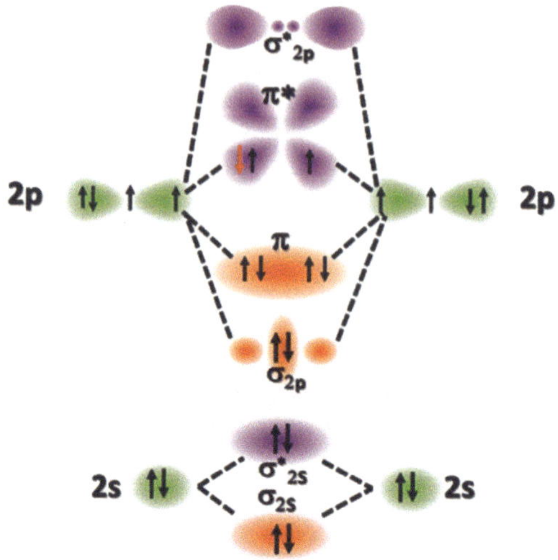

FIG. IV.39 Electronic configuration of superoxide anion O_2^-. The additional electron added to a π^* orbital is shown in *red*. Orbitals as depicted are symbolic and not meant to represent the actual atomic or molecular orbitals. 1 s orbitals are not shown.

energy orbital, the last two electrons occupy the two antibonding π^* orbitals of the same energy. According to Hund's rule, one electron each will occupy two antibonding π^* orbitals with parallel spins. This state is, thus, the ground state and is a triplet state, making O_2 a diradical.

Molecular oxygen is an oxidizing agent. During its oxidation reactions, it receives an electron(s) and becomes reduced. One electron reduction of molecular oxygen produces the superoxide anion, a reactive free radical (Fig. IV.39). The superoxide anion could undergo a further one-electron reduction to produce hydrogen peroxide, a powerful oxidizing agent. Another important free radical generated in cells from hydrogen peroxide through Fenton reaction is the hydroxyl radical.

$$Fe^{2+} + H_2O_2 \rightarrow Fe^{3+} + OH\bullet + OH^-$$

The most reactive of the oxygen metabolites is the hydroxyl radical (OH•). This radical reacts with DNA bases to create oxidized and ring-opened products. Hydroxyl radical is a powerful electrophile as it has a strong tendency to gain an electron and convert itself into OH^- ion. Guanines are most susceptible to attack by the hydroxyl radical, and addition to electron-rich 7–8 and 4–5 bonds are major reaction pathways. The addition of the hydroxyl radical to 7–8 double bond of the guanine ring and subsequent oxidation to 8-oxo-guanine are shown in Fig. IV.40.

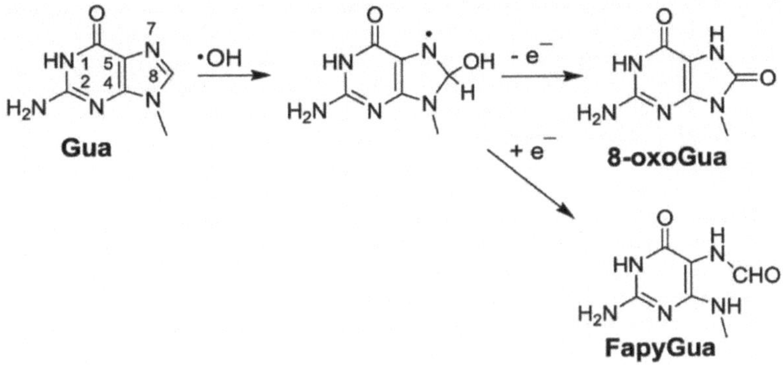

FIG. IV.40 Oxidation of guanine by hydroxyl radical. *Adapted from Cadet et al., 2017. Free Radic. Biol. Med. 107 13–34. Elsevier.*

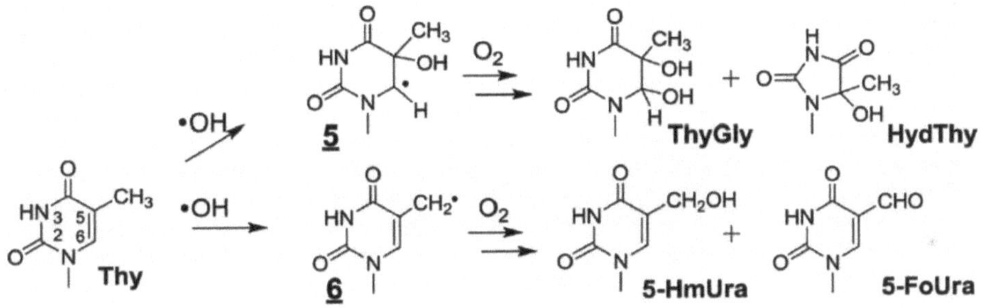

FIG. IV.41 Oxidation of thymine by hydroxyl radical. *Adapted from Cadet et al., 2017. Free Radic. Biol. Med. 107 13–34. Elsevier.*

Pyrimidines are also susceptible to attack by the hydroxyl radical. Addition to the electron-rich 5–6 bond of both thymine and cytosine is the most favored reaction that upon oxidation gives the 5,6 di-hydroxy products. The hydroxyl radical can extract a hydrogen atom from the 5-methyl group of thymine and 5-methyl cytosine; further oxidation leads to 5-hydroxymethyl and 5-formyl pyrimidines (Fig. IV.41).

IV.3.4 Free-radical reactions with the backbone

The hydroxyl radical attacks 2′-deoxyribose sugars of the phosphodiester backbone of the genomic DNA. Hydroxyl radicals abstract hydrogen from 2′-deoxyribose, creating radicals of deoxyribose sugar. The abstraction of a hydrogen atom from all five carbon atoms (Fig. IV.42) may occur, but an abstraction from 4′C is most favored in the double-stranded DNA, leading to a single-strand break (Fig. IV.43). Ultimately, the 2′-deoxyribose radical cation produces a degraded product, 2,3 dideoxy pentose-4-ulose.

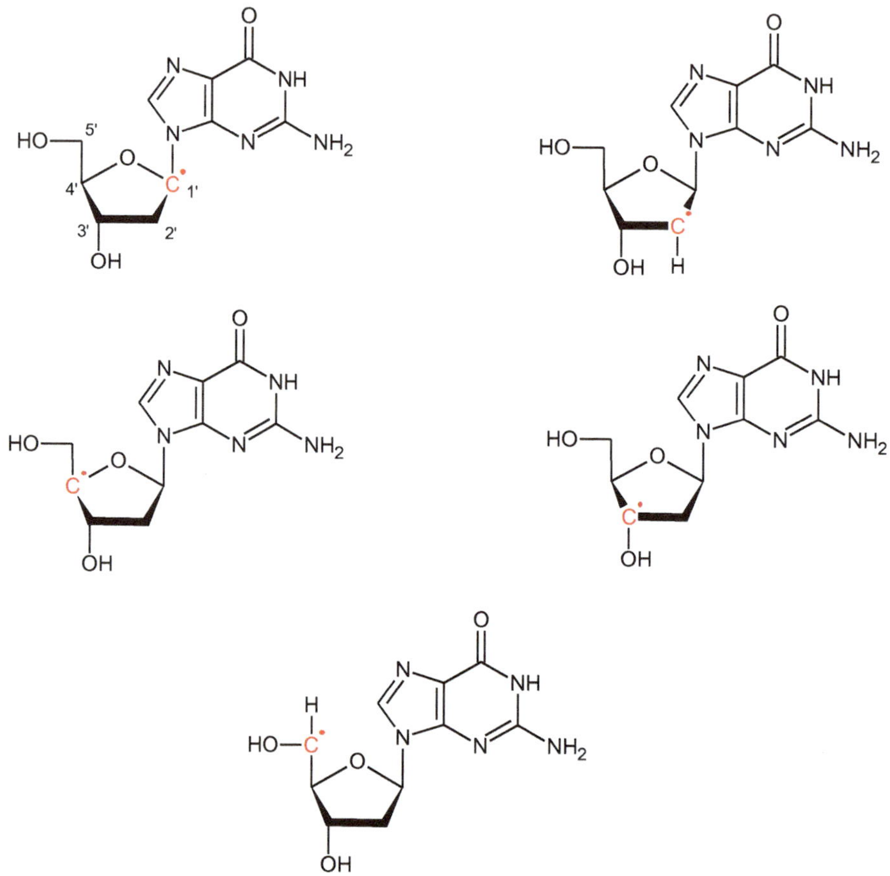

FIG. IV.42 Structure of 2′deoxyribose radical produced by hydrogen atom abstraction by the hydroxyl radical.

IV.3.5 Repair of modified bases

All living organisms possess DNA repair mechanisms. In mammalian cells, base excision repair systems remove and repair the base oxidation products, uracil produced from deamination of cytosine, and many other aberrant DNA bases. Two significant pathways exist, BER and nucleotide-excision repair (NER) (Fig. IV.44). BER primarily removes and repairs the nonbulky lesions, such as 8-oxo-guanine. NER primarily removes the bulky lesions. DNA-glycosylases play a key role in BER. In mammals, there are four superfamilies of DNA-glycosylases that remove oxidized bases from the DNA. They are uracil DNA-glycosylase superfamily, HhH superfamily, MPG (3-Methyl Purine Glycosylase), and NEIL

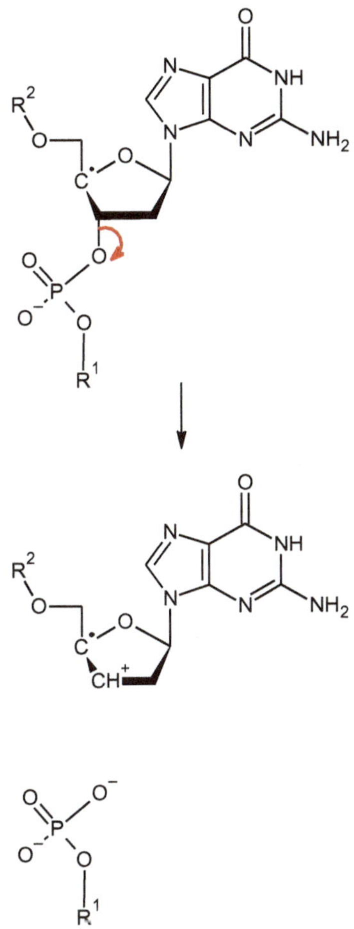

FIG. IV.43 Mechanism of single-strand break from C4′ radical.

superfamily. The glycosylases have partially overlapping base specificity. DNA-glycosylases from HhH and NEIL (Nei Like) superfamily remove some of the oxidation products of Guanine, including 8-oxo-Guanine. Similarly, the thymine glycol can be removed by multiple DNA glycosylases. A common mechanism of detecting and reading the oxidized base is through base flipping. Fig. IV.45 shows a base-flipped structure of thymine glycol containing DNA duplex, with the flipped base inserted into the active site of the glycosylase enzyme. The glycosyl bond is then cleaved by an activated water molecule attacking the already polarized glycosidic bond and hydrolyzing it.

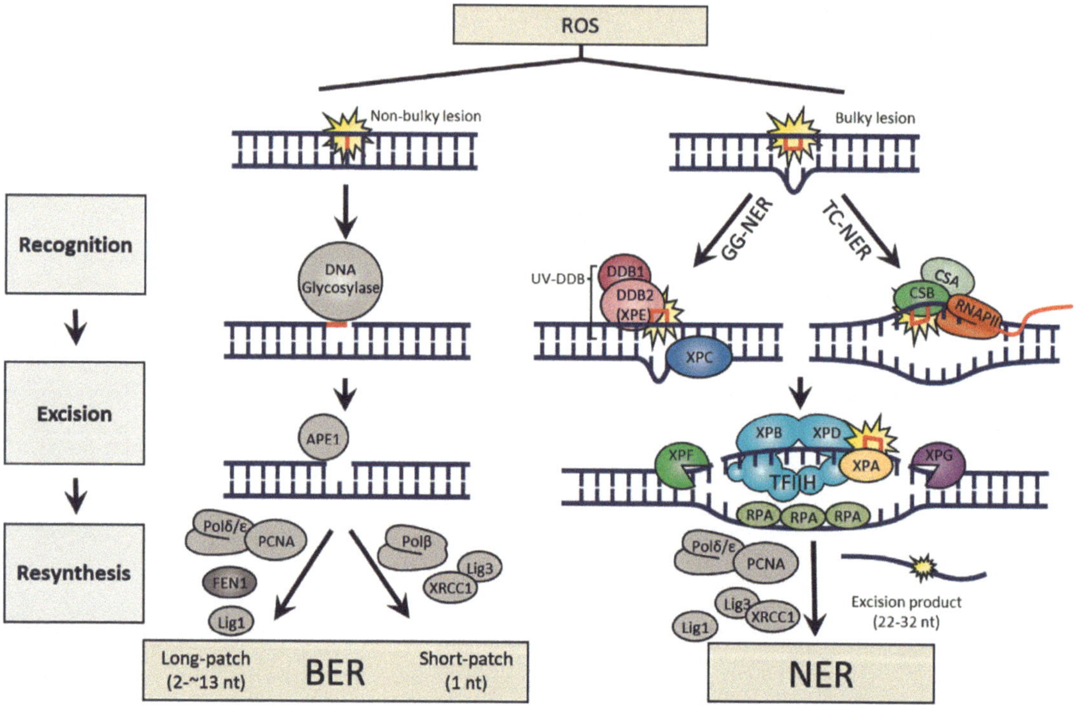

FIG. IV.44 DNA excision repair mechanisms for bulky or nonbulky DNA lesions caused by reactive oxygen species. *Reproduced from Int. J. Mol. Sci. 2019, 20 (23), 6092. doi:10.3390/ijms20236092. License CC-BY.*

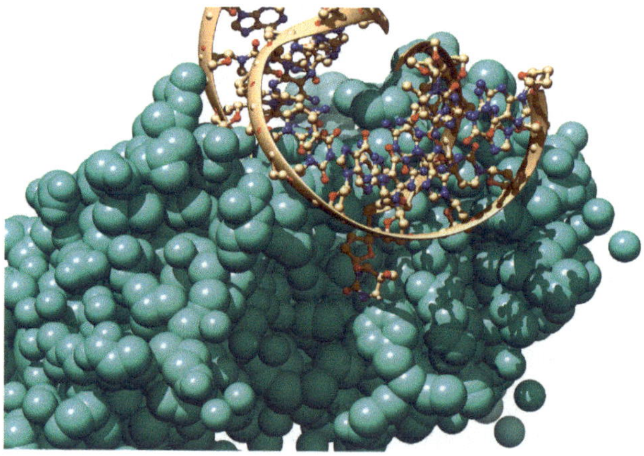

FIG. IV.45 Active site structure of NEIL1 with damaged base thymine glycol flipped into the active site. *The figure was produced from PDB5ITX.*

In the case of spontaneously deaminated base cytosine, the uracil in the DNA is detected and removed by the enzyme uracil DNA glycosylase, using a similar base-flipping mechanism. In the case of TET-mediated oxidation of 5-methyl cytosine, the oxidized bases are removed by TDG. The whole base excision repair is proposed to be highly processive and strongly coupled to the TET-mediated oxidation (Fig. IV.46).

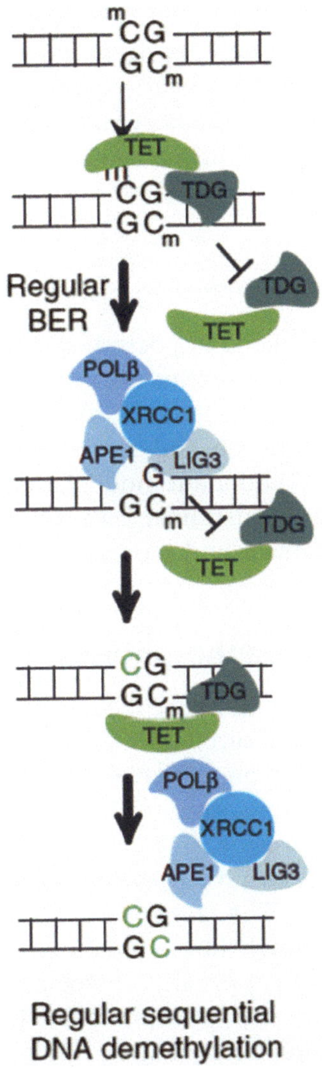

FIG. IV.46 Mechanistic model of ten eleven translocation-thymine DNA glycosylase-base-excision repair-mediated DNA demethylation. In the presence of all the necessary factors, DNA demethylation at fully methylated CpGs occurs in a coordinated and sequential manner to correctly reestablish the unmodified state (regular base-excision repair). *Reproduced from Weber, A., Krawczyk, C., Robertson, A. et al., 2016. Biochemical reconstitution of TET1—BER-dependent active DNA demethylation reveals a highly coordinated mechanism. Nat. Commun. 7, 10806. doi:10.1038/ncomms10806. CC-BY license.*

IV.4 Chemical modifications of proteins and enzymes that catalyze such modifications

Proteins undergo a large number of posttranslational modifications. Many of these modifications are catalyzed by enzymes and often in an amino acid sequence-selective manner. Chemical modifications that occur on the genome are usually dynamic. Enzymes that catalyze a modification are counteracted by enzymes that remove that modification. For example, kinases phosphorylate and phosphatases remove the phosphoryl group. We have shown before how modification-removal systems reach steady states (Section IV.1.4). Some modifications are of immense regulatory significance, such as histone posttranslational modifications. A few of these modifications are cellular compartment specific. Our focus will be on those modifications that are specific to the nucleus and occur on chromatin-associated proteins.

IV.4.1 Protein side chains undergoing chemical modifications

Activities of proteins are often regulated by posttranslational modifications of constituent amino acids. Nucleophilic groups on proteins are primary modification sites. In genome-associated proteins, chemical modifications of the ε-amino group of lysine, hydroxyl groups of serine, threonine, and tyrosine, and guanidine group of arginine have immense regulatory significance. Other less abundant modifications occur on the sulfhydryl group of cysteine, the imidazole group of histidine, carboxyl groups of glutamic and aspartic acids, and amide groups of glutamine and asparagine. We will explore the first group of amino acids in more detail.

IV.4.2 Phosphorylation and dephosphorylation

Phosphorylation of protein plays an immense regulatory role in many signaling and metabolic pathways. Phosphorylation of serine, threonine, and tyrosine is arguably the most crucial regulatory chemical modification that occurs in the cell. In this reaction, the γ-phosphoryl group of the ATP (adenosine triphosphate) is transferred to the hydroxy groups of serine, threonine, or tyrosine. In most kinase-catalyzed reactions, Mg^{+2}-ATP is the actual substrate, and the acceptor hydroxyl group attacks the γ-phosphate in a nucleophilic substitution reaction. The reaction occurs with the inversion of configuration at the phosphorus atom. Although the reaction bears all the signatures of an S_N2 reaction, recent results suggest that it is primarily dissociative, meaning the P—O—P bond breaks to a large extent, creating a metaphosphate like intermediate, before the nucleophile (OH^-) attacks the phosphorus atom (Fig. IV.47).

Phosphoprotein phosphatases or protein phosphatases remove phosphate groups from proteins. Serine or threonine phosphatases mostly follow different mechanisms than the tyrosine phosphatases. Serine or threonine phosphatases are metal-dependent enzymes. The mechanism involves the activation of an active site water molecule by

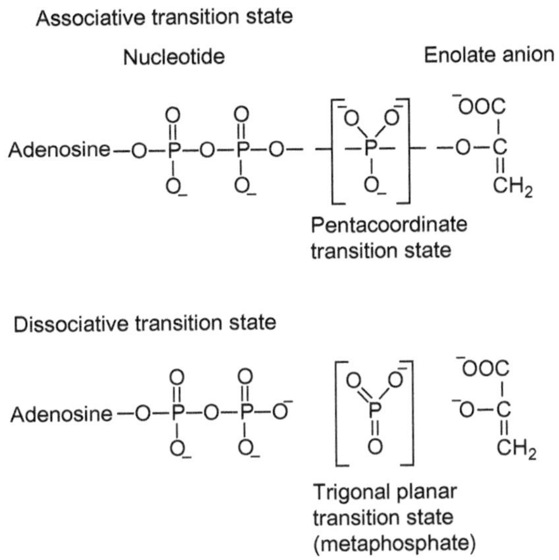

FIG. IV.47 Mechanisms and transition states for phosphoryl transfer reactions. The transition state for phosphoryl transfer may be described as one of two possibilities, either associative (S_N2-like) or dissociative (S_N1-like). In the associative case, the transition state is dominated by bond formation and involves a pentagonal bipyrimidal (phosphorane) structure with the axial ligands being the nucleophile of the attacking substrate and bridging oxygen atom between the β- and γ-phosphates, respectively. The dissociative case involves a planar, trigonal, metaphosphate (PO^{3-}) structure and is dominated by bond breaking. The two transition states are further defined by differences in charge distribution, with a net charge of −3 on the phosphoryl group in the associative case and −1 on metaphosphate in the dissociative case. *Adapted from Matte, et al., 1998. Structure 6 (4), 413–419. Elsevier.*

the metal atoms, followed by a direct attack on the phosphoserine or phosphothreonine moiety (see later). In tyrosine phosphatases, an active-site cysteine residue attacks the phosphotyrosine residue, transferring the phosphate to the cysteine. In the second step, an attack by a water molecule on this intermediate releases the phosphate, regenerating the cysteine thiol (Fig. IV.48).

IV.4.2.1 Protein kinases that phosphorylate histones

Phosphorylation of proteins at serine, threonine, and tyrosine are regulatory events of global significance. Kinases that phosphorylate other proteins also phosphorylate histones and other genomic proteins. We will highlight an important kinase that phosphorylates histones. Several phosphorylations on histones occur upon activation of DNA damage response. A crucial phosphorylation at serine 139 of the histone variant H2AX is catalyzed by the kinase ATM. Fig. IV.49 shows the active site of Tel1 (the yeast ortholog of ATM) in AMP-PNP or Mg^{2+} bound form. Three amino acid residues important for catalysis are shown in sea green. D2612 acts as a general base to deprotonate the substrate –OH, whereas H2614 and N2616 stabilize the transition state.

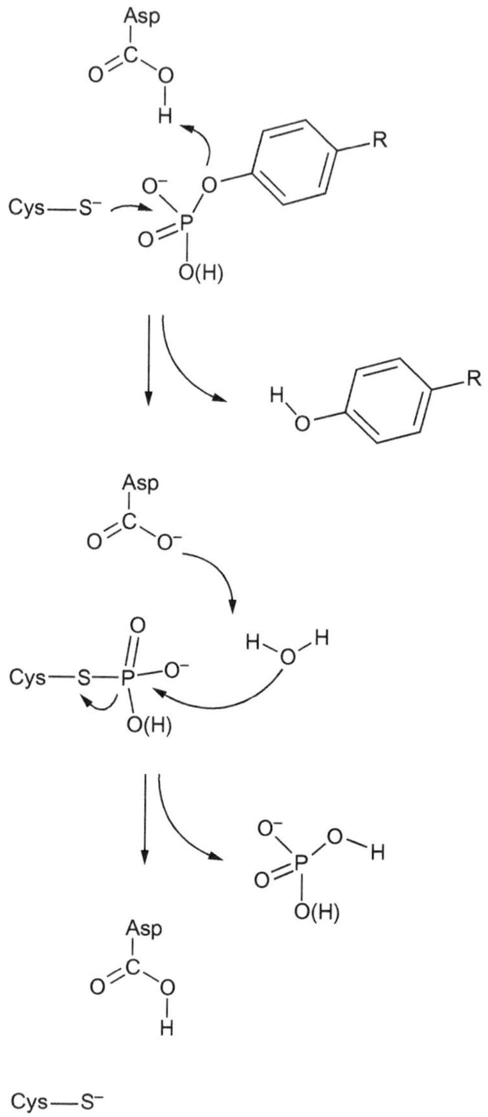

FIG. IV.48 The mechanism of reaction catalyzed by the phosphotyrosine phosphatases. The two proposed ionization states of the substrate are indicated. *Reproduced from Kolmodin, et al., 2001. FEBS Lett. 498 (2–3), 208–213. Elsevier.*

IV.4.2.2 Phosphoprotein phosphatases that act on histones

Like most chemical modifications in the genome, phosphorylations are also dynamic. Phosphorylations are catalyzed by protein kinases and removed by protein phosphatases. Many phosphatases act on phosphorylated histones. PP2A, an important phosphatase, acts on γH2AX. The crystal structure of PP2A with an inhibitor shows two active site

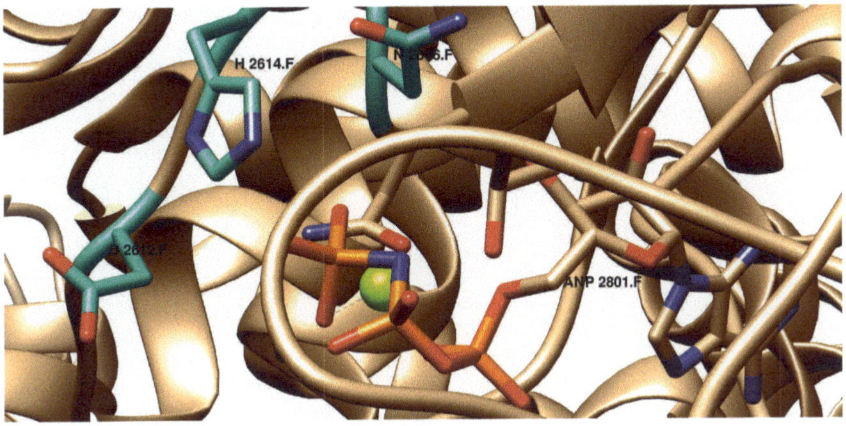

FIG. IV.49 Active site of Tel1 (yeast ATM ortholog) structure as obtained from Cryo-EM. *Figure generated from PDB6S8F.*

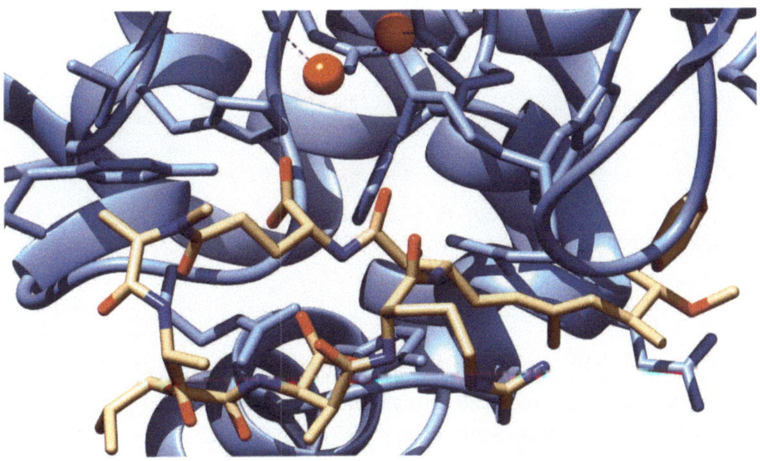

FIG. IV.50 Active site configuration of PP2A complexes with microcystin LR, a competitive inhibitor of the peptide substrate. The two Mn^{2+} atoms are shown as red spheres. *Structure generated from PDB2IAE.*

Mn^{2+} ions (Fig. IV.50). Mn^{2+} ions activate a water molecule that directly attacks the substrate phosphoserine, displacing the phosphate group in an S_N2-like mechanism.

IV.4.3 Acetyl-transfer reactions

Posttranslational acetylation of proteins is one of the significant chemical modifications of proteins that occur in vivo. Acetylation of protein residues occurs primarily at the N-terminal amino group and ε-amino group of lysine. The universal acetyl group donor in all such reactions is acetyl-Coenzyme A (Ac-CoA) (Fig. IV.51). The acetyl group in the

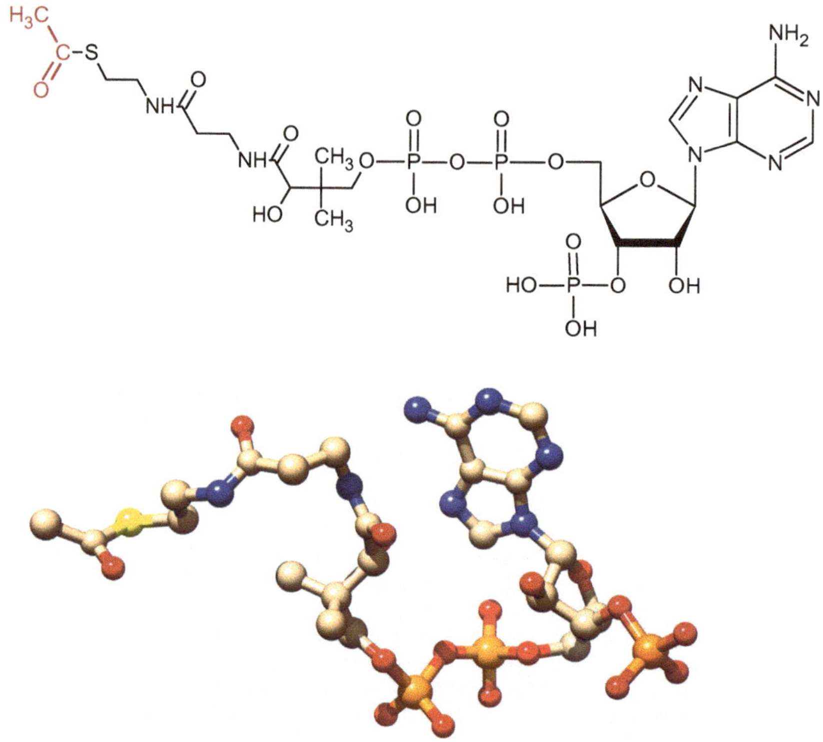

FIG. IV.51 Structures of acetyl coenzyme A. *Top*: chemical structure with acetyl group shown in red. *Bottom*: the crystal structure. *The crystal structure was generated from PDB4CSC.*

Ac-CoA is in the form of a thioester, and it is transferred to an amine group producing an amide functional group. In the reverse direction, the acetyl group is removed to generate the free ε-NH_2 group of the acetylated lysine (Fig. IV.52).

In the majority of the enzymes, the acetyl group is transferred directly, likely through a tetrahedral intermediate (Fig. IV.53). The first step in this reaction is a nucleophilic addition to the C=O bond of Ac-CoA, creating a tetrahedral intermediate. This tetrahedral intermediate then facilitates the exit of the CoA, resulting in acetyl lysine. Initially, for the MYST family acetyltransferases, a ping-pong mechanism was proposed which the Ac-CoA binds first and transfers the acetyl group to a cysteine residue on the enzyme and then leaves. The protein then forms a binary complex, and the acetylated cysteine residues transfer the acetyl group to the protein lysine group. However, later mutagenesis studies had not supported this ping-pong mechanism. It is still unclear about the exact mechanism followed by the MYST family of histone acetyltransferases (HATs).

Histone acetylation is a dynamic process. It is necessarily so, as otherwise turning genes off or on would be impossible. Histone deacetylases hydrolyze ε-acetyl-lysines in histones to acetate and lysine (Fig. IV.52). Mechanistically, the histone deacetylases fall

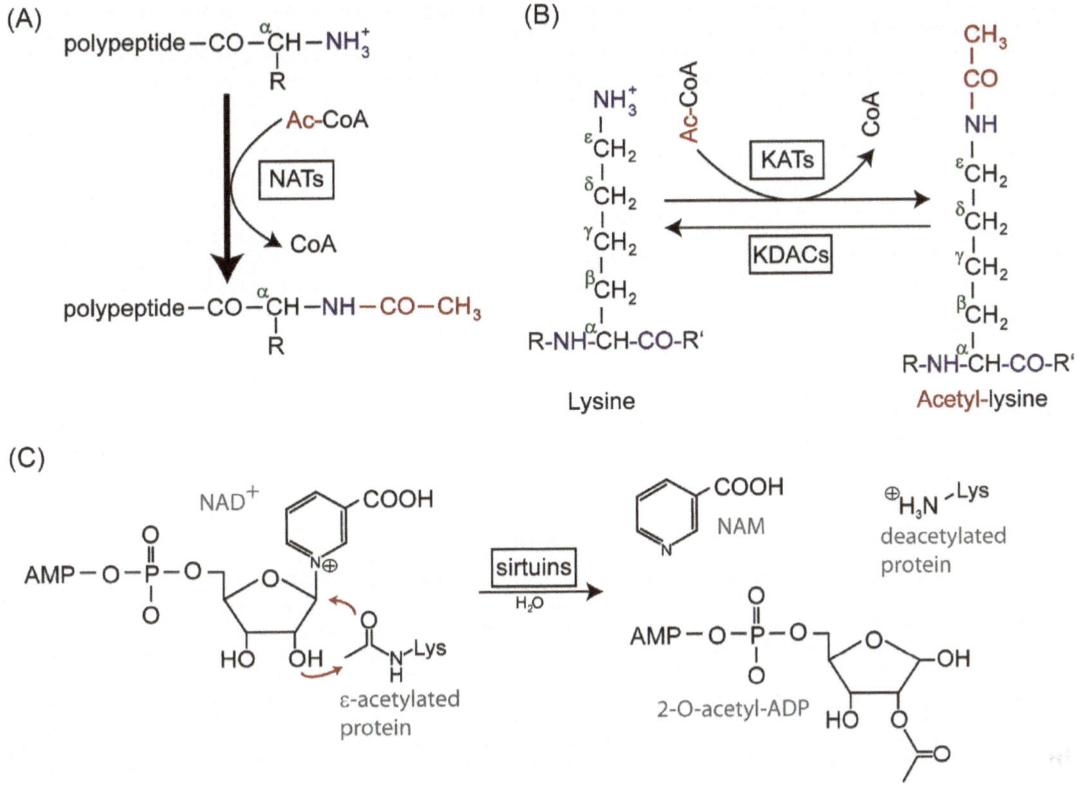

FIG. IV.52 Schematic overview of acetylation and deacetylation. (A) Enzymatic transfer of an acetyl group from acetyl-coenzyme A to the N-terminal amino group of a polypeptide, catalyzed by N-terminal acetyltransferases. The reaction is considered to be irreversible. (B) Reversible acetylation of the ε-amino group of a lysine residue by lysine acetyltransferases (KATs). The deacetylation of lysine residues is catalyzed by lysine deacetylases (KDACs) and (C) is NAD+-dependent in case of sirtuins. *NAD+*, nicotinamide adenine dinucletide; *NAM*, nicotinamide. *Reproduced from Drazic, et al., 2016. Biochim. Biophys. Acta 1864 (10), 1372–1401. Elsevier.*

into two categories, the zinc dependent and the NAD^+-dependent. In the former mechanism, zinc atom supports the nucleophilic addition of a hydroxide ion to the carbonyl group of the acetylated lysine and perhaps stabilizes tetrahedral oxyanion intermediate (Fig. IV.54). The oxygen atom of a water molecule is a weak nucleophile, especially when compared with the hydroxide ion, OH^-. Thus an active site-bound water molecule is first deprotonated to produce OH^-, a powerful nucleophile that can attack the carbonyl group of acetyl lysine. The attack of the hydroxide ion creates a tetrahedral oxyanion intermediate that then collapses to form acetate and lysine.

The NAD^+-dependent histone deacetylases (Sirtuins) use NAD^+ as the cofactor to deacetylate N-ε-acetyl lysines on histones. The carbonyl oxygen on the acetyl group of the lysine performs a nucleophilic substitution on the C1″ (called C1″ to distinguish it from

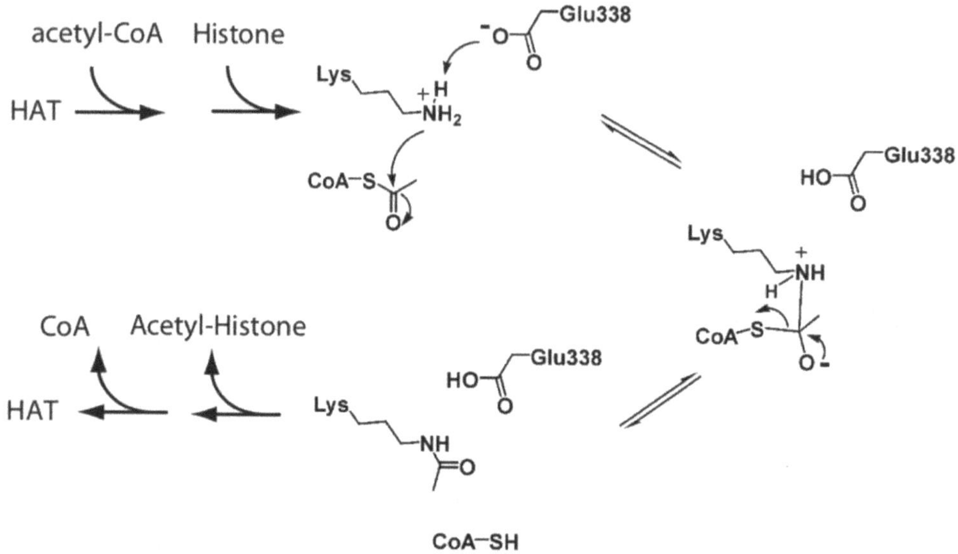

FIG. IV.53 Proposed sequential mechanism of acetylation for histone or protein acetyltransferases. Lysine attacks the carbonyl carbon of the acetyl moiety of acetyl-CoA forming a tetrahedral intermediate that then collapses to form CoA and acetylated product. *Reproduced from Berndsen, et al., 2008. Curr. Opin. Struct. Biol. 18 (6), 682–689. Elsevier.*

the adenine attached ribose numbering. Henceforth, we will number any ribose distal from adenine, such as in NAD+ and ADP-ribose, this way) position of ribose attached to the nicotinamide ring. Whether the mechanism of this substitution is associative or dissociative (i.e., S_N2 or S_N1 like, respectively) remains unresolved. The neighboring OH group (2″) then attacks the carbonyl carbon of the acetyl group to produce a cyclic structure, which then reacts with water to produce the free lysine and 2″-*O*-acetyl ADP-ribose (Fig. IV.55).

IV.4.4 Other acylations

Although acetylation is the most well-known acylation reaction known, many other acylations occur on histones. They include formylation, propionylation, crotonylation, 2-hydroxyisobutyrylation, β-hydroxybutyrylation, succinylation, malonylation, glutarylation, benzoylation, and long-chain fatty acylations (Fig. IV.56). These acylations may have distinct effects on gene expression. Many of these acylations occur through respective Acyl-CoAs. Even though we do not know either the enzymes involved or reaction mechanisms, the mechanistic path is likely to be similar to acetylation. At least in some cases, CBP or p300 catalyzes these acylations. If the same enzymes catalyze both acetylation and other acylations, then the regulation will depend on different Acyl-CoA concentrations. It appears that this is precisely the mechanism through which the relative abundance of different acyl groups on histone is regulated.

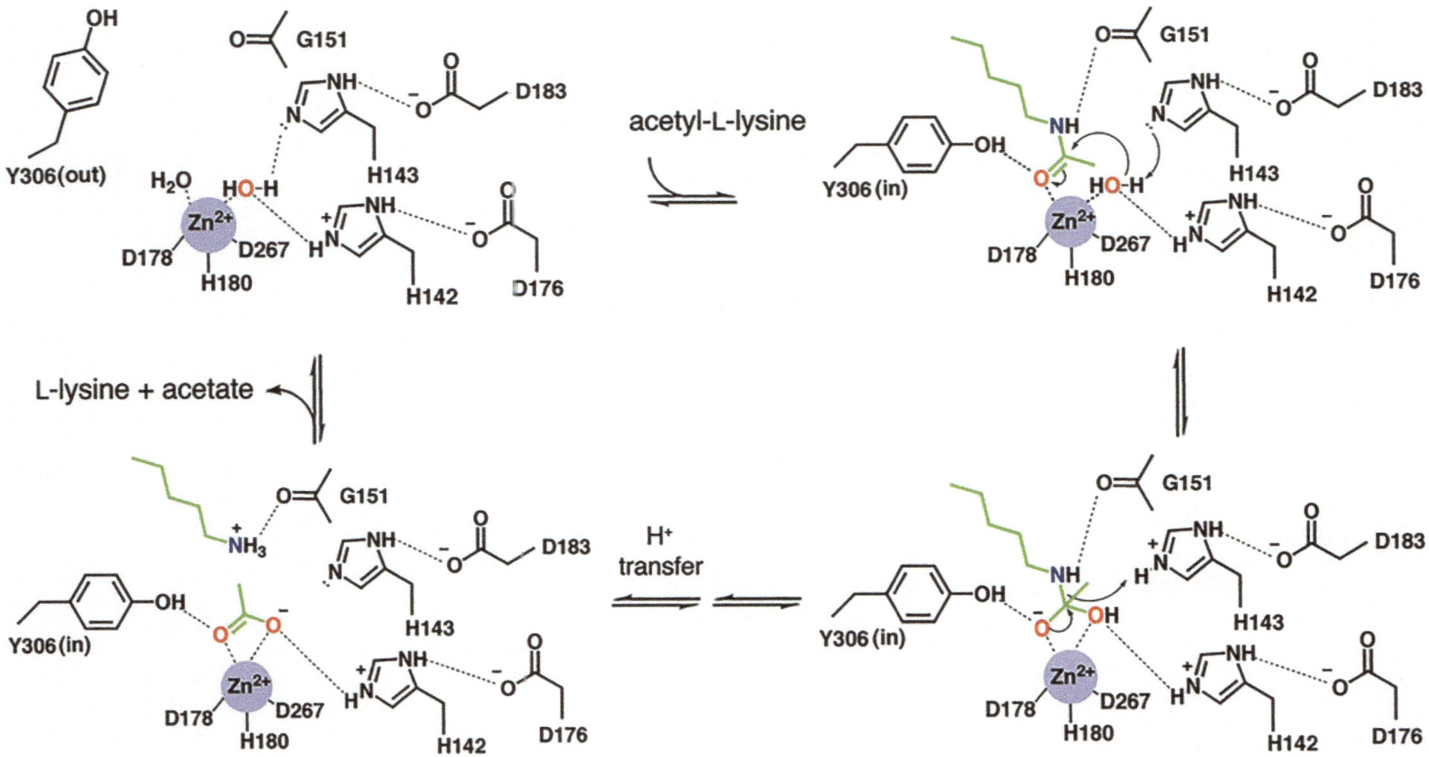

FIG. IV.54 Proposed mechanism of histone deacetylases and histone deacetylase-related enzymes. The active site transition metal ion of histone deacetylase 8 (Zn^{2+} or Fe^{2+}) and general base H143 promote the nucleophilic attack of a metal-bound water molecule at the metal-coordinated C═O group of the acetyl-L-lysine substrate (for clarity, only the side chain of acetyl-L-lysine is shown). As drawn, the nucleophilic lone electron pair on the metal-bound water molecule becomes available only upon proton abstraction, for example, the electron pair of the breaking O—H bond could add to the π^* orbital of the substrate carbonyl, although other molecular orbital explanations are possible. The oxyanion of the tetrahedral intermediate and its flanking transition states are stabilized by metal coordination and hydrogen bond interactions with Y306, H143, and H142. H143 serves as a general acid catalyst to facilitate the collapse of the tetrahedral intermediate to form acetate and L-lysine after an intervening proton transfer (possibly mediated by H143). We speculate that the side chain of Y306 may undergo a conformational transition from an "out" conformation to an "in" conformation to accommodate substrate binding and catalysis, based on the conformational mobility of the corresponding res due in related enzymes (Y976 in H976Y histone deacetylase 4 and Y323 in APAH). If so, this could suggest an induced-fit substrate-binding mechanism reminiscent of that described for Y248 in carboxypeptidase A. *Reproduced from Lombardi, et al., 2011. Curr. Opin. Struct. Biol. 21 (6), 735–743. Elsevier.*

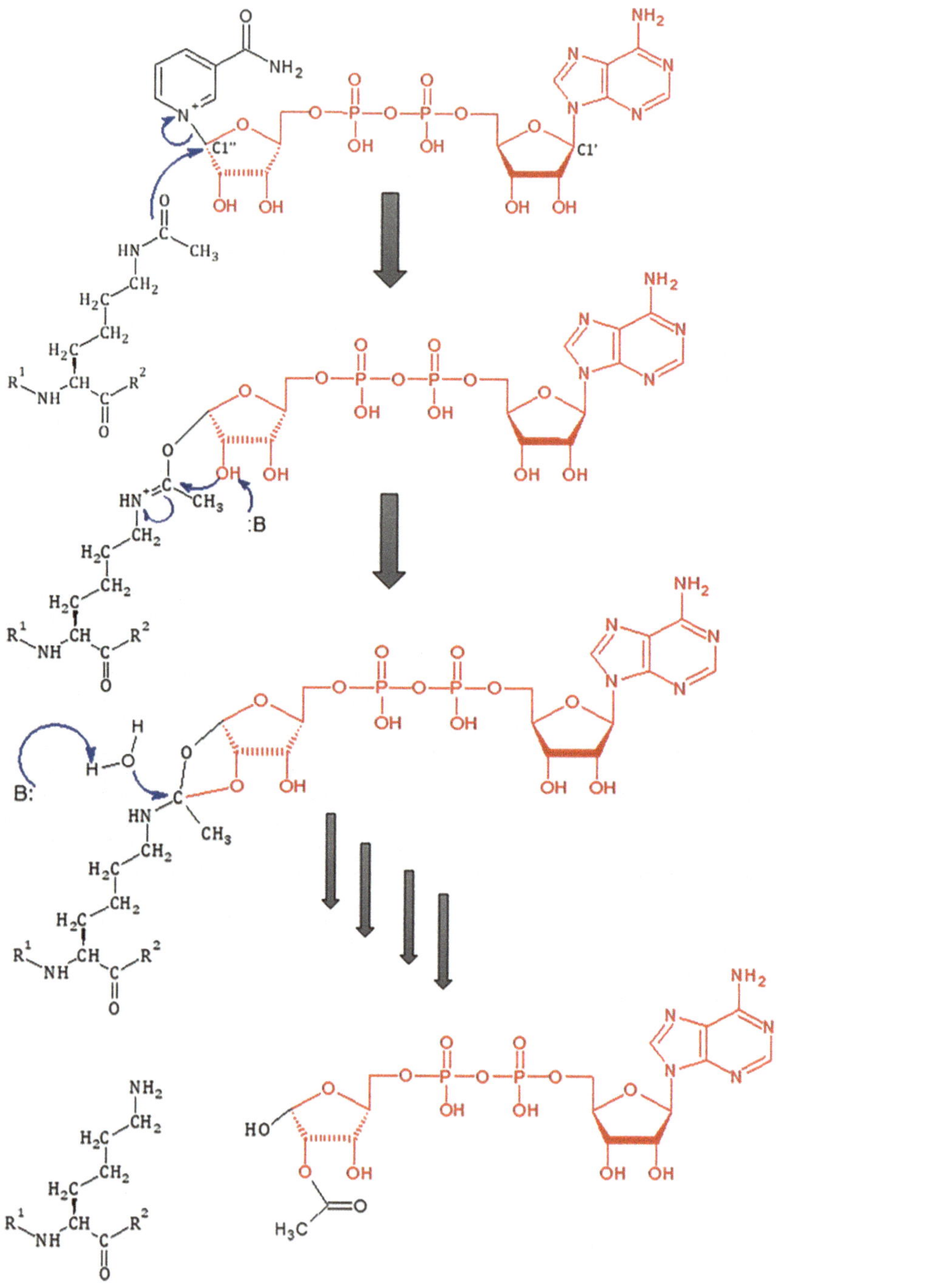

FIG. IV.55 Mechanism of sirtuins (NAD$^+$-dependent lysine deacetylases). Multiple gray arrows indicate presence of more than one step.

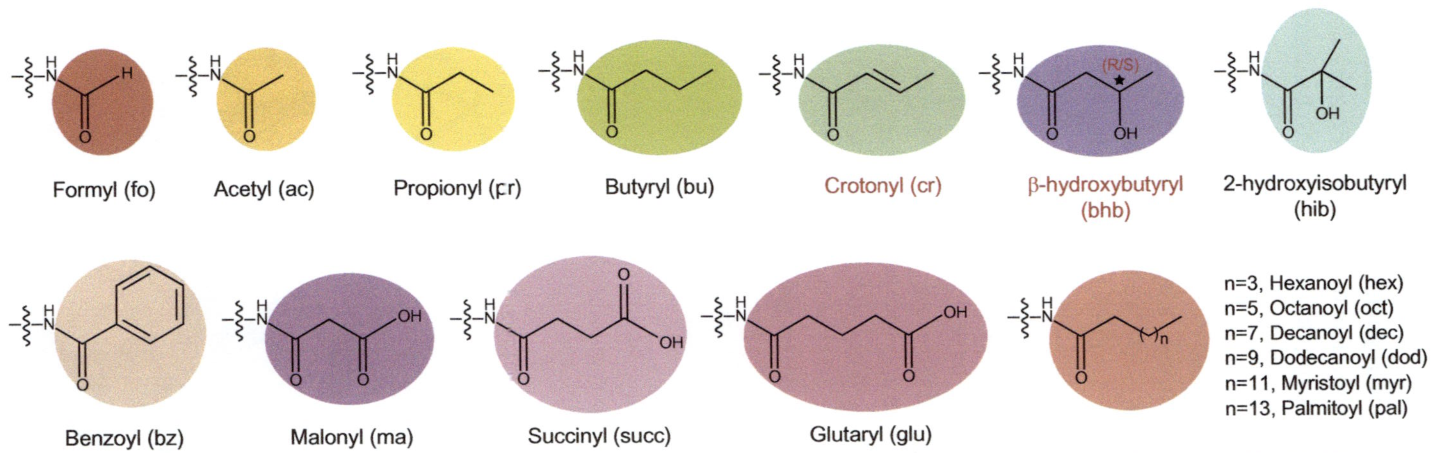

FIG. IV.56 Chemical structures of histone or nonhistone lysine acylations. *Reproduced from Zhao, et al., 2018. Curr. Opin. Struct. Biol. 53, 169–177. Elsevier.*

IV.4.4.1 Acetyltransferases

HATs are critical elements of gene regulatory systems of eukaryotes. In general, the acetylation of histones loosens the nucleosomes and increases gene expression. Although their primary function is transferring the acetyl group to lysine residues on histones, they are known to acetylate nonhistone proteins as well. HATs that are prevalent in the metazoan lineage are broadly classified into four families. HAT1 is the type B histone acetyltransferase that acetylates the histone H4 and H2A in the cytoplasm. Others are the GNAT (GCN5-Related N-Acetyltransferases) family, MYST family, and p300 family. The role of HAT1 in cellular processes is less well defined than that of the members of the other three families. HATs of the other families have been studied more intensively.

The catalytic and kinetic mechanisms of the GNAT and the MYST family are similar. It proceeds through a sequential kinetic mechanism in which the Ac-CoA and the protein bind to form a ternary complex. The ε-amino group of the lysine to be acetylated binds in the protonated form. A general base on the enzyme, often a conserved glutamate residue (E575 in human GCN5) in its ionized form, deprotonates the lysine in the first step (Fig. IV.57). The deprotonated amino group attacks the C=O group of the thioester and forms a tetrahedral intermediate. The CoA then leaves, and the tetrahedral intermediate collapses to form ε-acetyl-lysine. In the case of p300 or CBP, the ternary intermediate is very short-lived so that it is not detectable in kinetic measurements, thus showing the Theorell-Chance kinetics.

IV.4.4.2 Histone deacetylases

Histone deacetylases (HDACs) have been grouped into four classes, I, II, III, and IV, based on their structural and evolutionary relationships. Classes, I, II, and IV use zinc to facilitate

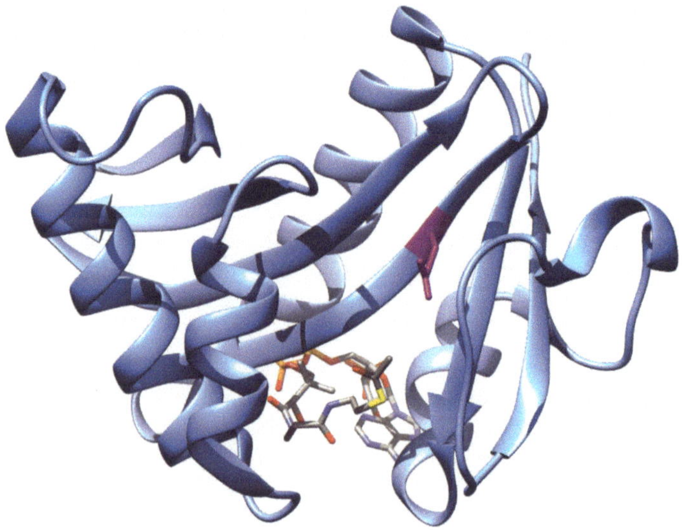

FIG. IV.57 Active site of human GCN5 (acetyl CoA-binding domain). The structure shown here was generated from PDB1Z4R. The conserved glutamate residue E575 is shown in purple. Bound AcCoA is shown in the stick model.

the hydrolysis reaction, whereas the class III seems to be mechanistically and evolutionarily unrelated. In the zinc-dependent deacetylases, during the formation of the oxyanion intermediate and transition states, negative charges develop and disappear (Fig. IV.54). A histidine-aspartic acid charge relay system can mitigate some of the unfavorable charge distributions. H142 and H143 play a crucial role as the general acid and base in the catalysis (Figs. IV.54 and IV.58). Class III enzymes, also known as Sirtuins, use NAD^+ for removing the acetyl group. Class III HDACs or Sirtuins, on the other hand, have a completely different enzymatic mechanism. They use NAD^+ as a cofactor in the deacetylation process in which the NAD^+ is consumed. In SirT2, the catalytically important general base histidine 187, which acts to remove a proton from ribose 2″-O, is seen close to ribose 2″-O and 3″-O in the crystal structure (Figs. IV.55 and IV.59).

IV.4.5 ADP-ribosylation

ADP-ribosylation of proteins is a common posttranslation modification that occurs in both prokaryotic and eukaryotic domains of life. It involves the transfer of the ADP-ribose moiety from NAD^+ to an amino acid side chain in a protein. Two types of ADP-ribosylations are known: the reaction that transfers a single ADP-ribose moiety (mono ADP ribosylation or MARYlation) and the reaction that transfers two or more ADP-ribose moiety (poly ADP-ribosylation or PARYlation) (Fig. IV.60). The most common acceptors of the ADP-ribose moiety are the guanidino group of arginine, hydroxyl group of serine, and

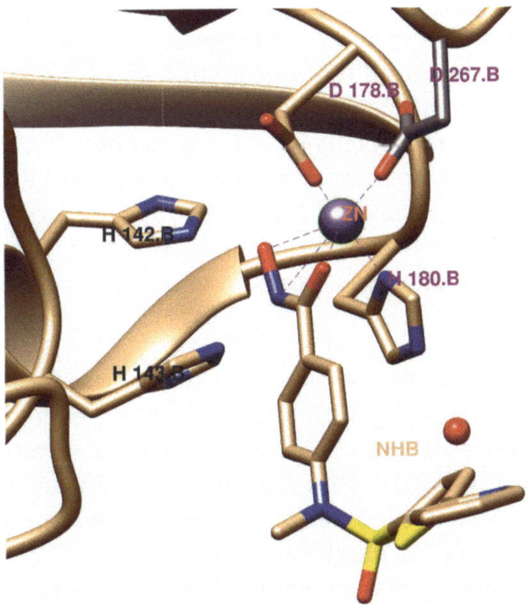

FIG. IV.58 Active site architecture of human histone deacetylase 8. Refer to Fig. IV.36 for catalytic roles of residues shown.

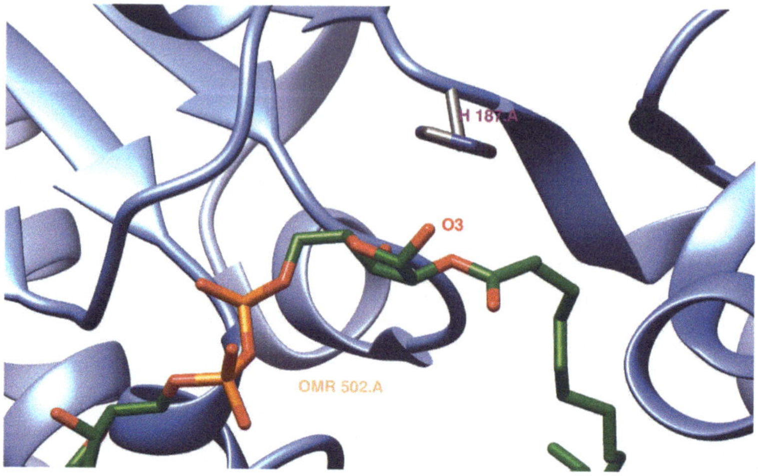

FIG. IV.59 Active site architecture of SirT2 bound with 2-O-myristoyl-ADP-ribose. Structure was generated PDB4Y6Q. The catalytically important histidine 187 is shown above the O3. This histidine acts as the general base. See Fig. IV.37.

carboxyl groups of aspartic and glutamic acids. Other known acceptors are tyrosine, lysine, and cysteine. The mechanism is a nucleophilic substitution at the C1″ carbon of the ribose attached to the nicotinamide (Fig. IV.61). Although initially an S_N2 type mechanism was proposed, a recent report suggests an S_N1 type mechanism in which the N-glycosidic bond of the nicotinamide portion is cleaved first to produce an oxocarbenium ion that after some changes reacts with the acceptor amino acid.

The reversal of ADP-ribosylation requires hydrolysis of the bond between the ribose (C1″) and the connected amino acid. It involves an attack by a water molecule at the C1″ position. Two possible mechanisms have been discussed, and both are nucleophilic substitution reactions. In one type, the activated water molecule directly attacks the C1″ carbon and displaces the amino acid side chain (the leaving group) in an S_N2 type mechanism. The other mechanism involves S_N1 type dissociation first, freeing the conjugated amino acid side chain and creating an oxocarbenium intermediate.

IV.4.5.1 (Poly)ADP-ribose polymerase

Attachment of ADP-ribose groups to proteins is catalyzed by a group of enzymes called (poly)ADP-ribose polymerase (PARP). This group has many members and diverse domain structure. However, conserved ADP-ribosyl transfer domains are present in all the family members. PARP1 is the best-studied enzyme in this group. Three signature amino acid residues, H862, Y896, and E988, play crucial roles in NAD^+ recognition and catalysis (Fig. IV.62). H862 forms a hydrogen bond with 2′-OH group of the adenine proximal ribose. Y896 stacks against the nicotinamide ring of the NAD^+. E988, in addition to being involved in substrate positioning, plays an important role in catalysis.

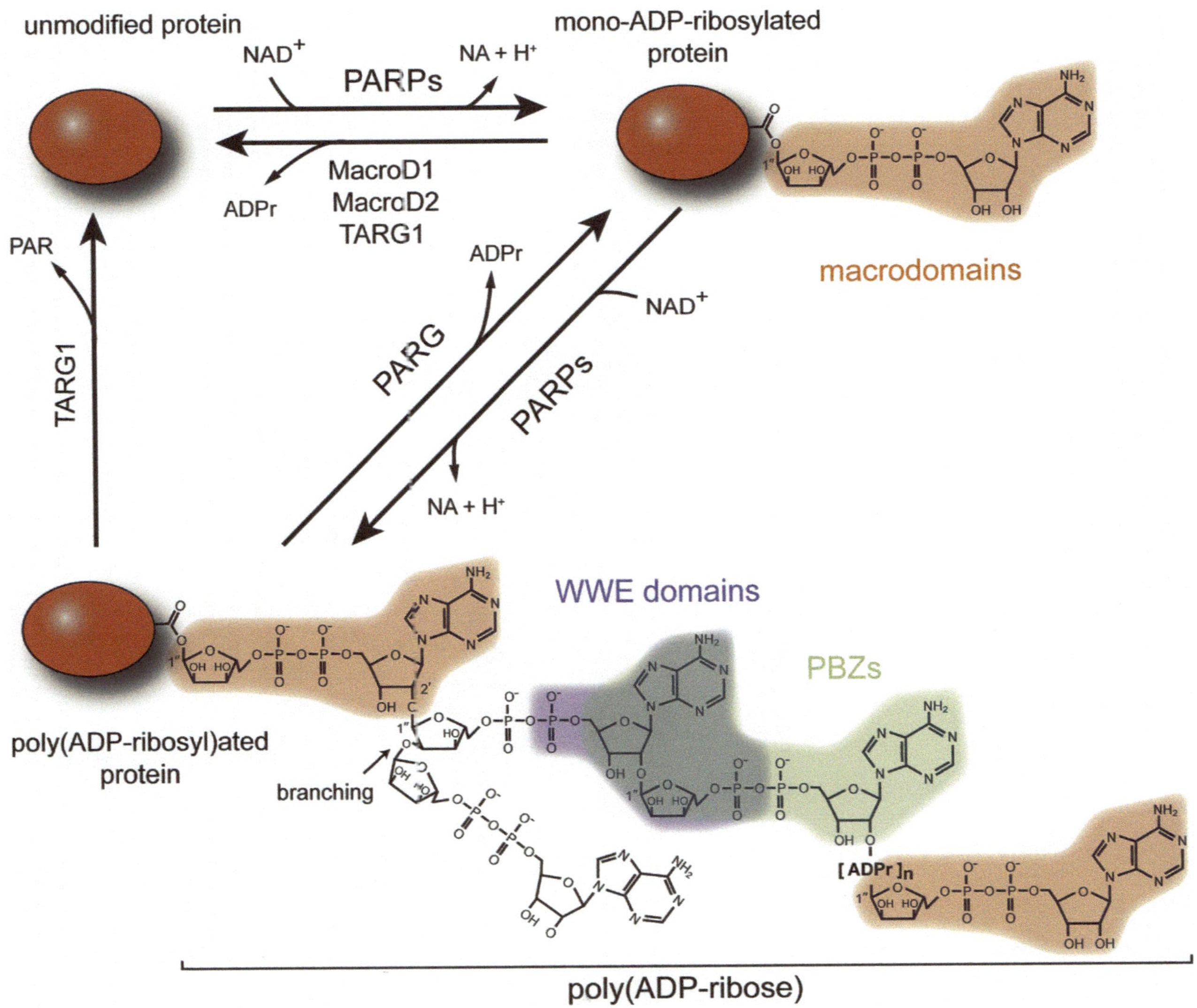

FIG. IV.60 Poly ADP-ribose polymerase-dependent ADP-ribosylation cycle. Proteins are ADP-ribosylated by poly ADP-ribose polymerases using NAD^+ as a cofactor, releasing nicotinamide. Specific poly ADP-ribose polymerases can ADP-ribosylate previous ADP-ribose units, which results in poly(ADP-ribose) formation with occasional branching. Different domains recognize mono-ADP-ribose or poly(ADP-ribose). Macrodomain recognition parts of mono-ADP-ribose and poly(ADP-ribose) are indicated in orange. PAR (poly ADP-ribose) recognition parts of WWE domain are highlighted in violet and those of PBZs (PAR-binding Zinc fingers) in green. Poly(ADP-ribosyl)ated proteins can be substrates of TARG1, which can remove whole PAR chain from the targets, or of poly ADP ribose glycohydrolase, which degrades PAR, leaving mono-ADP-ribosylated protein. Mono-ADP-ribose can be removed by the action of MacroD1, MacroD2, or TARG1, which complete the ADP-ribosylation cycle. *Reproduced from Barkauskaite, et al., 2015. Mol Cell 58, 935–946. Elsevier.*

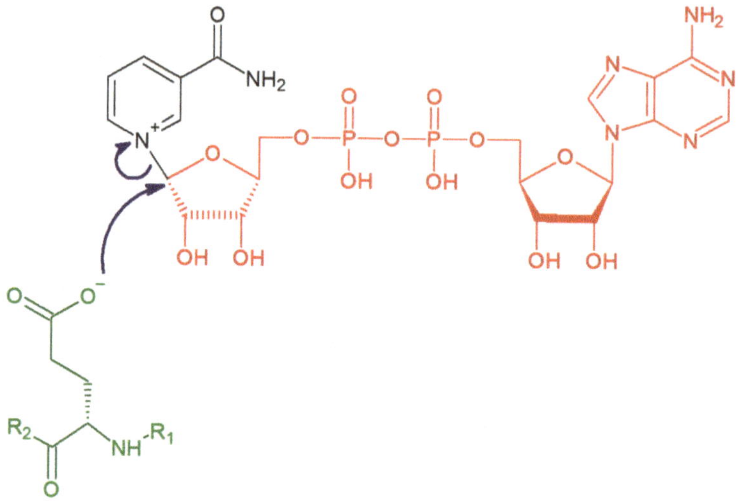

FIG. IV.61 ADP-ribosyl transfer to a protein glutamic acid. ADP-ribose part is shown in red, glutamic acid in green, and the nicotinamide group is in black. The substitution occurs in the C1″ of the ribose attached to the nicotinamide moiety.

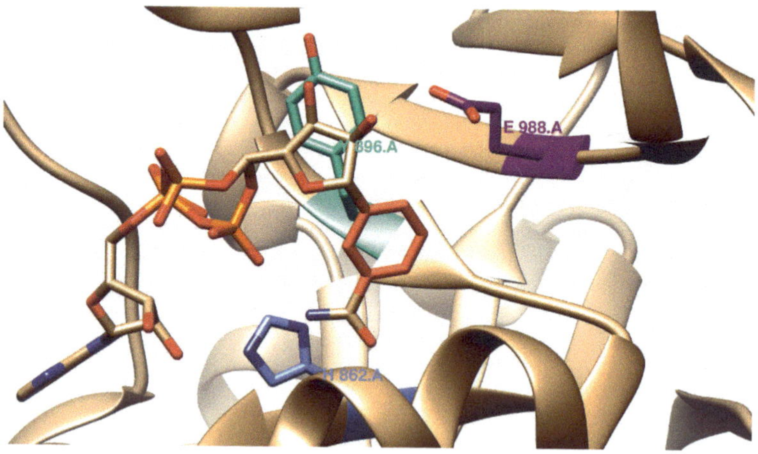

FIG. IV.62 Active site architecture of PARP1. Three important signature residues, H862, Y896, and E988, are shown in *cornflower blue*, *sea green*, and *purple*, respectively. The nicotinamide ring is shown in salmon color. *The figure was generated from PDB6BHV.*

IV.4.5.2 ADP-ribose glycohydrolases

The enzymes that remove ADP-ribose from proteins are a large group consisting of two major subclasses, macrodomain containing hydrolases and ADP-ribosyl hydrolases (Fig. IV.63). Each enzyme may have a somewhat different substrate specificity that can be overlapping. It appears that poly ADP-ribose glycohydrolase (PARG) may be the principal enzyme in human that removes the poly (ADP-ribose). However, its catalytic activity

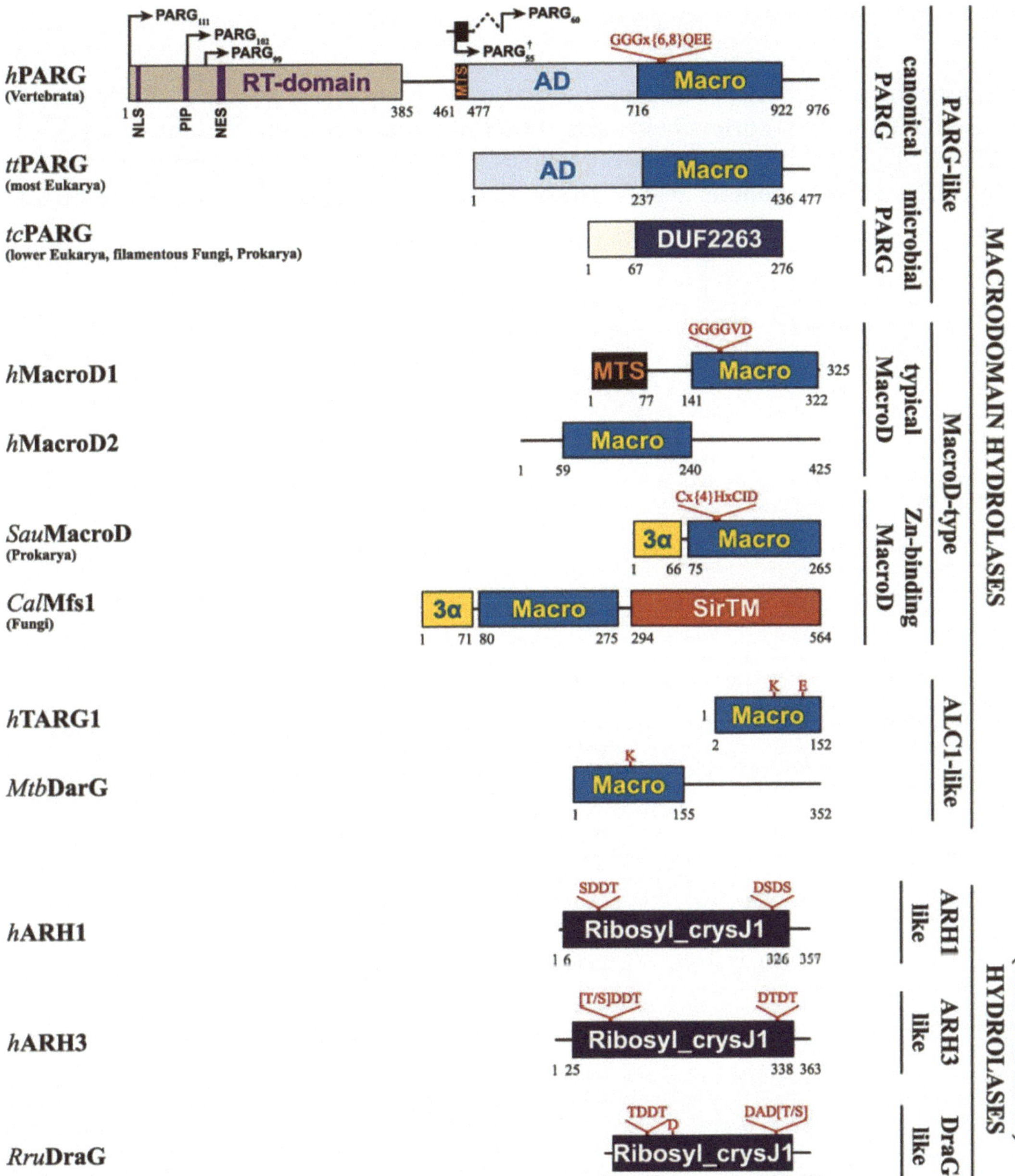

FIG. IV.63 Domain structure of macrodomains and (ADP-ribosyl)hydrolases. The hydrolytic domains are Macro (macrodomain), DUF2263, (microbial poly ADP ribose glycohydrolase), and Ribosyl_crysJ1 (ADP-ribosylation or Crystallin J1 fold), respectively. Subtype-specific sequence motifs are given above the first domain structure (red) of its type. Canonical poly ADP ribose glycohydrolases contain an accessory domain (AD). In vertebrata, the AD contains a mitochondrial-targeting signal, and the N terminus is extended by a regulatory and targeting domain (RT domain) that holds the nuclear localization and export signal (NLS and NES, respectively) and a PCNA-interacting protein (PIP) box. Other domains: 3α, 3-α-helical bundle; SirTM, sirtuin of M class. Alternative splicing of the single poly ADP ribose glycohydrolase gene in humans is indicated earlier as poly ADP ribose glycohydrolase. Note that the poly ADP ribose glycohydrolase 60 transcript involves splicing of exons 1 and 4 and the exclusion of exon 5 leading to an altered N-terminal sequence but including the mitochondrial-targeting signal. The arrow indicates the position from which the primary sequence corresponds to the other splice variants. (†) poly ADP ribose glycohydrolase 55 derives from the usage of an alternative start codon in the poly ADP ribose glycohydrolase 60 transcript. *Reproduced from Rack, et al., 2020. Genes Dev. 34, 263–284. CC BY 4.0 license.*

decreases for shorter chains, suggesting that other hydrolases, possibly ARH3, may be involved in removing shorter chains. Fig. IV.65 shows the architecture of the catalytic site of hPARG. The human PARG probably follows an S_N1-like dissociative mechanism (Fig. IV.64). E756 acts as a general acid to protonate the C1″ oxygen. Deprotonated E756 also

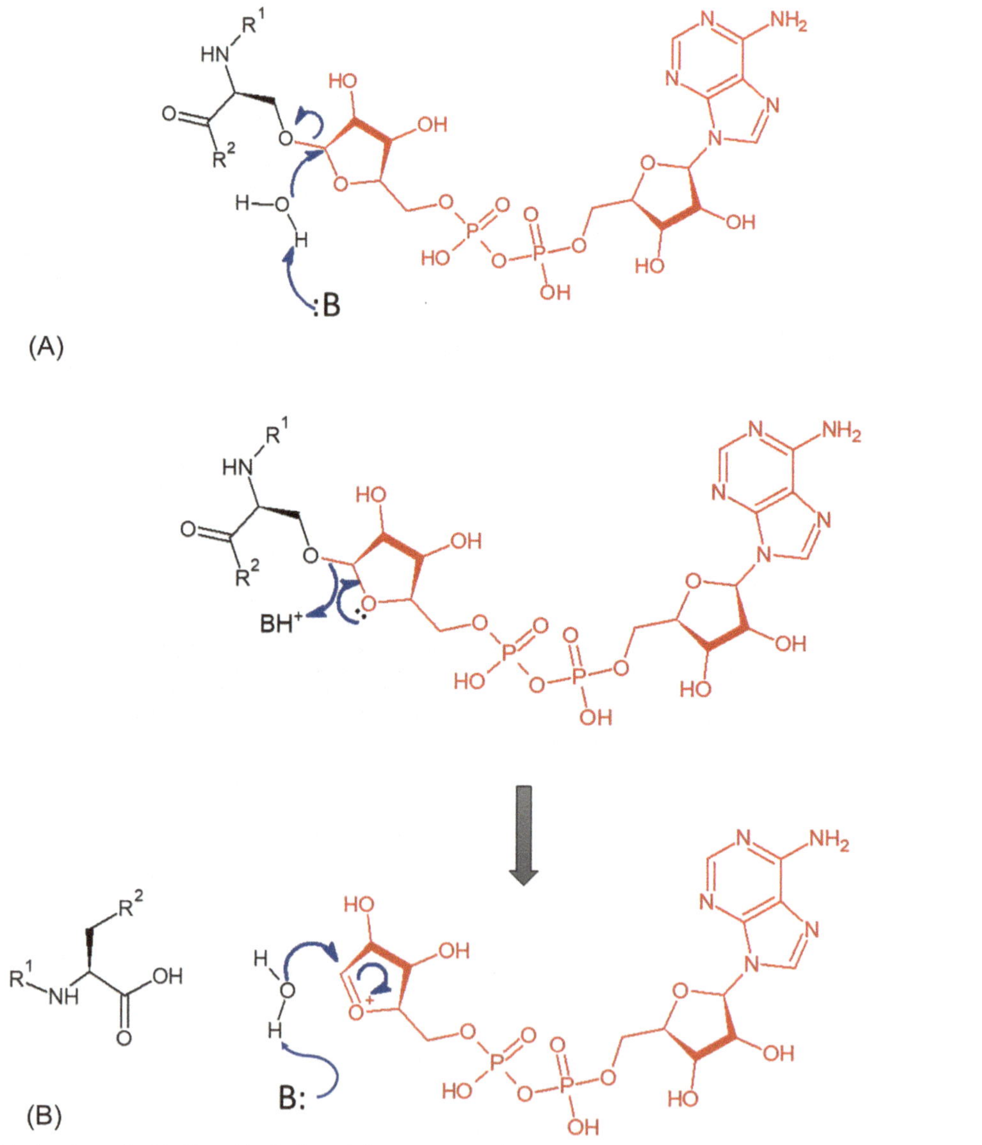

FIG. IV.64 Mechanism of ADP-ribosyl hydrolysis. (A) S_N2 type mechanism of hydrolysis of ADP-ribosyl group. (B) S_N1 type mechanism of hydrolysis of ADP-ribosyl group.

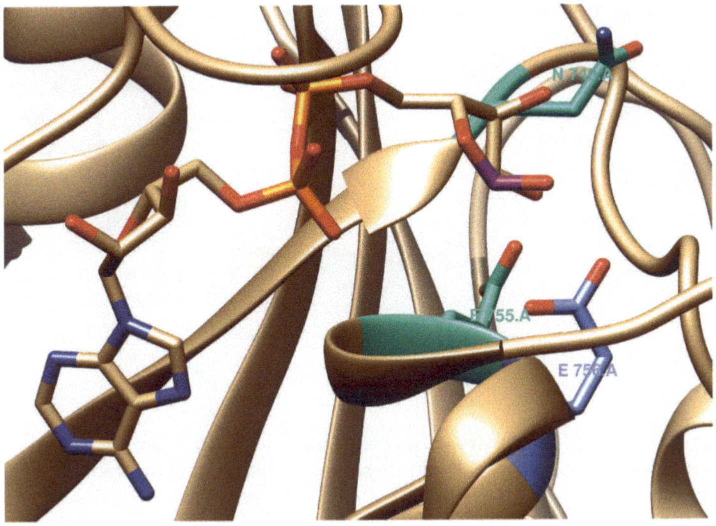

FIG. IV.65 Active site of poly ADP ribose glycohydrolase catalytic domain in complex with ADP-ribose. Three important catalytic residues are shown in stick model. The C1″ of the distal ribose is shown in *purple*. *Structure was generated from PDG 4B1H.*

acts as the general base in the next step to pick off a proton from the water to general OH^-, which is a powerful nucleophile. E755 and N740 form hydrogen bonds with hydroxyl groups of C2″ and C3″, respectively (Fig. IV.65).

IV.4.6 Methylation and demethylation of lysine

Methylation of histones plays a crucial role in regulating many vital processes, including gene expression. Arginine and lysine are the two amino acids that are extensively methylated. Methylation of lysine is an important regulatory modification of histones and crucial for epigenetic regulation of genes. S-adenosyl methionine is the methyl group donor in all methylations. Most methylations occur through direct attack of the incoming nitrogen lone pair on the methyl group of S-adenosyl methionine and proceed through an S_N2 mechanism. The transition state appears to depend on the methyltransferase; both product like and substrate-like transition states have been predicted (Fig. IV.66). Methylations on lysine can occur multiple times, producing mono-, di-, and tri-methyl lysines. In each case, the reaction presumably occurs by the same pathway.

Lysine demethylases catalyze the demethylation of methylated lysines in histones. Lysine demethylases are grouped into two major subtypes based on their reaction mechanism. One group catalyzes the removal of methyl groups through FAD (Flavin Adenine Dinucleotide)-dependent dehydrogenation of the CH_3-(εNH_2)- of lysine. The dehydrogenation produces the corresponding imine that is hydrolyzed to the N–CH_2OH group, ultimately leading to a free NH_2 group and formaldehyde (CH_2O). The dehydrogenation reaction, like other dehydrogenases, is likely to be a direct hydride transfer to FAD. In

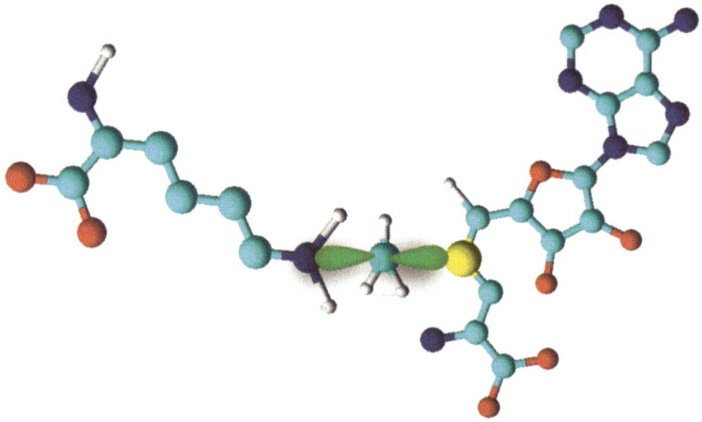

FIG. IV.66 S_N2 mechanism of methyl transfer from S-adenosyl methionine to unprotonated lysine. The lysine and the S-adenosyl homocysteine part are shown in ball and stick model. The trigonal bipyramidal methyl group in the transition state is shown with two lobes of *green* orbital indicating partial bond formation and breakage.

the other subtype of lysine demethylase, the methyl group is converted into the hydroxymethyl group through nonheme iron-dependent oxidation using α-ketoglutarate as a cofactor (Fig. IV.67). This reaction is mechanistically similar to that of the 5hmC demethylases that we discussed before.

IV.4.6.1 Histone lysine methyltransferases

Human arginine and lysine methyltransferases are a large family of enzymes consisting of more than 50 enzymes (Fig. IV.68). Except for a few, all the lysine methyltransferases contain the catalytic SET (Su(var)3-9 and "Enhancer of zeste" and Trithorax) domain. Many of them are specific for individual histone marks. The crystal structure of SET7/9, a member of the protein family that methylates H3K4, shows that H3K4 inserts into a deep tunnel. The methyl group S-adenosyl methionine is present at the other end of the tunnel (Fig. IV.69). The protonated lysine (—NH_3^+) must be deprotonated before methyl transfer reaction. The mechanism of this crucial step remains unresolved.

IV.4.6.2 Histone lysine demethylases

Histone lysine demethylase family also contains multiple enzymes in humans. They can be grouped into two broad classes, the FAD-containing lysine-specific demethylases and the Jumonji-containing α-ketoglutarate or Fe-dependent dioxygenases. The histone demethylating enzyme, LSD1, belongs to the former class. LSD1 demethylates mono and di-methylated H3K4 and H3K9. The crystal structure of LSD1, in complex with the cofactor FAD and an H3 derived peptide, which is substituted at the K4 position (K4M), shows the proximity of the methyl group of M4 and the flavin ring at the active site (Fig. IV.70). The proximity is consistent with the direct hydride transfer mechanism. JMJD2A is a representative demethylase of the second class. It demethylates di and trimethylated H3K9 and H3K36. The active site (Fig. IV.71) shows the Fe in coordination with

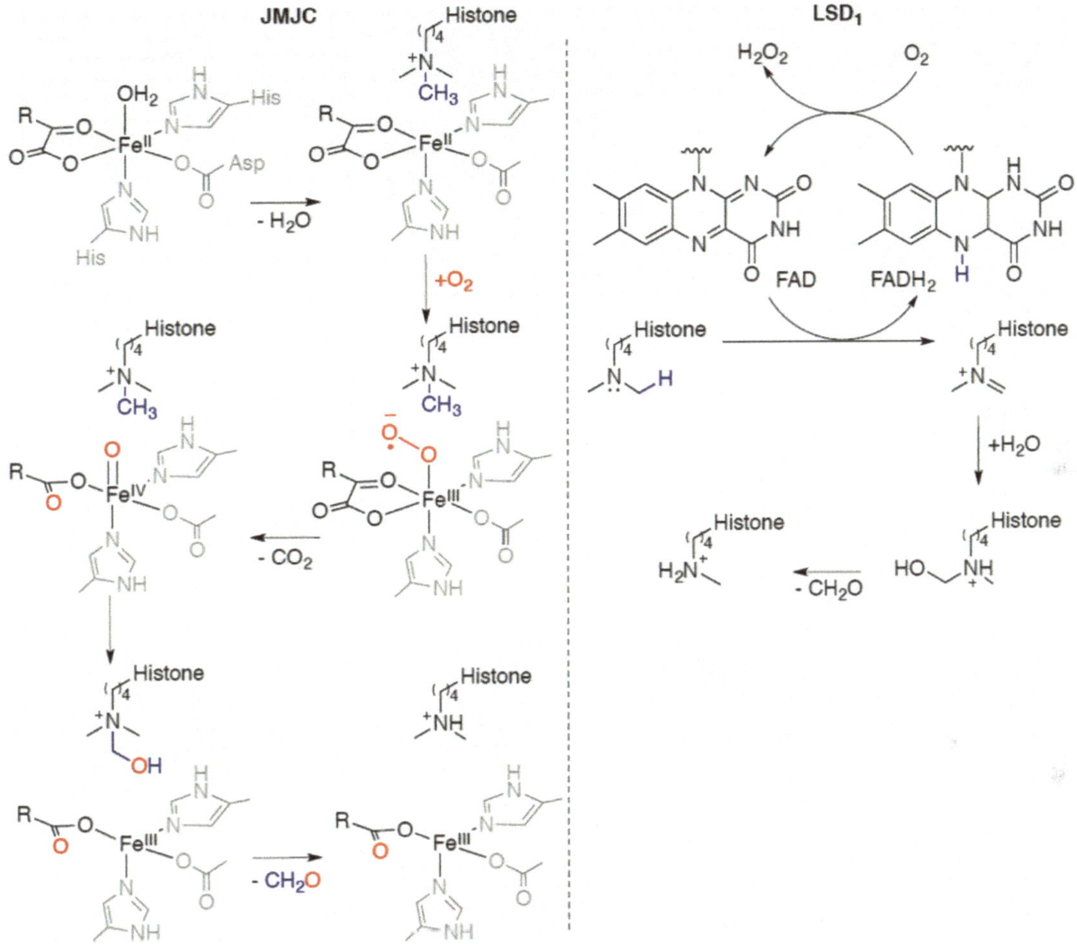

FIG. IV.67 Mechanism of two classes of lysine demethylases. *Reproduced from Cortopassi, et al., 2016. J. Mol. Graph. Model. 67, 69–84. https://doi.org/10.1016/j.jmgm.2016.04.011. Elsevier.*

histidine residues 276 and 188, glutamic acid 190, along with an analog of α-ketoglutarate. The Fe-bound O_2 molecule is in proximity to the H3K36.

IV.4.7 Methylation and demethylation of arginine

Arginine can be methylated on both of the unsubstituted nitrogen atoms of the guanidinium group (Fig. IV.72). Due to its high stability, the guanidinium cation is not easily deprotonated. Thus nucleophilic substitution occurs in its protonated state. The lone pair on each nitrogen can participate as an incoming nucleophile in an S_N2-like concerted reaction. The reaction appears to be similar to the lysine methylation described in a previous section. The demethylation of arginine remains controversial. No consensus has emerged whether there is an active demethylation process in a cell under normal

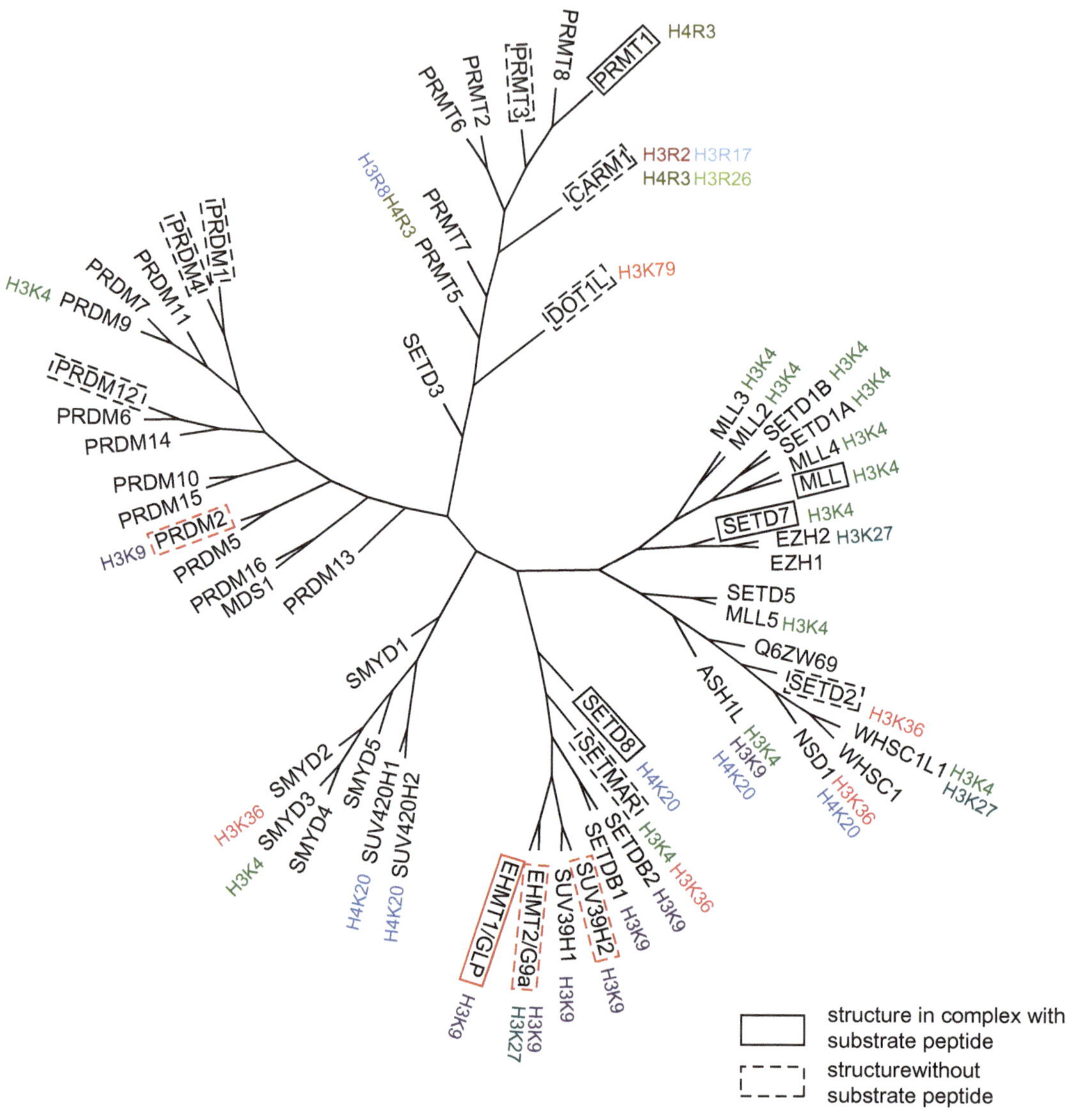

FIG. IV.68 Phylogenetic tree of human histone methyltransferases. Phylogeny is based on a multiple sequence alignment of the methyltransferase domain including the N-SET, Pre_SET, SET, I-SET, and PostSET motifs. Substrate selectivity was extracted from Kouzarides. Enzymes with solved structure are highlighted by a *frame*, *dotted* if no peptide substrate complex is available. Structures solved in the present work are framed in *red*. *Reproduced from Wu, H., Min, J., Lunin, V.V., Antoshenko, T., Dombrovski, L., Zeng, H., et al., 2010. Structural biology of human H3K9 methyltransferases. PLoS ONE 5 (1), e8570. https://doi.org/10.1371/journal.pone.0008570. CC-BY license.*

conditions. A few reports suggest that the demethylation of histone arginine residues may occur through the Jumonji domain-containing α-ketoglutarate- or Fe-dependent enzymes. These enzymes are similar in action to the α-ketoglutarate- or Fe-dependent lysine demethylases and may even have overlapping activity.

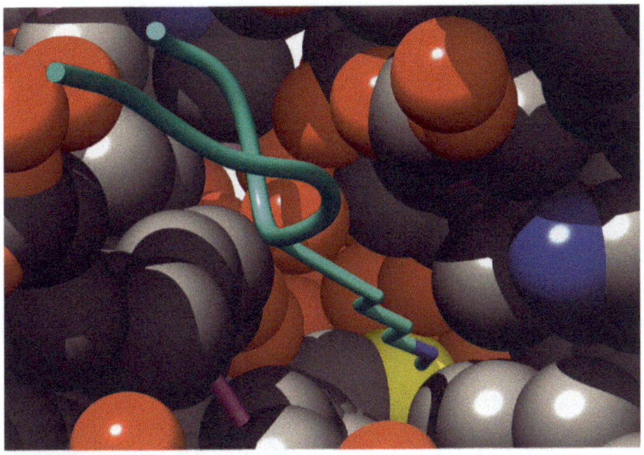

FIG. IV.69 Active site of SET7/9 (KMT7). The histone-derived peptide is shown in *sea green*, and the sulfur atom of the S-adenosyl homocysteine is seen in *yellow* at the end of the tunnel. *The structure was generated from PDDB1O9S.*

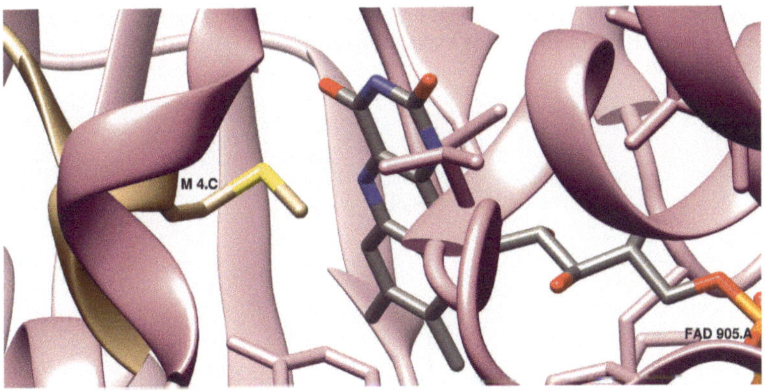

FIG. IV.70 Histone lysine demethylase LSD1 active site, complexed with the cofactor FAD and a H3-derived peptide carrying a K4M substitution. The proximity of M4 and the FAD suggests optimized geometry for direct hydride transfer. *The structure was generated from PDB5H6Q.*

IV.4.7.1 Arginine methyltransferases and demethylases

Arginine methyltransferases are a family of enzymes and have been named as PRMT 1-9. PRMT1 is a major arginine methyltransferase in the cell, and its structure and catalytic mechanism have been studied in detail. It carries out asymmetrical dimethylation of H4R3 and other nonhistone arginine residues. The proximity of the guanidino group, and the sulfur atom of the S-adenosyl methionine in the crystal structure, suggests a direct transfer of the methyl group in an S_N2-like concerted transfer (Fig. IV.73). E144 transfers a proton from the guanidine group upon the first methylation. It also likely plays a role in charge redistribution during the methyl transfer process. It appears that PRMTs are not processive enzymes. Upon transfer of the first methyl group, the monomethylated

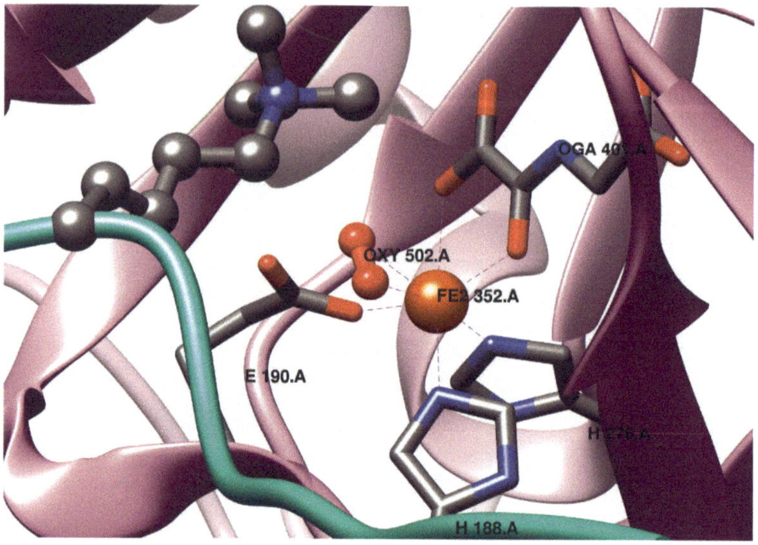

FIG. IV.71 Active site of the Jumonji domain containing histone lysine demethylase, JMJD2A, in complex with an analog of α-ketoglutarate and H3 derived peptide that is trimethylated at H3K36 (ball and stick model). *The structure was generated from PDB2P5B.*

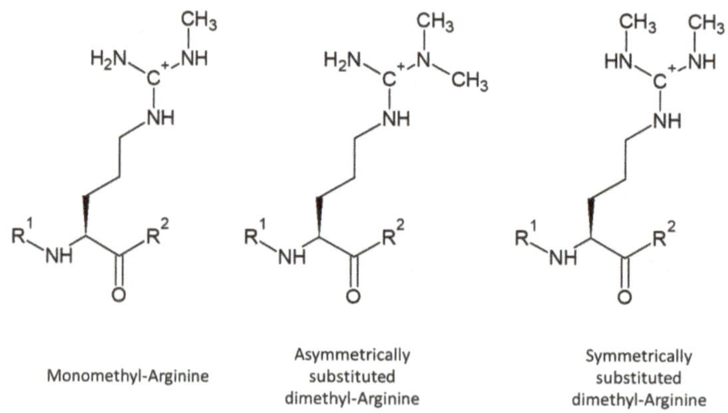

FIG. IV.72 Structure of mono and dimethyl arginines.

arginine is released, S-adenosyl homocysteine is exchanged with S-adenosyl methionine, followed by the monomethylated arginine substrate. The second methyl transfer then occurs.

Although consensus regarding the existence of an active arginine demethylation process in the cell is lacking, a couple of enzymes are capable of demethylating methylated arginine. They are two Jumonji domain-containing proteins, JMJD6 and JMJD1B. Both are α-ketoglutarate- or Fe-dependent hydroxylases.

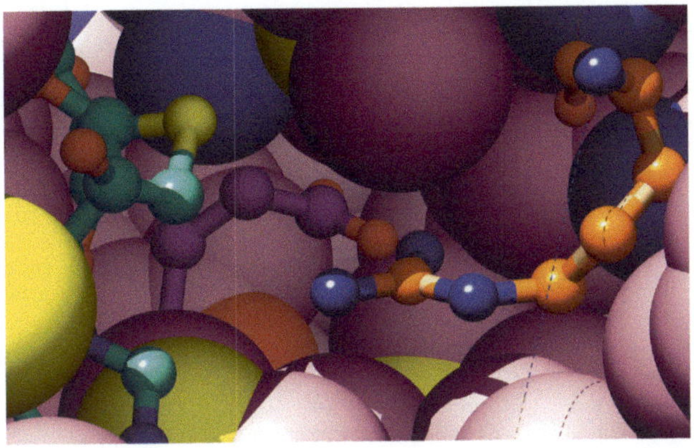

FIG. IV.73 Active site of PRMT1. The important components are shown in ball and stick models. The arginine to be methylated is shown in *orange* backbone, and S-adenosyl homocysteine is shown in *sea-green* backbone. Heavy atoms are in standard color. The arginine is close in space to the sulfur atom of S-adenosyl homocysteine. Glutamic acid 144 (shown in ball and stick and colored *purple*) plays a crucial role in the catalytic mechanism. *Figure prepared from PDB1OR8.*

IV.4.8 Thiol oxidations and di-sulfide formation

Cysteine is a highly reactive amino acid found in proteins. The thiol group of cysteine participates in many reactions in which it acts as a powerful nucleophile. Under physiological conditions, this reactivity is due to several factors. The thiol group of cysteine has a pK_a of around 8.0 in proteins, unless some local environmental factors influence the pK_a substantially. Thus, at physiological pH, the —SH group of the cysteine is substantially ionized, making it a powerful nucleophile. Cysteines are also prone to oxidations. Apart from oxidation to cystine, cysteine can be oxidized to produce several other oxidation products (Fig. IV.74). The nucleophilicity of cysteine is either modified or lost in all these oxidation products.

IV.4.8.1 Regulation of protein activities by redox signaling

The oxidation potential of intracellular milieus and that of the extracellular spaces vary widely and may change according to the change of cellular states. The presence and concentrations of many species, such as peroxides, oxidized and reduced glutathione, change significantly, thus, altering the ability of cellular environments to oxidize different species present. Under normal cellular conditions, three amino acids are reversibly oxidized or reduced. These are cysteine, methionine, and selenocysteine. Oxidation of these amino acids may profoundly alter their function in proteins. Some of these oxidations are important for the regulation of biological activities of certain proteins.

One well-known example is the reversible oxidation of the catalytic cysteine 215 in the active site of the protein tyrosine phosphatase 1B (PTP1B), a phosphotyrosine phosphatase. PTP1B plays a crucial role in insulin signaling, balancing the tyrosine kinase activity of the insulin receptor (IR). Binding of insulin stimulates the tyrosine kinase activity of IR. A burst of hydrogen peroxide is then released, oxidizing the catalytic cysteine 215 to its sulfenic acid form and ultimately to sulfenamide form (Fig. IV.75). This inactivation is

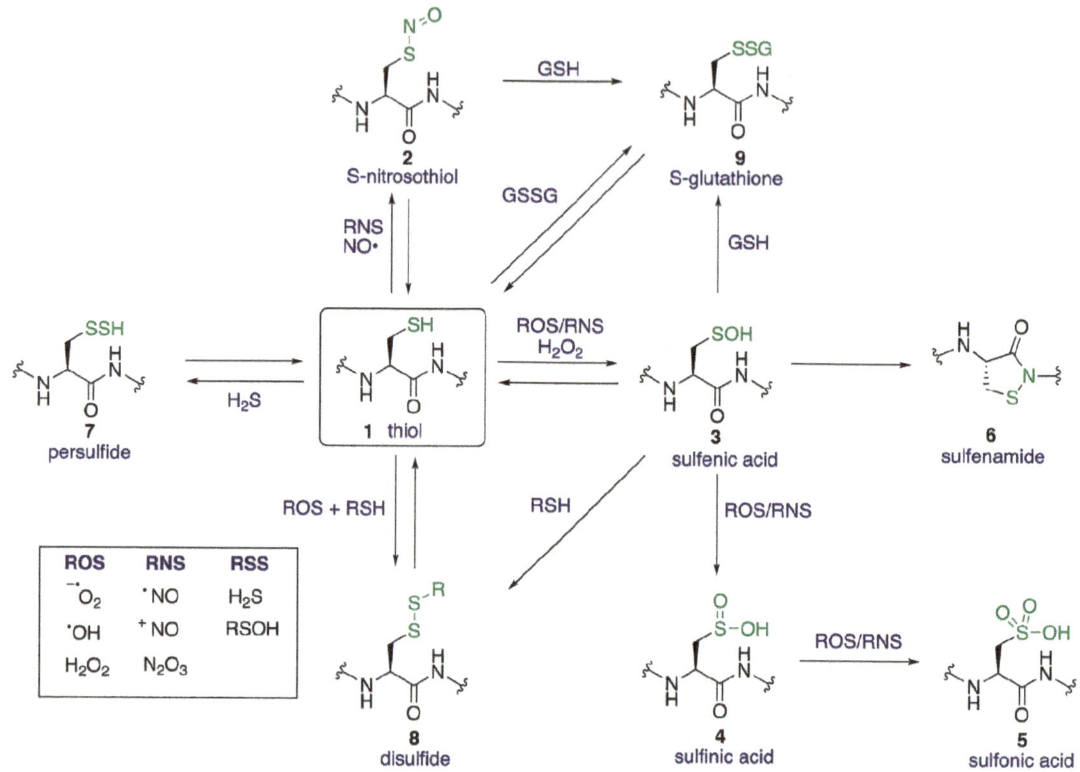

FIG. IV.74 A selection of biologically relevant oxidative posttranslational modifications of cysteine. *Reproduced from Alcock, L.J., Perkins, M.V., Chalker, J.M., 2018. Chem. Soc. Rev. 47, 231. https://doi.org/10.1039/C7CS00607A Under CC-BY 3.0 license.*

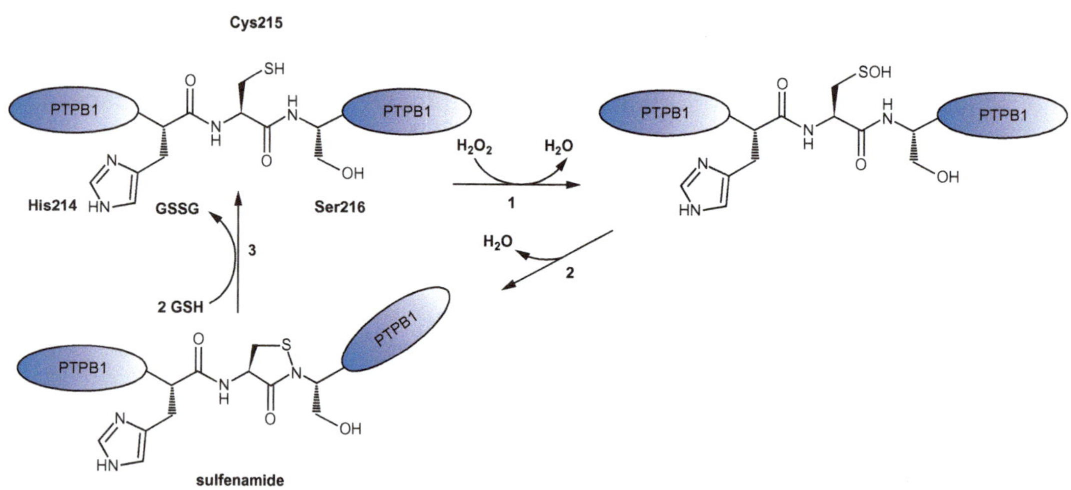

FIG. IV.75 Mechanism of sulfenamide formation in protein tyrosine phosphatase 1B (1) Exposure of the catalytic cysteine 215 (RSH) of protein tyrosine phosphatase 1B to H2O2 leads to the formation of a sulfenic acid (RSOH), which undergoes (2) intramolecular cyclization with a nearby amino group, yielding in sulfenamide (RSNR). (3) Reduction of the sulfenamide is mediated by forming a mixed disulfide formation with glutathione, which is subsequently resolved by glutaredoxin. *Reproduced from Groitl and Jakob, Biochim. Biophys. Acta.*

transient and reversible. One crucial issue is that this oxidation occurs only at cysteine 215 and not at other cysteines present in the protein. The selective oxidation occurs because the microenvironment of cysteine 215 lowers it pK_a from around 8.0 to 5.5, thus, ensuring that a significant fraction of this cysteine thiol is unprotonated and reactive at the pH of the microenvironment. In contrast, the other cysteines remain protonated and unreactive to oxidation.

IV.4.9 Deimination or citrullination

Proteins, including histones, undergo enzymatic deimination of arginine to yield citrulline. The deimination of arginine destroys the positive charge on the arginine to make it neutral. Side chain of arginine has a guanidine group that has a pK_a of greater than 12. Under cellular environments, where the pH is near neutral, it is fully protonated. The deimination process is catalyzed by a family of enzymes, called peptidylarginyl deiminase. The chemical mechanism of deimination starts with an attack of a cysteine thiol on the central carbon of the guanidine group and producing a tetrahedral intermediate. The tetrahedral intermediate then undergoes an elimination addition process, in which the NH_3 group leaves, and an incoming water molecule provides the hydroxyl group to produce another tetrahedral intermediate, distinct from the previous tetrahedral intermediate. The thiol is then eliminated from the tetrahedral intermediate to produce citrulline.

Chapter summary

Genomes undergo massive amounts of modifications, both in the DNA and the associated proteins. A large number of enzymes catalyze these modifications. This chapter focuses on the chemistry behind a myriad of modifications that occur on the genome. The diversity of both the chemical mechanisms and the enzymes that catalyze the chemical reactions are breathtaking.

- In the first section, we have reviewed the principles behind chemical reactions that occur in the genome. Some readers may find this redundant, but this section aims to refresh the memory of the readers who do not deal with chemistry on a day-to-day basis. We have described the electronic structures and properties of amino acid side chains of proteins and bases present in the DNA. We have reviewed some basic concepts of chemical equilibrium and kinetics. We have briefly described some aspects of enzyme catalysis that we have used in the following sections.
- In the second section, we have described mechanisms of common types of reactions that occur on the genome. The major ones are the nucleophilic substitutions and additions, β-elimination, and free-radical reactions.
- In the third section, we have discussed the reactions that modify the DNA. Some of these are enzymatic modifications, but some others are reactions that occur without the involvement of enzymes. The most prominent reactions discussed are methylation and demethylation of the DNA bases, strand breaks, deamination of cytosine, and repair of modified bases.

- In the fourth section, we have discussed chemical modifications of proteins and enzymes that catalyze such modifications. Major chemical modifications that were covered include phosphorylation, acetylation, ADP-ribosylation, and methylation of lysines and arginines. We have also discussed the oxidation of cysteine and redox regulation of enzyme activities. For each modification, we have discussed the chemical mechanisms and the enzymes that catalyze such modifications. In each case, we have also discussed the reverse reactions and enzymes that catalyze the reverse reaction.

Further reading

Berndsen, C.E., Albaugh, B.N., Tan, S., Denu, J.M., 2007. Catalytic mechanism of a MYST family histone acetyltransferase. Biochemistry 46, 623–629.

Bilyard, M.K., Becker, S., Balasubramanian, S., 2020. Natural, modified DNA bases. Curr. Opin. Chem. Biol. 57, 1–7.

Black, J.C., Van Rechem, C., Whetstine, J.R., 2012. Histone lysine methylation dynamics: establishment, regulation, and biological impact. Mol. Cell 48, 491–507.

Cadet, J., Davies, K.J., Medeiros, M.H., Di Mascio, P., Wagner, J.R., 2017. Formation and repair of oxidatively generated damage in cellular DNA. Free Radic. Biol. Med. 107, 13–34.

Cho, U.S., Xu, W., 2007. Crystal structure of a protein phosphatase 2A heterotrimeric holoenzyme. Nature 445, 53–57.

Fitch, C.A., Platzer, G., Okon, M., Garcia-Moreno, E.B., McIntosh, L.P., 2015. Arginine: its pKa value revisited. Protein Sci. 24, 752–761.

Hammes, G., 2012. Enzyme Catalysis and Regulation. Elsevier.

Kaypee, S., Sahadevan, S.A., Patil, S., Ghosh, P., Roy, N.S., Roy, S., Kundu, T.K., 2018. Mutant and wild-type tumor suppressor p53 induces p300 autoacetylation. Science 4, 260–272.

Langelier, M.-F., Eisemann, T., Riccio, A.A., Pascal, J.M., 2018. PARP family enzymes: regulation and catalysis of the poly (ADP-ribose) posttranslational modification. Curr. Opin. Struct. Biol. 53, 187–198.

Lee, K.K., Workman, J.L., 2007. Histone acetyltransferase complexes: one size doesn't fit all. Nat. Rev. Mol. Cell Biol. 8, 284–295.

Lombardi, P.M., Cole, K.E., Dowling, D.P., Christianson, D.W., 2011. Structure, mechanism, and inhibition of histone deacetylases and related metalloenzymes. Curr. Opin. Struct. Biol. 21, 735–743.

Lyko, F., 2018. The DNA methyltransferase family: a versatile toolkit for epigenetic regulation. Nat. Rev. Genet. 19, 81.

Maiti, A., Michelson, A.Z., Armwood, C.J., Lee, J.K., Drohat, A.C., 2013. Divergent mechanisms for enzymatic excision of 5-formylcytosine and 5-carboxylcytosine from DNA. J. Am. Chem. Soc. 135, 15813–15822.

Marmorstein, R., Zhou, M.-M., 2014. Writers and readers of histone acetylation: structure, mechanism, and inhibition. Cold Spring Harb. Perspect. Biol. 6, a018762.

McMurry, J., Begley, T.P., Begley, T., 2005. The Organic Chemistry of Biological Pathways. Roberts and Company Publishers.

Mondal, S., Thompson, P.R., 2019. Protein arginine deiminases (pads): biochemistry and chemical biology of protein citrullination. Acc. Chem. Res. 52, 818–832.

O'Sullivan, J., Ferreira, M.T., Gagné, J.-P., Sharma, A.K., Hendzel, M.J., Masson, J.-Y., Poirier, G.G., 2019. Emerging roles of eraser enzymes in the dynamic control of protein ADP-ribosylation. Nat. Commun. 10, 1–14.

Palazzo, L., Mikoč, A., Ahel, I., 2017. ADP-ribosylation: new facets of an ancient modification. FEBS J. 284, 2932–2946.

Pidugu, L.S., Flowers, J.W., Coey, C.T., Pozharski, E., Greenberg, M.M., Drohat, A.C., 2016. Structural basis for excision of 5-formylcytosine by thymine DNA glycosylase. Biochemistry 55, 6205–6208.

Shi, Y., Whetstine, J.R., 2007. Dynamic regulation of histone lysine methylation by demethylases. Mol. Cell 25, 1–14.

Smith, B.C., Denu, J.M., 2009. Chemical mechanisms of histone lysine and arginine modifications. Biochim. Biophys. Acta Gene Regul. Mech. 1789, 45–57.

Sykes, P., 1986. A Guidebook to Mechanism in Organic Chemistry. Pearson Education India.

Tanner, J.J., Parsons, Z.D., Cummings, A.H., Zhou, H., Gates, K.S., 2011. Redox regulation of protein tyrosine phosphatases: structural and chemical aspects. Antioxid. Redox Signal. 15, 77–97.

Tewary, S.K., Zheng, Y.G., Ho, M.-C., 2019. Protein arginine methyltransferases: insights into the enzyme structure and mechanism at the atomic level. Cell. Mol. Life Sci. 1–16.

Vossenaar, E.R., Zendman, A.J., van Venrooij, W.J., Pruijn, G.J., 2003. PAD, a growing family of citrullinating enzymes: genes, features and involvement in disease. BioEssays 25, 1106–1118.

Wang, X., Chu, H., Lv, M., Zhang, Z., Qiu, S., Liu, H., Shen, X., Wang, W., Cai, G., 2016. Structure of the intact ATM/Tel1 kinase. Nat. Commun. 7, 1–8.

Wang, Z., Cole, P.A., 2014. Catalytic mechanisms and regulation of protein kinases. In: Methods in Enzymology. Elsevier, pp. 1–21.

Wu, X., Zhang, Y., 2017. TET-mediated active DNA demethylation: mechanism, function and beyond. Nat. Rev. Genet. 18, 517.

Xu, W., Parmigiani, R., Marks, P., 2007. Histone deacetylase inhibitors: molecular mechanisms of action. Oncogene 26, 5541–5552.

Zhao, S., Zhang, X., Li, H., 2018. Beyond histone acetylation—writing and erasing histone acylations. Curr. Opin. Struct. Biol. 53, 169–177.

V

Chemical principles of DNA sequence recognition and gene regulation

V.1 Introduction

The nucleus consists of highly organized supramolecular structures of proteins, nucleic acids, and other classes of molecules. Genomic DNA itself exists as massive protein-DNA complexes, folded and condensed in the form of chromatins. Although highly organized, the chromatins are also dynamic entities, with constant movements of proteins and localized changes in the DNA structure. The chromatin packing is nonrandom. In resting cells, each chromatin has a definite position. Its positions may change significantly depending on the cellular state. Prokaryotic genomes appear to be less densely packed than their eukaryotic counterparts. However, prokaryotic genomes are also highly organized. In eukaryotes, packaging limits the accessibility of DNA sequences to proteins and other classes of molecules. Accessibility of proteins for a specific gene regulatory sequence plays a crucial role in the regulation of gene expression in eukaryotic cells. Accessibility of regulatory sequences, although necessary, is not believed to play as crucial a role in the prokaryotic gene regulation as other factors play crucial roles. We have described the genome organization in Chapter I and focus on other topics here.

V.2 Review of DNA structure

V.2.1 B-DNA, major and minor grooves

Double-stranded DNA can adopt different types of structures depending on the base sequence, solution conditions, and the presence of other molecules, such as proteins. The B-form of DNA is the most stable structure under ordinary cellular conditions. However, other higher-energy structures, such as G-Quadruplexes, may occur transiently that can interact with proteins. We will mostly focus on the B-DNA structure and its variations as these are primary protein-recognition targets.

The double-stranded B-DNA is cylindrical, with two grooves running parallel to each other throughout the length of the DNA. The wider of the two grooves is termed the major groove (MG), whereas the other one is called the minor groove (mG) (Fig. V.1). The major and minor groove widths are approximately 12 Å and 6 Å, respectively. The base pairs are

Chemical Biology of the Genome. https://doi.org/10.1016/B978-0-12-817644-3.00005-2

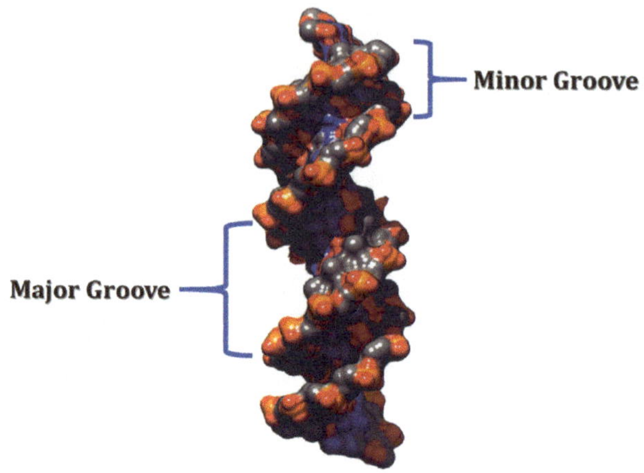

FIG. V.1 Structure of B-DNA. *The structure was generated using freeware Chimera 1.14.*

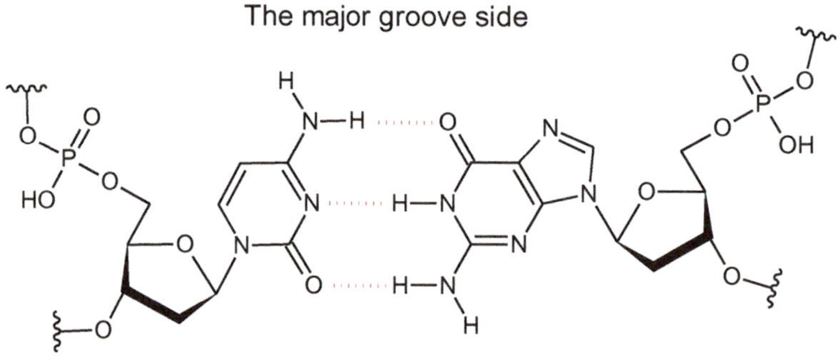

FIG. V.2 Orientation of a base pair in the B-form of DNA showing the sides exposed to the major and minor grooves.

exposed to solvents through the grooves. The purine C6-N7 side and the pyrimidine C4-C5 side face the major groove (Fig. V.2), whereas the purine N3 side and the pyrimidine C2 side face the minor groove. Sequence-specific DNA-binding proteins interact with the DNA through the major groove with occasional exceptions.

Duplex DNA can attain different conformational states in the isolated form depending upon solution conditions. The most common alternate form is the A-DNA. Both the A- and the B-DNA are right-handed helices. However, several conformational parameters are different. For example, there are 11 base pairs per turn of the helix in A-DNA, whereas there are approximately 10.5 base pairs per turn in the B-DNA. The main feature of A-DNA is its narrow major groove as opposed to wide major groove of B-DNA and wide and shallow minor groove of A-DNA. The *Z*-DNA, on the other hand, is a left-handed helix having 12 base pairs per turn. The biological roles of A- and Z-DNA are not as widespread as those of the B-DNA, and we will primarily focus on the B-form of DNA.

V.2.2 Base pairing in the B-DNA and its structural parameters

The first B-DNA structure solved by Watson and Crick from fiber diffraction gave an impression of a conformationally homogeneous helix, independent of the base sequence. With the elucidation of high-resolution structures of defined oligonucleotides, it became apparent that the B-DNA helix is not conformationally homogeneous, and there are significant deviations from the canonical B-DNA structure and these deviations are sequence dependent. These deviations are often referred to in the literature as conformational micro-heterogeneity or local shape variation. In an idealized B-DNA helix, each base pair may be considered as planar, with both bases in a pair lying on the same plane. This planarity makes the complementary base pairing most stable. In an idealized B-DNA helix, two adjacent base pairs are stacked above one another and rotated around the vertical (z-axis) about 35 degree for maintaining the helical pitch of about 10.5 base pairs per turn. However, sequence-dependent deviations from this ideal situation occur frequently and are thought to play a significant role in some DNA-protein interactions.

High-resolution structures of oligonucleotides made it clear that within a range, the base pair conformations in the B-DNA are significantly different. We will define some of the important conformational deviations from the average structure here. The shift is a displacement of a base pair along the x-axis, whereas the slide is a displacement along the y-axis. When a base-pair plane rotates around the x-axis relative to its neighboring base-pair plane, it is called the tilt, whereas the roll is the rotation of the base-pair plane along the y-axis (Fig. V.3). Base pairs can also move relative to a neighboring base pair

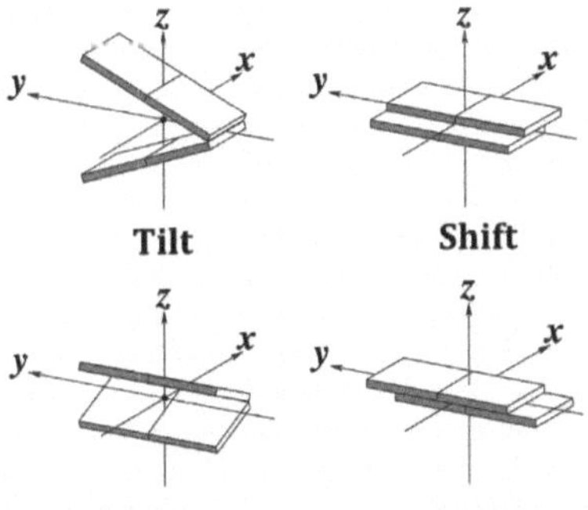

FIG. V.3 Base pair movements (deviation from planarity) between two adjacent base pairs involving rotation of one with respect to the other. *Reproduced from website http://x3dna.org/. Lu, et al., 2003. 3DNA: a software package for the analysis, rebuilding and visualization of three-dimensional nucleic acid structures. Nucleic Acids Res. 31 (17), 5108–5121; Lu, et al., 2008. 3DNA: a versatile, integrated software system for the analysis, rebuilding and visualization of three-dimensional nucleic-acid structures. Nat. Protoc. 3, 1213–1227. The licensor has permitted in writing for reproduction in this book.*

while retaining the parallel orientation. Fig. V.4 shows a couple of these movements, the twist, and the slide. Fig. V.5 shows the buckle and the propeller twist, examples of movements of a base in a base pair relative to the other. In addition to sequence-dependent variation, the geometry of a base pair may vary with protein binding and change in environmental conditions.

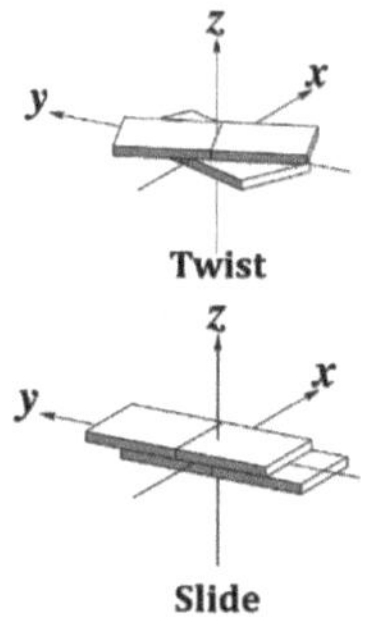

FIG. V.4 Base pair movements, retaining the parallel orientation, between two adjacent base pairs. *Reproduced from website http://x3dna.org/. Lu, et al., 2003. 3DNA: a software package for the analysis, rebuilding and visualization of three-dimensional nucleic acid structures. Nucleic Acids Res. 31 (17), 5108–5121; Lu, et al., 2008. 3DNA: a versatile, integrated software system for the analysis, rebuilding and visualization of three-dimensional nucleic-acid structures. Nat. Protoc. 3, 1213–1227. The licensor has permitted in writing for reproduction in this book.*

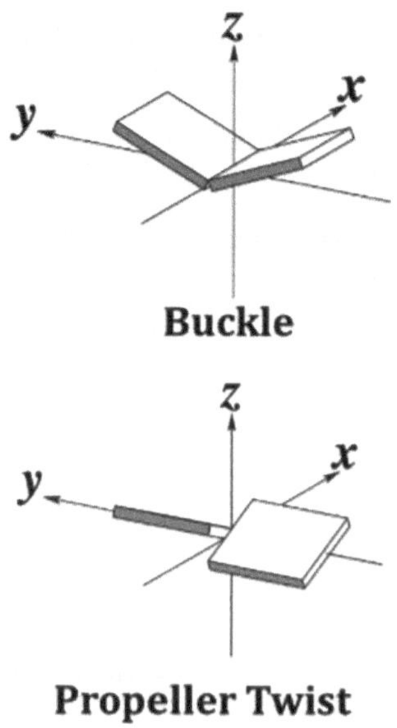

FIG. V.5 Base movements within a base pair. *Reproduced from website http://x3dna.org/. Lu, et al., 2003. 3DNA: a software package for the analysis, rebuilding and visualization of three-dimensional nucleic acid structures. Nucleic Acids Res. 31 (17), 5108–5121; Lu, et al., 2008. 3DNA: a versatile, integrated software system for the analysis, rebuilding and visualization of three-dimensional nucleic-acid structures. Nat. Protoc. 3, 1213–1227. The licensor has permitted in writing for reproduction in this book.*

V.2.3 Sequence-dependent variation of B-DNA shape

Previously, we have discussed sequence-dependent conformational heterogeneity of B-DNA. These deviations include changes in values of groove widths, twist, roll, propeller twist, and other parameters. Such deviations result in changes of shape, which are recognized by proteins as variant shapes, without directly reading the base sequence responsible for such variations. This shape or indirect recognition by proteins will be discussed in more detail in a later section. A frequent change in shape that occurs due to specific nucleotide sequences is the change in the width of the minor groove. This variant shape is now well characterized and conclusively demonstrated to be essential for sequence recognition by certain classes of transcription factors.

The minor groove width is defined as the smallest phosphorus-phosphorus separation across the groove, after subtracting 5.8 Å for the diameter of a phosphate group. DNA minor grooves vary significantly with the sequence of the base, and generally, AT-rich sequences tend to have narrower minor grooves. The B-DNA backbone conformation is characterized by the torsion angles, α, β, γ, δ, ε, and ζ. In B-DNA, correlated transitions in the backbone ε(C4′-C3′-O3′-P) and ζ(C3′-O3′-P-O5′) occur. The transitions in ε and ζ occur from the more frequent t, g- to g-, t conformation, and the two structures are often referred to as BI and BII conformational substates, respectively (Fig. V.6). These changes in torsion angles also affect roll, slide, and twist of the structures to adopt unusual values. The dinucleotide steps having GC, CG, CA, TG, and TA have a high preference for BII conformation; on the other hand, TT, TC, CT, and CC dinucleotide steps are rarely found in

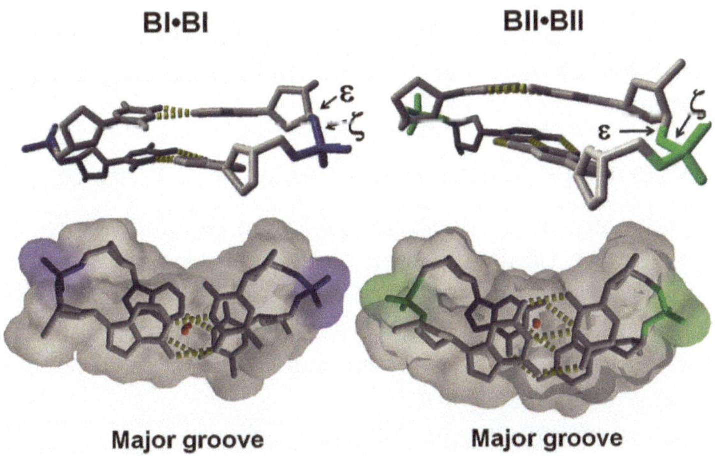

FIG. V.6 Illustration of the relationship between BI- or BII-rich regions and DNA grooves. The BI-rich (PDB code 1EHV, *left*) and BII-rich (PDB code 3GGI, *right*) decamer X-ray structures are in the same orientations. BI and BII phosphate groups are in *blue* and *green*, respectively. *Top*: the lateral view shows that the minor groove is considerably enlarged in the BII-rich structure *(right)*. *Bottom*: view along the major groove, showing its reduced concavity in the BII-rich structure *(right)*. *Reproduced from Oguey, C., Foloppe, N., Hartmann, B., 2010. Understanding the Sequence-Dependence of DNA Groove Dimensions: Implications for DNA Interactions. PLoS ONE 5 (12), e15931. https://doi.org/10.1371/journal.pone.0015931.*

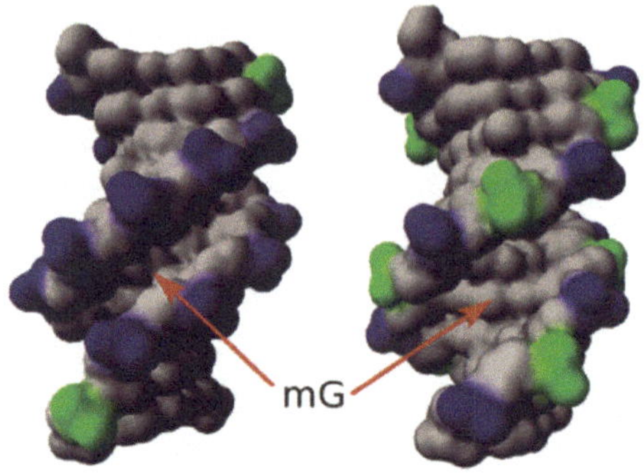

FIG. V.7 *Left*: BI–BI conformation with the narrow minor groove. *Right*: BI–BII substate. *Oguey, C., Foloppe, N., Hartmann, B., 2010. Understanding the sequence-dependence of DNA groove dimensions: implications for DNA interactions. PLOS ONE 5 (12), e15931. https://doi.org/10.1371/journal.pone.0015931. https://journals.plos.org/plosone/article?id=10.1371/journal.pone.0015931.*

this conformation. Dinucleotide steps having BII substates in both strands rarely occur. The presence of the BI substate in both strands leads to the narrowing of the minor groove (Fig. V.7).

V.3 Thermodynamics of DNA-protein interaction

Recognition of specific DNA sequences by DNA-binding proteins, such as transcription factors, is a result of a complex balance of inter and intramolecular forces. Measurement of equilibrium thermodynamic quantities, such as entropy, enthalpy, and heat capacity, provides us ways to assess the role of different kinds of intermolecular forces responsible for such interactions and characterize structural and dynamical changes that may occur upon complex formation.

V.3.1 Enthalpy, entropy, Gibbs free energy, and heat capacity

The critical thermodynamic quantities relevant for understanding protein-DNA interactions are enthalpy, entropy, Gibbs free energy, and heat capacity. In any thermodynamic measurements, one is always interested in changes of thermodynamic parameters, rather than absolute values. Our focus will be on this as well. We will recapitulate only the aspects that will be used later in this book. Interested readers may consult textbooks on physical chemistry for more detailed description.

Enthalpy of a closed system may be defined as follows:

$$H = U + pV,$$

where H is the enthalpy, U is the internal energy, p is the pressure, and V is the volume. Biological systems mostly operate at constant pressure and within a narrow temperature range. Under such conditions, the change in enthalpy is a good measure of the change in internal energy of a system. We are interested in enthalpy changes that occur upon molecular interactions. Hence, a more useful and measurable parameter is the enthalpy change or ΔH. Change in enthalpy upon complex formation leads to heat flow to or from the surrounding, depending on whether the reaction is exothermic or endothermic, respectively. By convention, a negative sign precedes the exothermic change and a positive sign precedes the endothermic change. Either Calorie or Joule is the unit of enthalpy. An important thing to remember is that many thermodynamic state properties, such as the enthalpy, entropy, and Gibbs free energy, are characterized with respect to a reference state, called the standard state. It is denoted by a superscript "o" symbol as a suffix. By convention, the standard enthalpy of formation (ΔH^o) of all elements in their most stable form is zero.

The concept of entropy can be approached from more than one direction. We will try to explain it from a probabilistic viewpoint. Entropy is a direct reflection of disorder in a system. The disordered state has a higher probability of occurrence, and hence, a system tends to move in that direction. We are concerned with systems here that include the protein, the DNA, and their immediate environment consisting of solvent and ions. Interaction of macromolecules may influence the entropy of the system in many ways. The protein and the DNA are both flexible molecules, and their intra and intermolecular dynamics may change upon complex formation. Dynamic molecules are more disordered as they can visit multiple states and, hence, are entropically more favorable. Solvent molecules and ions that are in the neighborhood of the macromolecule also can change their mobility upon complex formation. Change in dynamic character of molecules influences entropy as the degree of disorder changes. Thus, in many cases, the complex formation leads to change in entropy. The SI unit of entropy is the Joules. degree^{-1}. Kelvin^{-1}.

Gibbs free energy, (G), can be conceptualized in different ways. For chemical reactions, it is the most useful predictor of the spontaneity of a chemical reaction. Change in free energy between two states (ΔG) is usually measured and related to other thermodynamic quantities by the following equation:

$$\Delta G = \Delta H - T\Delta S$$

ΔG must be negative for a spontaneous change to occur.

We will try to explain the above relation in terms of entropy. Any spontaneous change will cause an increase in the total entropy of the universe, which is the sum of entropies of a system and the surroundings. Surroundings encompass the rest of the universe. For a closed system, the entropy of the system may even decrease in a spontaneous reaction, as long as the entropy of the universe increases. The ΔS term in the above equation is the entropy of the system. The ΔH term actually hides the change in entropy of the surroundings. In isothermal and constant pressure conditions, ΔH reflects the heat flow between the system and the surroundings, which will change the entropy of the surroundings. Thus a condition of a negative ΔG is equivalent to the condition of the increase in entropy of the universe.

Heat capacity of a substance is the amount of heat required to raise the temperature by one unit, usually 1°K. Heat capacity is dependent on mass and usually defined per unit mass. Thus the unit of C_p is Joule. Kelvin^{-1}. Kilogram^{-1}. External variables may also affect the heat capacity. In biological systems that essentially operate under constant pressure, the heat capacity value at constant pressure (C_p) is the most useful parameter. During protein-DNA complex formation or protein association, C_p often changes (ΔC_p). As currently understood, a change in heat capacity (ΔC_p) during these processes is primarily due to a change in solvent-exposed apolar surface.

V.3.2 Intermolecular forces important for sequence-specific binding

Like other macromolecular complexes, protein-DNA complexes are also characterized by ionic interactions between positive and negative charges, hydrogen bonds, and van der Waals interaction. Ionic interactions between positively charged protein side chains and negatively charged phosphate groups of the DNA are commonly found in protein-DNA complexes. Fig. V.8 shows an example of ionic interaction in λ-repressor-O_L1 complex. Hydrogen bonds are a commonplace in protein-DNA interactions. Fig. V.9 shows hydrogen bonding in a protein-DNA complex. In addition to these intermolecular forces, hydrophobic interactions play an important role in sequence-specific DNA recognition. In Chapter I, we have discussed the basic features of different intermolecular forces.

V.3.3 Energetics of sequence-specific recognition

Interactions of several DNA-binding domains with their respective target DNA have been examined. In almost all cases, the entropy change upon complex formation is positive (favorable), whereas enthalpy change could be either positive (unfavorable) or negative (favorable). This variation of enthalpy occurs because of the varied nature of the complexes and accompanying diversity of atomic interactions. In addition, when a protein binds to DNA, both the protein and the DNA may change their structure. This structural

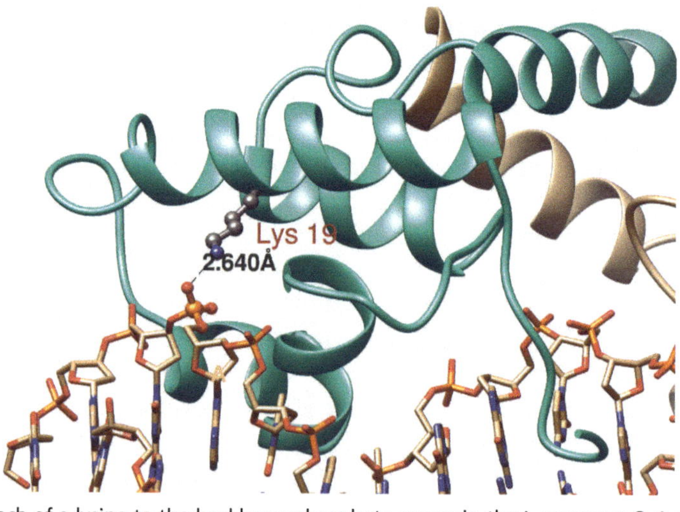

FIG. V.8 Close approach of a lysine to the backbone phosphate group in the λ-repressor-O_L1 complex (pdb1LMB).

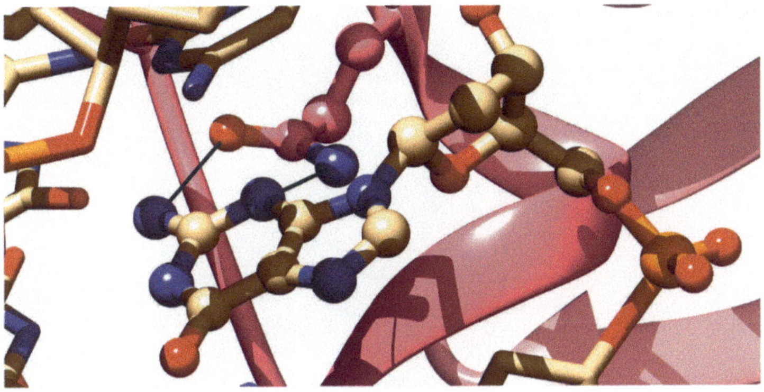

FIG. V.9 An example of hydrogen bonds *(black line)* between a glutamine residue of a protein and a guanine base through the minor groove. The glutamine and the guanine are shown in ball and stick models. *The structure is derived from PDB ID 4ux5.*

deformation from the lowest energy conformation present in the free state requires the loss of some interactions, leading to loss of enthalpy.

V.3.4 Release of water and ions during complex formation

Both proteins and DNA are polyelectrolytes. Counter ions and water molecules surround them. In proteins, the water molecules may be either tightly bound or loosely associated. The type of atom with which the water molecule interacts may determine its character. In general, both classes of water, particularly the tightly associated ones, lose a significant degree of freedom of movement. The water surrounding the DNA plays an essential role in maintaining the structure and stability of the double helix. In B-DNA, the major and the minor grooves appear to have different patterns of hydration (Fig. V.10). The minor groove in the DNA often has a "spine of hydration" at AT-rich sequences. These water molecules are less dynamic than the water molecules present in other areas of the DNA and may have an ice-like character.

The release of any macromolecule-bound species leads to enhanced mobilities of the released molecules and, hence, a source of increased entropy. Thus disruption of the spine of hydration upon binding of proteins maybe a source of the favorable entropy change, in some protein-DNA complexes. The favorable entropic component may also originate from displaced water molecules and ions that are associated with the protein and other parts of the DNA. The loss of mobility of both the protein and the DNA that usually accompany complex formation may attenuate the favorable entropy change mentioned above.

In solution, DNA is a negatively charged polyelectrolyte surrounded by counterions of which a fraction is tightly associated with negatively charged phosphate groups of the DNA. Binding of a protein, a polyelectrolyte itself, displaces some tightly associated cations and transfers them to the bulk solvent. The process increases the entropy of the system. Fig. V.11 depicts this binding equilibrium in a pictorial form. As the ion concentration in the bulk solution increases, the equilibrium will shift toward dissociation of the

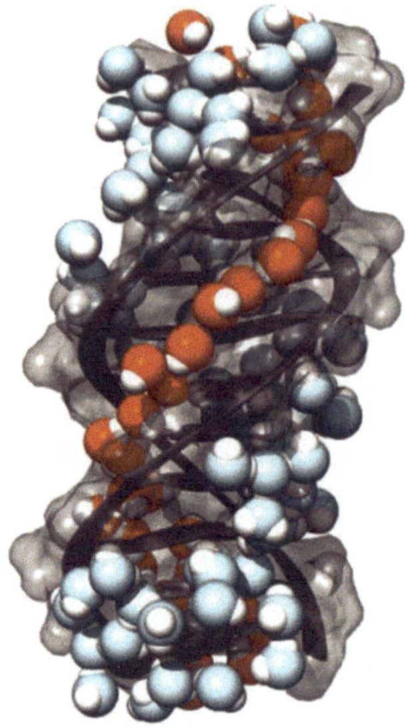

FIG. V.10 Water structure in the minor groove *(red)* and major groove *(blue)* of the double-stranded CGCAAATTTGCG oligonucleotide in a MD (Molecular Dynamics) simulation. Water in the minor groove is substantially more ordered with respect to the DNA axis than water in the major groove. *Reproduced from McDermott et al., 2017. ACS Cent. Sci. 3 (7), 708–714. With permission from ACS https://pubs.acs.org/doi/full/10.1021/acscentsci.7b00100. Further permissions related to the material excerpted should be directed to the ACS.*

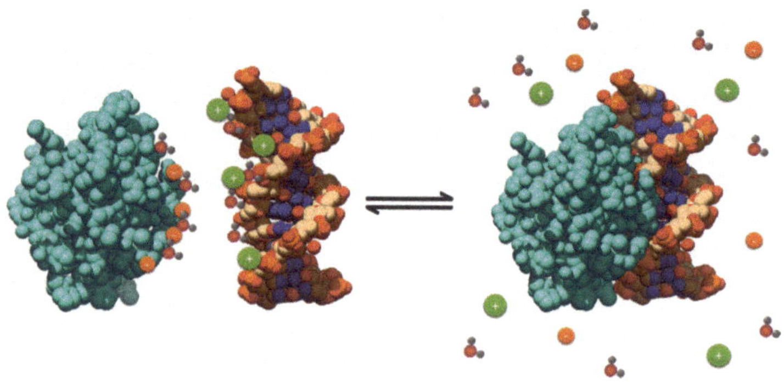

FIG. V.11 A model of ion and water release upon DNA-protein complex formation. Only the ions and water molecules are shown that are released upon complex formation.

protein-DNA complex. The amount of ion released upon protein-DNA complex formation can be estimated from the dissociation constant of the protein-DNA complex as a function of the ionic strength of the solution. Similarly, the water-release stoichiometries have been estimated for various protein-DNA interactions from the variation of dissociation constants as a function of osmolytes concentrations in the solution (Fig. V.11).

The *E. coli* lac repressor is a tetrameric protein that binds to several sites on the *E. coli* genome and regulates the expression of several genes responsible for lactose metabolism.

It was one of the earliest DNA-binding proteins purified and later studied intensively. Estimates of ion release, obtained from ionic strength dependence of the binding constant of lac repressor to DNA, indicate the release of 8 ± 1 and 11 ± 2 monovalent cations upon the formation of specific and nonspecific complexes, respectively. When osmolyte dependence of the binding constant of lac operator with DNA was studied, the release of 260 ± 32 and 165 ± 43 water molecules was observed upon specific and nonspecific complex formation, respectively. Other DNA-binding proteins show qualitatively similar results. We may conclude that the nonspecific complex has significantly higher ionic interactions. A lower amount of water released during nonspecific complex formation in comparison to the specific complex suggests that the protein-DNA interface in the nonspecific complex is less extensive than the specific complex. We will see later that NMR (Nuclear Magnetic Resonance) experiments bear this out.

V.4 Origin of sequence specificity

A picture emerges from the previous discussion that the sequence-specific binding of a protein is a complex function of many structural and solution properties and is unlikely to be explained by a simple code. Many different classes of proteins bind to DNA. In broad terms, there exist two classes of proteins, sequence-specific DNA-binding proteins and non-specific DNA binding proteins. The former class of proteins includes most transcription factors. The latter class includes many architectural proteins involved in chromatin condensation, such as histones. Many, if not all, sequence-specific DNA-binding proteins form both specific and nonspecific complexes, albeit with different affinities. We have to understand the specificity at two levels. At the first level, we have to understand how the binding affinity and interactions differ for the target and unrelated DNA sequences. At the second level, we have to understand the specificity for the target sequence in the backdrop of numerous unrelated sequences and some related sequences present in the genome.

V.4.1 Structural differences between sequence-specific and nonspecific complexes

Although nonspecific protein-DNA complexes share some of the same qualitative characteristics with those of the specific complexes, substantial differences exist. For example, binding of DNA-binding proteins to both specific and nonspecific DNA sequences leads to ion release, water release, and significant favorable free-energy changes. However, the values of some of these solution parameters are significantly different for specific and nonspecific complexes. For nonspecific complexes in general, the water release is lower, and ion release is higher than that of the specific complexes. These characteristics suggest that the protein primarily interacts with the phosphodiester backbone in the nonspecific complex. The following section compares structural aspects of the specific and nonspecific complexes of the lac repressor.

In specific complexes, many DNA-binding proteins interact with base atoms in the major groove. In nonspecific complexes, interactions occur primarily with the DNA backbone. Lac repressor headpiece (amino acid residues 1–62), which is the DNA-binding domain of the lac repressor, and its complexes with both specific and nonspecific DNA have been studied in great detail by NMR spectroscopy. Comparison of lac repressor (1–62) interactions with a nonspecific DNA and O1 (one of the operators in the lac operon) highlights many differences. In the specific complex, the C-terminal tail region folds into an α-helix and interacts with the minor groove situated between the two major grooves of the symmetrically related halves (Fig. V.12). In the nonspecific complex, the C-terminal tail part remains unstructured and does not interact with the minor groove in the same way (Fig. V.13). In general, many proteins and protein domains are dynamic in the free state, sampling many conformations. This ensemble of conformations collapses to a distinct conformation in the specific complex, forming many base-protein noncovalent

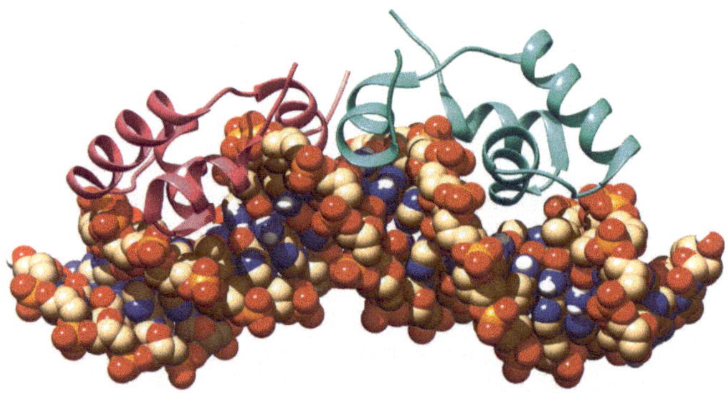

FIG. V.12 Lac repressor headpiece specific complex with operator O1. *Modeled from Pdb1l1m using Chimera.*

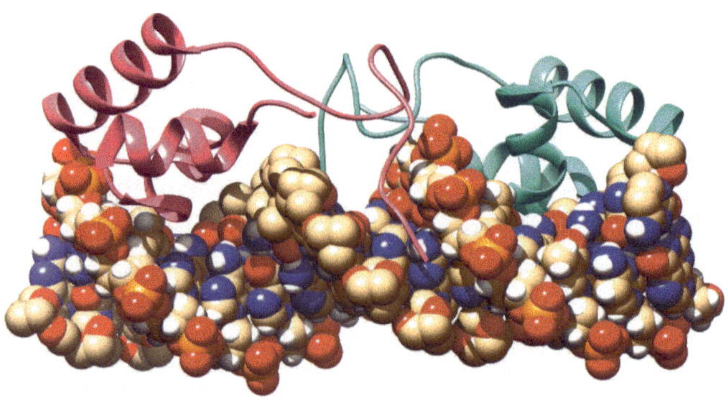

FIG. V.13 Lac repressor headpiece nonspecific complex. *Modeled from Pdb1osl using Chimera.*

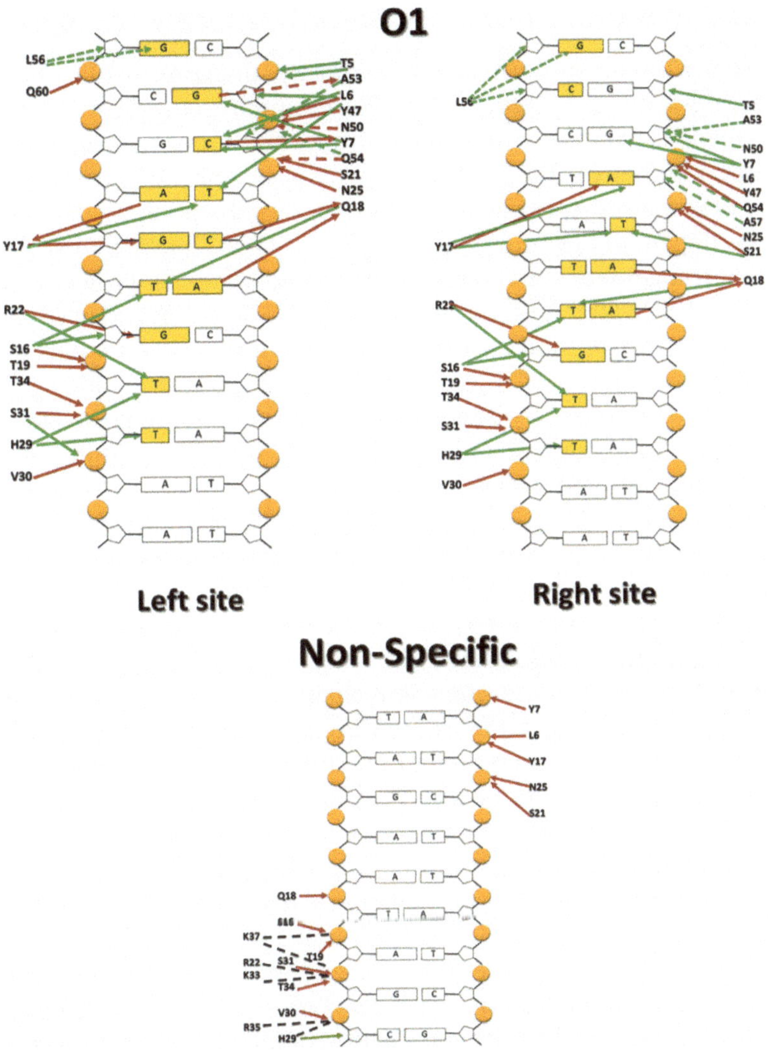

FIG. V.14 Protein-DNA contacts in the HP62V52C–O1 *(top)* and nonspecific complexes *(bottom)*. The bases that are recognized specifically are shown in *yellow*, and continuous and broken lines indicate interactions observed in the majority (50%) of the structures in the final ensembles in the major and minor grooves, respectively. *Red* and *green* lines indicate hydrogen bonding and hydrophobic contacts, respectively. *Redrawn with modifications (Figs. 6 and 8) from Romanuka, et al., 2009. J. Mol. Biol. 390, 478–489 (Elsevier).*

bonds. In the nonspecific complex, the protein retains its native-like flexibility, without forming base-specific hydrogen bonds. Thus hydrogen bonds between protein and base atoms of the DNA and consequent restriction of conformational mobility characterize specific DNA-protein interaction (Fig. V.14).

V.4.2 Bonding in sequence-specific and nonspecific complexes

The B-DNA structure is conformationally mostly uniform, with the exception of sequence-dependent micro-heterogeneity. The dominant contribution to specificity in the sequence-specific complex comes from the formation of hydrogen bonds and Van der Waals interactions between protein atoms and base atoms. In the major groove, where a vast majority of sequence-specific interactions occur, the polar atoms of the bases present a sequence-dependent geometric arrangement. Thus a particular sequence of bases carries a unique arrangement of polar and some nonpolar atoms. Typically, N7 atom of the purine ring, exocyclic O6 atom of guanine, exocyclic N6 amino group of adenine, exocyclic N4 amino group of cytosine, and exocyclic O4 atom of thymine are part of this pattern. Besides, the C5-methyl group of thymine is also present in the major groove. The N7, O6, and O4 atoms are excellent hydrogen bond acceptors, whereas the N6 and N4 amino groups are hydrogen bond donors. Hydrogen bonds between base atoms and protein side chains occur in many different ways.

In some cases, a single hydrogen bond forms between an amino acid side chain and an atom in the base. Fig. V.15 shows such an example between Lys4 of λ repressor and a GC base pair. Another interaction pattern involves the formation of two hydrogen bonds between N7 and O6 atoms of the same guanine residue and two guanidine nitrogen atoms of an arginine side chain (middle panel, Fig. V.15). A variation of this motif is hydrogen bonding between two atoms of the same side chain to atoms on two different base pairs (bottom panel, Fig. V.15). These types of bonding are examples of direct readout.

As mentioned earlier, in the lac repressor headpiece (1–62)-DNA complexes, the interactions with backbone atoms are observed in both the specific and nonspecific complexes. However, in the sequence-specific complex in addition to the backbone contacts, direct interactions with base atoms were observed. Thus we may conclude that the DNA backbone-protein interactions contribute toward affinity, and the base-protein interactions confer specificity as well. The backbone interactions in this example consist of hydrogen bond formation between neutral amino acid side chains with phosphate groups, van der Waals interaction between deoxyribose sugars with hydrophobic side chains such as valine, and electrostatic interaction of the phosphate group with positively charged amino acids such as lysine and arginine.

V.4.3 Indirect readout of sequence information

Initial elucidation of the DNA structure by fiber diffraction provided an average of conformations of the base pairs and backbone. The determination of high-resolution structures of synthetic oligonucleotide duplexes later showed significant sequence-dependent conformational differences between base pairs and backbone units. We have given examples of such conformational heterogeneity at the beginning of this chapter. Some DNA-binding proteins recognize these conformational variations or their consequences, instead of reading out the base sequence through direct noncovalent interactions with the base atoms. This mode of recognition is called the shape recognition or indirect readout.

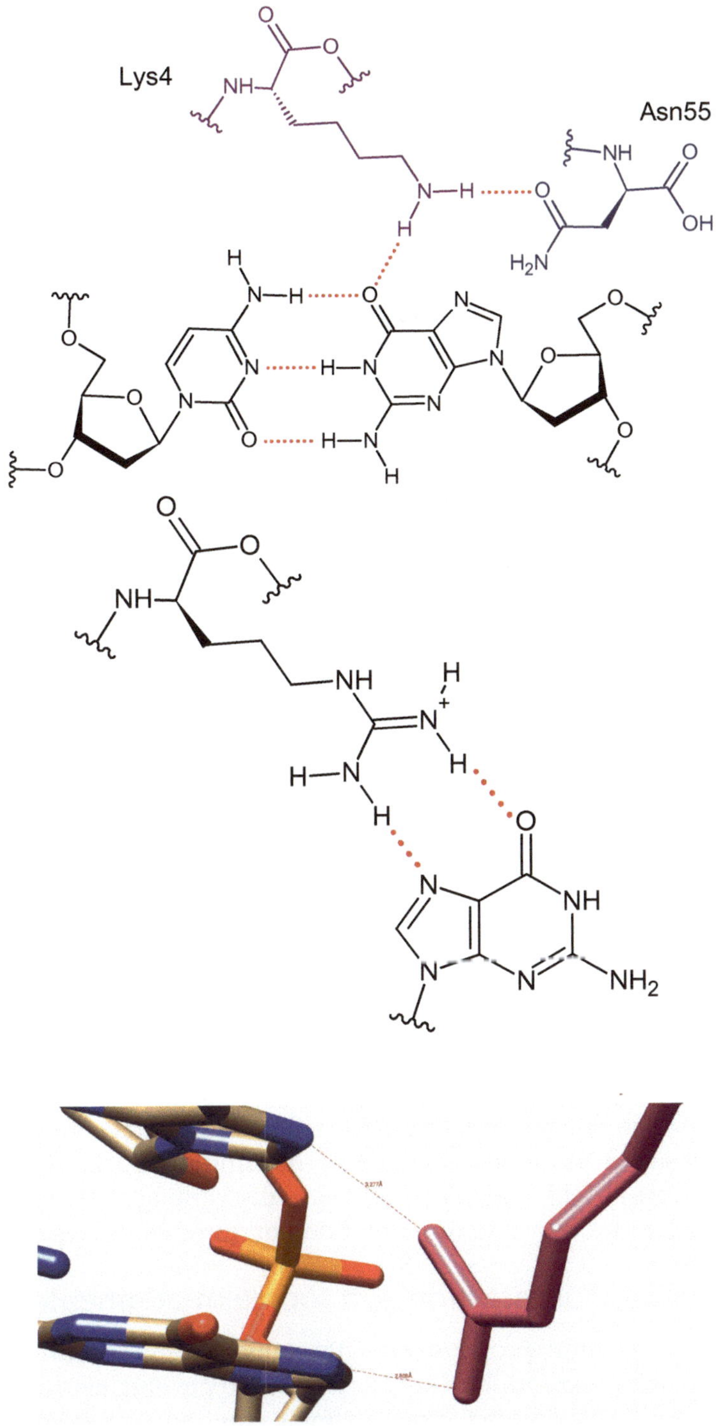

FIG. V.15 Examples of hydrogen bond patterns between proteins and DNA. The examples shown above are all interactions through the major groove. *Top panel* is adapted from λ-repressor N-terminal domain or O_L1 complex (1lmb), the *middle panel* is an example of how arginine residue forms a bidentate bond to a Guanine base, and the *bottom panel* is an example of a bidentate hydrogen bond across two adjoining base pairs by an arginine.

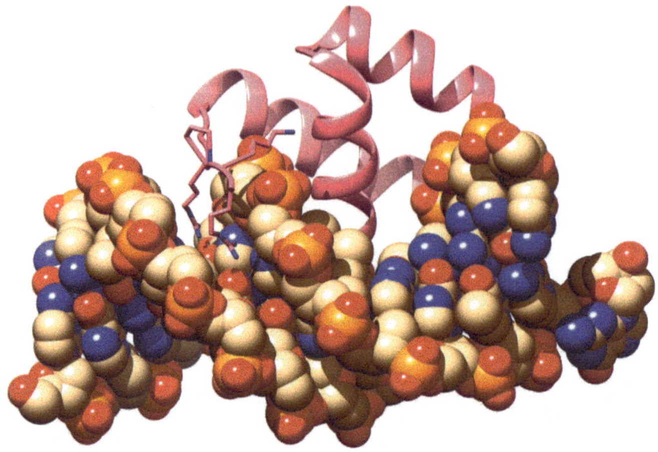

FIG. V.16 The Dlx5 homeodomain in complex with its target DNA. The N-terminal arm can be seen in hot *pink* in the minor groove with two arginine positioned to sense the enhanced electrostatic field of the narrow minor groove. *Figure prepared from PDB4rdu.*

The indirect readout mechanism is fairly widespread. This mode of recognition may not always replace direct readouts but supplement them, at least in some cases, to enhance sequence specificity. As will be discussed in a later section, a combination of multiple recognition elements is a theme utilized in many DNA-binding proteins for enhancing binding specificity. The homeodomain-containing transcription factors are a large group of eukaryotic transcription factors. They utilize the mechanism of indirect readout of the minor groove width of AT-rich sequences (Fig. V.16). AT-rich sequences that lack a TpA dinucleotide step give rise to narrow minor grooves. The presence of a narrow minor groove enhances the electrostatic potential (negative) in the minor groove. Positively charged arginine and lysine residues are often present in these narrow minor grooves without directly bonding with the bases present there. The electrostatic interaction between the enhanced negative electrostatic potential and positively charged arginine and lysine residues provides additional stabilization. Fig. V.17 shows this mode of interaction for the homeodomain-containing protein, Scr. This protein binds specifically to the fkh250 binding site but not to the fkn250* (three base pairs substituted that widen the minor groove) site. In sequences that lack narrow minor groove, such enhanced electrostatic potentials are absent, making it possible to discriminate between sequences that produce narrow minor grooves and those that do not, through electrostatic interactions.

V.4.4 Protein-protein interaction as a source of augmented specificity

In eukaryotes, gene regulatory proteins generally cooperate and combine with each other for their function. In the next couple of sections, we will try to elucidate a physical basis for cooperation between transcription factors. A simple equilibrium model, as shown below,

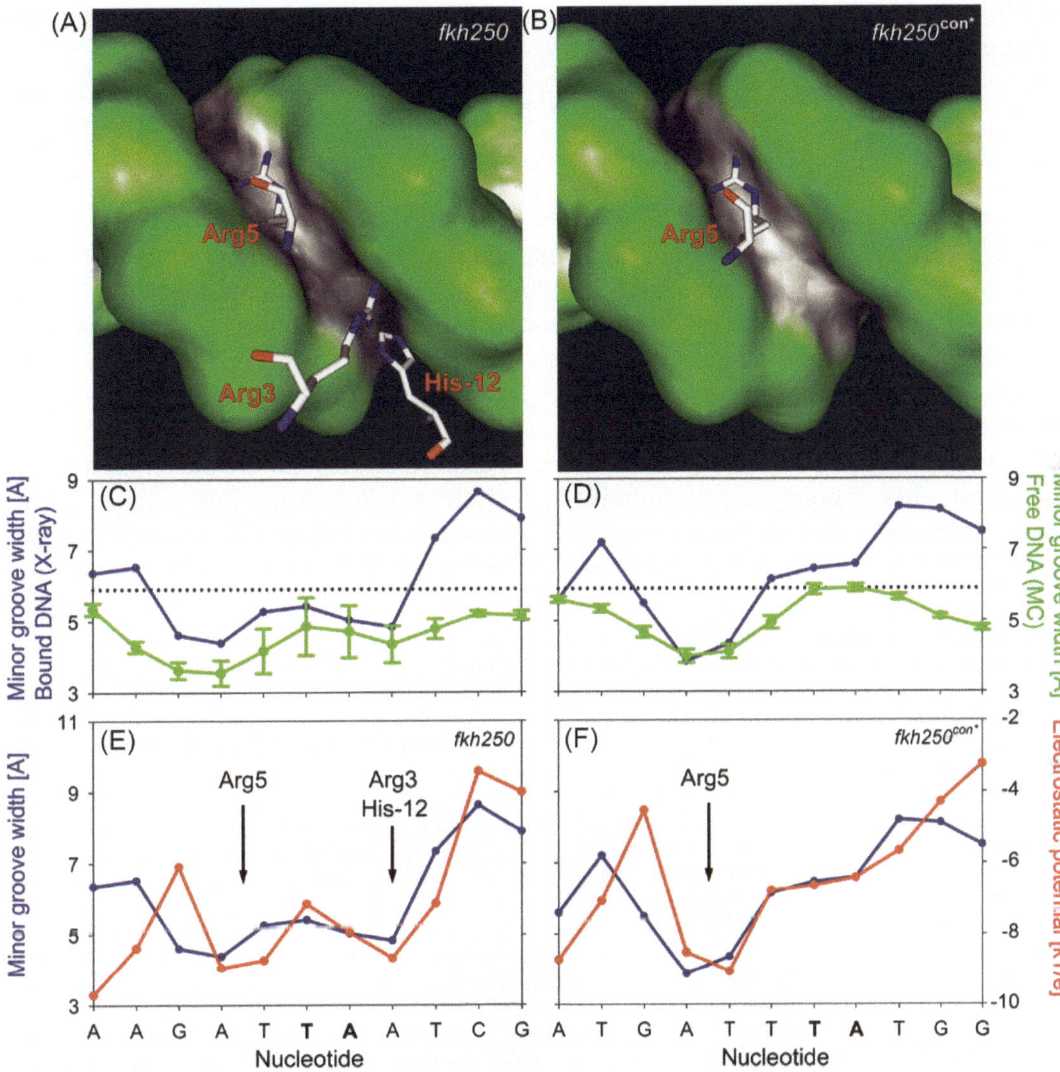

FIG. V.17 Differences in minor groove geometries and electrostatic potentials between fkh250 and fkh250con*. (A and B) The surfaces of the (A) fkh250 and (B) fkh250con* DNAs are color coded according to shape using GRASP (Nicholls et al., 1991). *Black* or *gray* and *green* surfaces represent concave and convex surfaces, respectively. Residues Arg5 (A and B) and Arg3 or His-12 (A, only) with side chains entering the minor groove are shown. (C and D) Graphs comparing minor groove widths seen in the crystal structures *(blue curves)* with those predicted by the MC (Monte Carlo) simulations *(green curves)* for fkh250 (C) and fkh250con* (D). The sequences of the two binding sites are shown below; TpA steps are shown in *bold* lettering. The error bars represent standard deviations of five independent MC predictions (for details see Supplemental Data). (E and F) Graphs comparing minor groove widths (Å) and electrostatic potential (kT/e) for fkh250 (E) and fkh250con* (F), based on the crystal structures. The positions of Arg5 and Arg3 or His-12 minor groove insertion are indicated. *Reproduced from Joshi, et al., 2007. Cell 131 (3), 530–543 Elsevier.*

describes the binding of a ligand (A) to a protein or DNA (B) with a single binding site and binding constant of K_{eq}.

$$A + B \rightleftharpoons A.B$$

$$K_{eq} = \frac{[A.B]}{[A][B]}$$

However, when a protein or the DNA binds multiple ligands (same or different), binding of the first ligand may influence the affinity of subsequent ligand bindings. If the binding of the first ligand results in a higher affinity for subsequent ligands, we call it positive cooperativity; if the binding affinity of the subsequent ligands weakens upon binding of the first ligand, we call it negative cooperativity. Many DNA-binding proteins have multiple functional binding sites in the genome. The proteins bound to these sites may interact with each other, provided geometric and spatial relationships are satisfied, and the interaction energy between two proteins is sufficient to overcome unfavorable spatial relationships (see later sections for details). This interaction often occurs with looping out or bending of the intervening DNA and protein-protein interaction between two proteins (Fig. V.18). The protein-protein interaction energy largely compensates for the energy required to bend the DNA.

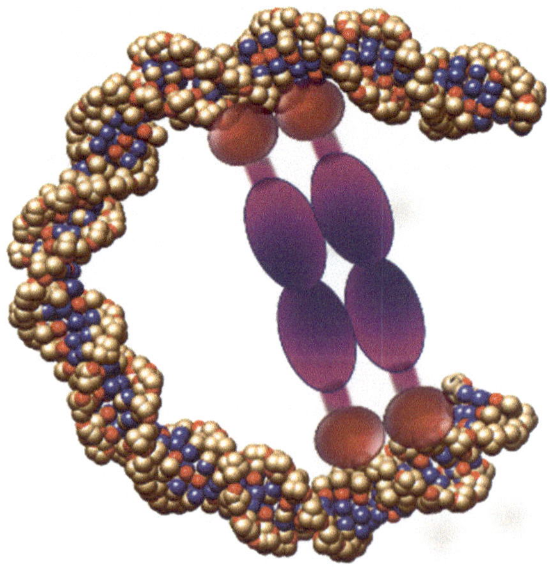

FIG. V.18 A model of a looped complex.

Energetics of formation of multiprotein gene regulatory complexes

Protein-protein association in the DNA-bound state and accompanying formation of a looped or bent DNA appear to be a commonplace mechanism for the formation of multiprotein gene regulatory complexes. Elucidation of the underlying energetics of the looped complex may give us a more in-depth insight into many gene regulatory processes. DNA looping is driven by the interaction energy of the two proteins in the DNA-bound state.

Assessing the importance of different factors in the loop formation can be done by writing a thermodynamic cycle, as shown in Fig. V.19. We can decompose the net loop formation free energy, ΔG_{loop} into three free-energy terms: The interaction energy of two isolated protein molecules in solution, while remaining bound to isolated binding sites (ΔG_{int}), the free-energy cost of converting the complex from the undistorted state to the distorted state that is present in the looped complex (ΔG_{dis}), and the free energy lost due to loss of translational and rotational entropy when two isolated molecules interact in solution (ΔG_{prox}). When two proteins associate in solution, they lose translational and rotational entropies. However, the interaction between DNA-bound proteins avoids this loss of translational and rotational entropies as the two molecules are already present on the same piece of DNA.

In principle, all these energies are measurable or estimable from known properties of DNA.

$$\Delta G_{loop} \cong \Delta G_{int} + \Delta G_{prox} + \Delta G_{dis}$$

The ΔG_{int} is primarily dependent on the character of interaction surfaces of the two proteins. ΔG_{prox} is likely to be insensitive to the orientation of the bound proteins. However, ΔG_{dis} is

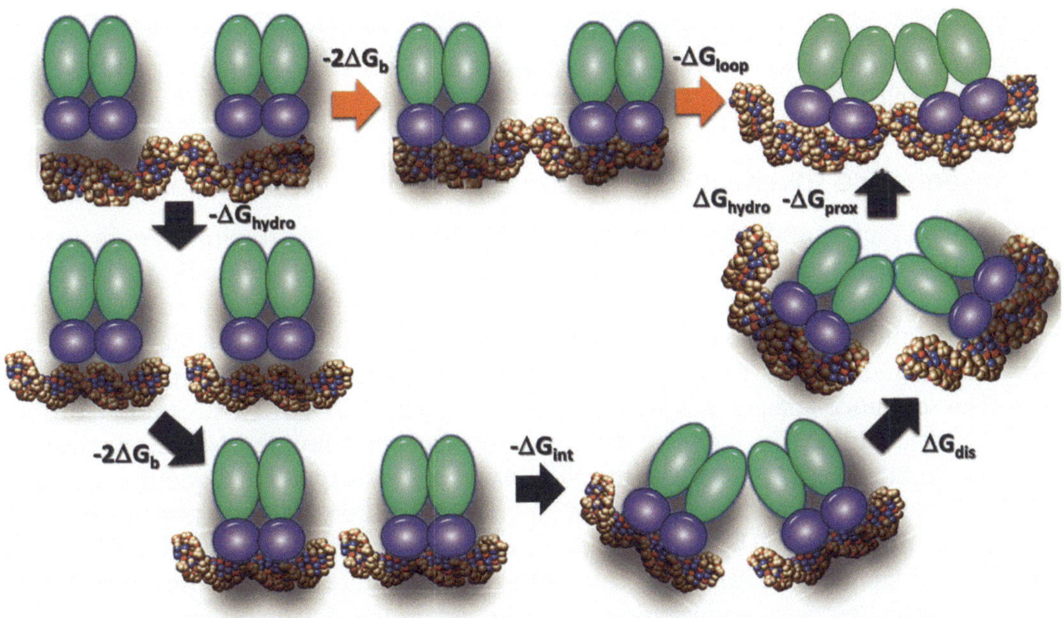

FIG. V.19 Analysis of energetics of looped complex formation using a thermodynamic cycle. As per the convention, the favorable free-energy changes are negative. ΔG_b = binding free energy; ΔG_{hydro} = hypothetical energy required to hydrolyze and separate two DNA fragments.

dependent on the orientation and distance between the protein molecules. In general, twisting the DNA requires much more input of energy than bending it. At specific orientations, for example, when the two proteins are bound to the opposite faces of the DNA, ΔG_{dis} becomes so large that $\Delta G_{int} + \Delta G_{prox} + \Delta G_{dis}$ is >0, and no loop formation takes place.
We use the case study of λ repressor for estimating different energy terms. In a regulatory complex, λ repressor binds to adjacent sites and directly interacts resulting in distortion or looping out of the intervening DNA. In the looped λ repressor regulatory complex, experiments suggest that ΔG_{loop} is approximately 2–3 kcal/mol, and ΔG_{int} is approximately 7 kcal/mol. From known mechanical properties of DNA, an estimate of ΔG_{dis} around 6–7 kcal/mol is obtained when the proteins are bound to the same face of the DNA. Thus ΔG_{prox} can be inferred to be approximately 2 kcal/mol, not far from what is expected from the loss of rotational and translational entropies. We may, thus, conclude that favorable spatial orientation and strong protein-protein interaction are needed for the cooperative binding to occur. If multiple proteins bind to the DNA and form a single regulatory complex, several protein-protein interaction energies may add up, giving very high stability and specificity to the regulatory complex.

Although direct protein-protein interaction accompanied by looping of the intervening DNA is the canonical model of binding cooperativity for DNA-binding proteins having multiple DNA-binding sites, recently, a nonlooping model without the direct protein-protein interaction has been proposed. In the nonlooping model, the binding of a protein to the first site influences binding to the neighboring site without protein-protein contact. The mechanism is believed to be DNA mediated. The looping model can only account for the positive cooperativity, whereas the nonlooping model can account for negative cooperativity as well.

V.4.5 Sequence specificity in the context of the whole genome

Transcription factors and other sequence-specific DNA-binding proteins operate inside a bacterium or a nucleus of a eukaryotic cell. In this setting, the DNA-binding proteins are exposed to an enormous number of base pairs other than the target sequence. The number of base pairs in the prokaryotic genome is typically in the range of a few million base pairs, whereas eukaryotic genomes are much larger, often reaching several billion base pairs. The sequence-specific DNA-binding proteins generally have a substantial affinity for the nonspecific DNA sequences. Thus the vast excess of genomic DNA acts as competitors for the target DNA. Although the affinity for the nonspecific sequence is several orders of magnitude lower than that for the specific sequence, the concentrations of the nonspecific DNA could be anywhere from a million to a billion times higher than those of the specific sequence. Because of the complexity of the state of the genome, it is difficult to estimate the degree of competition from the genomic DNA quantitatively. However, significant competition is very probable.

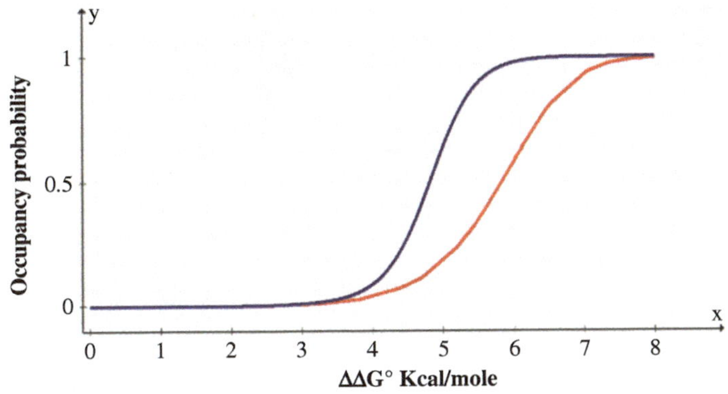

FIG. V.20 Simulation of transcription factor (TF) occupancy probability of the target site in a genome of 4×10^6 base pairs. The x-axis represents the difference between the binding free energies for the specific and nonspecific sites. The *blue line* represents a stabilization of 3 kcal/mol due to TF-TF interaction, and the red line represents no stabilization on account of protein-protein interaction. *The model is derived from the one described in Mazumder, et al., 2017. Phys. Chem. Chem. Phys. 19 (22), 14781–14792.*

Statistical mechanics-based models are useful tools for estimating the occupancy of a specific site against competition from nonspecific DNA. Based on models described in these studies, we analyze and compare two specific situations. In one situation, the transcription factor binds to specific and non-specific sites with no cooperativity. In the other situation, the protein binds to a bi-partite specific site with positive cooperativity, while showing no cooperativity for non-specific sites. We assume that like real-life situations, the transcription factor has a significant uniform affinity for non-specific DNA. A model-based simulation was performed to calculate the occupancy of the specific site as a function of the difference of binding energies for the specific and non-specific sites ($\Delta\Delta G_o$). Fig. V.20 shows that when positive cooperativity is present, full occupancy of the specific site is reached at a lower $\Delta\Delta G_o$ value. Thus, for the same $\Delta\Delta G_o$ value, transcription factors with positive cooperativity would require a lower concentration to reach the same level of occupancy at the specific site. This should provide an evolutionary advantage to the cell.

In the previous section, we have seen how cooperative binding can augment the sequence specificity of a transcription factor in the backdrop of the whole genome. This mechanism is frequently observed in the formation of transcription regulatory complexes of the eukaryotes. A well-known example is that of the multiprotein complex that regulates the interferon β gene. Multiple gene regulatory proteins bind to contiguous regions of the promoter showing strong cooperativity giving the protein complex, the enhanceosome, very high stabilization energy and, thus, providing a high degree of augmented specificity needed for selective binding and regulation (Fig. V.21).

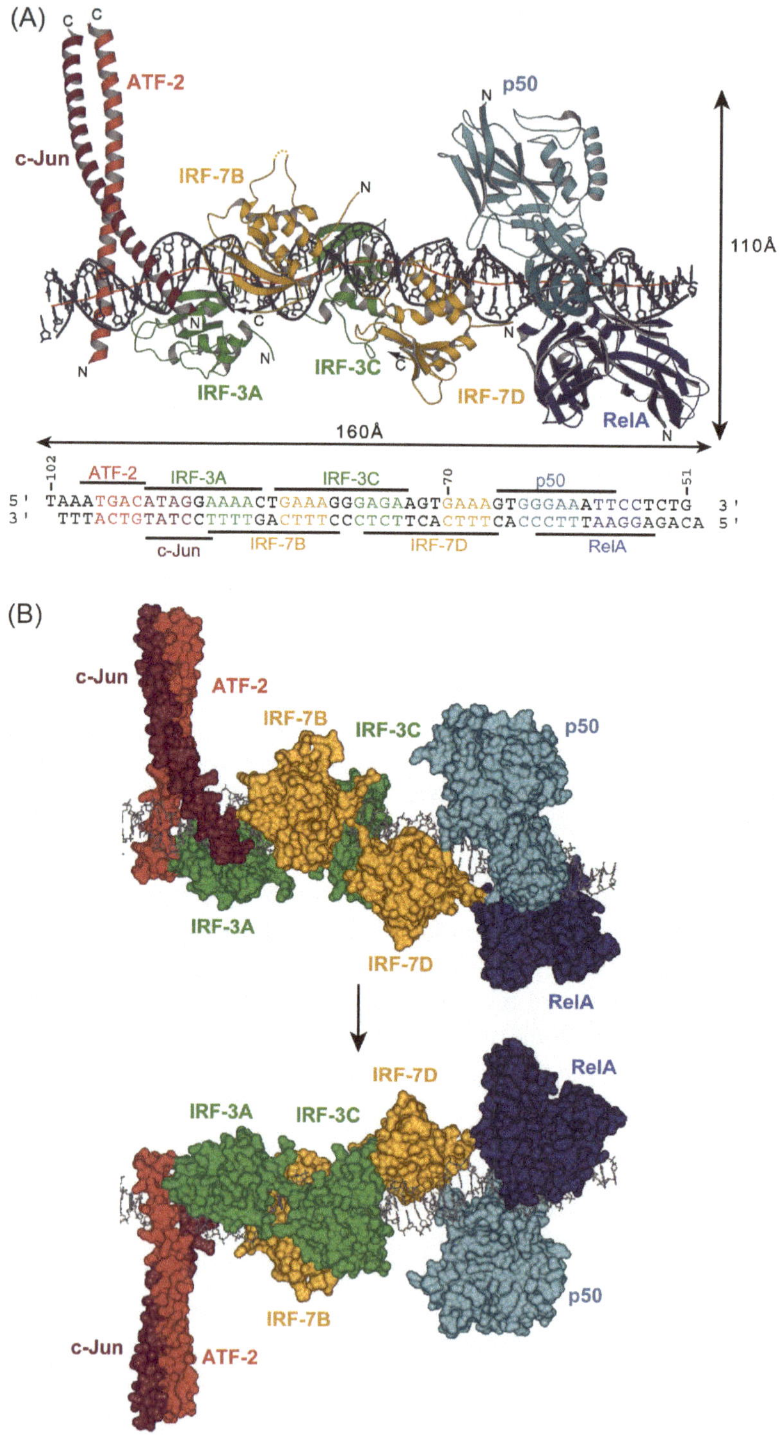

FIG. V.21 Overall structure of the IFN-β enhanceosome. (A) *Side view* of the complex shows bending of the DNA around the four IRF domains. The *red line* shows the local DNA helical axis as calculated with the program Curves (Lavery and Sklenar, 1988). The RHR (Rel Homology Region) of p50 is in *light blue* and that of RelA in *dark blue*. IRF-7B and D are in *yellow*, and IRF-3A and C are in *green*. ATF-2 is in *light red*, and c-Jun in *dark red*. The two DNA strands are in *gray*. (B) Molecular surface representation of the IFN-β enhanceosome showing that the eight polypeptides form a composite surface for DNA recognition. The two views, related by a 180 degree rotation as suggested by the *arrow*, are from opposite sides of the complex. *Reproduced from Panne, et al., 2007. Cell 129 (6), 1111–1123, Elsevier.*

V.5 DNA recognition motifs

V.5.1 Helix as the major DNA binding motif

One of the most abundant secondary structural elements in proteins is the α-helix. The width of an α-helix is such that it fits into the major groove of the DNA well (Fig. V.22). Most of the sequence-specific DNA-binding proteins have evolved to utilize the α-helix as the primary sequence recognition element. Most of the extant DNA binding motifs can be grouped into a few classes, depending upon the architecture of the recognition elements. A vast majority of them use α-helix as the primary recognition element. Four major classes are the helix-turn-helix, the homeodomain, the zinc fingers, and the b-helix-loop-helix motif-containing proteins. However, proteins belonging to each class are not structurally or conformationally homogeneous and variation occurs around the canonical motif. These variations are crucial for the recognition of diverse DNA sequences. We will not describe these motifs or their variations in detail. Multiple sources are available that deal with them. We aim to provide the chemical principles of how such variations are used to recognize diverse DNA sequences.

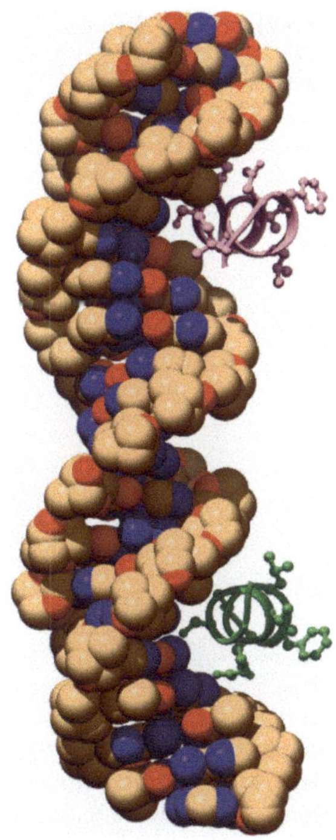

FIG. V.22 Fit of α-helices into the major groove of DNA. *The figure was generated from the crystal structure 1lmb.*

V.5.2 Combination of multiple recognition elements to increase recognition specificity

A single helix by itself is an ineffective and inadequate DNA binder, both with respect to affinity and specificity. As discussed above, the challenge to selectively bind the target DNA site in the genome is immense. This challenge is met by a combination of different strategies. One central theme appears to be the use of a combination of multiple recognition elements. Let us consider two motifs (M_1 and M_2) that bind to two particular DNA sequences, D_1 and D_2, with binding free energies of ΔG_1 and ΔG_2. When connected by an idealized linker, the construct M_1-linker-M_2 will bind to a sequence D_1–D_2 (with or without a spacer sequence in between) with a free energy of $\Delta G_1 + \Delta G_2$ (Fig. V.23). The ideal linker is such that M_1-linker-M_2 will neither lose any free energy due to linker distortion nor would it cause a DNA distortion. The additivity is not exact, as the linking injects two additional factors in the equation: one, a favorable entropic contribution due to the proximity of the second element and second, the loss of entropy of the linker upon binding as its freedom of movement is typically restricted in the DNA-bound state when compared with the free state. If the spatial relationships are not perfect for good binding, some distortions in either protein or DNA or both may be necessary to make a lock-and-key type fit. Such distortion energy would also contribute unfavorably toward binding (see the thermodynamic cycle for the DNA looping for a similar argument). However, if the linker is optimized, as may be the case for evolved proteins, the unfavorable contributions will be minimized, and the gain of free energy due to the combination of recognition elements will be significant and lead to a much tighter binding.

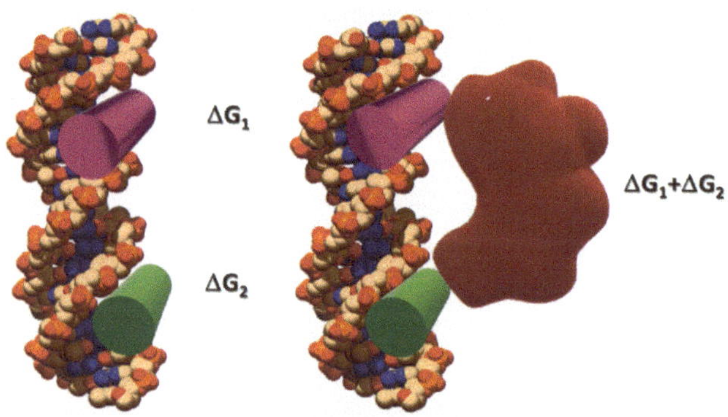

FIG. V.23 Cartoon representation of additivity of free energy of binding of two recognition elements. The two cylinders represent the two recognition elements, and the *brown figure* indicates the connector that does not distort upon binding. The *left figure* represents binding of two elements separately, whereas the *figure on the right* represents binding of the same two elements in the same mode while attached to a linking structure. The entropic advantage of having the two elements on the same connector is ignored here.

V.5.3 Use of helix in major DNA binding motifs

Helix-turn-helix is an abundant motif present in proteins. The helix-turn-helix motif is present in all domains of life. It is one of the most ubiquitous motifs present in transcription factors. Proteins containing the helix-turn-helix motif are involved in many processes, including DNA repair, RNA metabolism, and protein-protein interaction. A large number of prokaryotic repressors belong to the helix-turn-helix family of proteins. One of the earliest reports of DNA-protein cocrystal structures was the λ repressor with its cognate operator, O_L1 (Fig. V.24). In its elementary form, the motif consists of two α-helices separated by a turn (Fig. V.25). The second helix from the N-terminus in this motif is the principal recognition element (the recognition helix). This helix often interacts with the major groove bases. However, amino acid residues from beyond the recognition helix may make supplemental contacts with additional elements of DNA, making significant contributions to the binding affinity and specificity. In proteins, it is usually present with other structural elements. A common form is a three-helix bundle (Fig. V.26), with the third helix from the N-terminus acting as the recognition helix. The helix-turn-helix motif also occurs as a tetrahelical bundle and multihelical bundle. In the tetra-helical version, an additional helix is added at the C-terminus. The basic three-helix bundle gives rise to

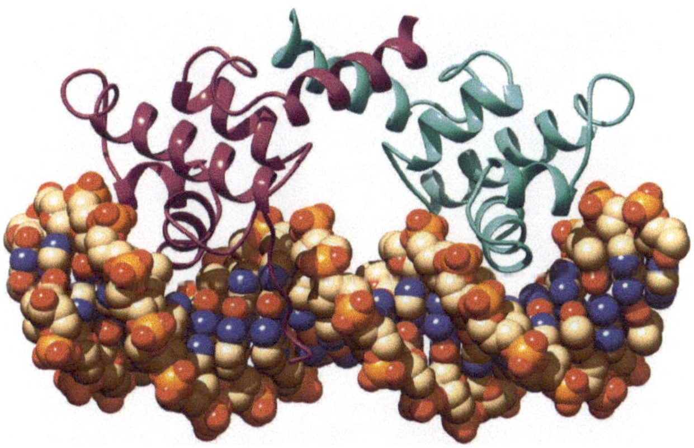

FIG. V.24 Crystal structure of λ-repressor N-terminal domain (Pdb1lmb).

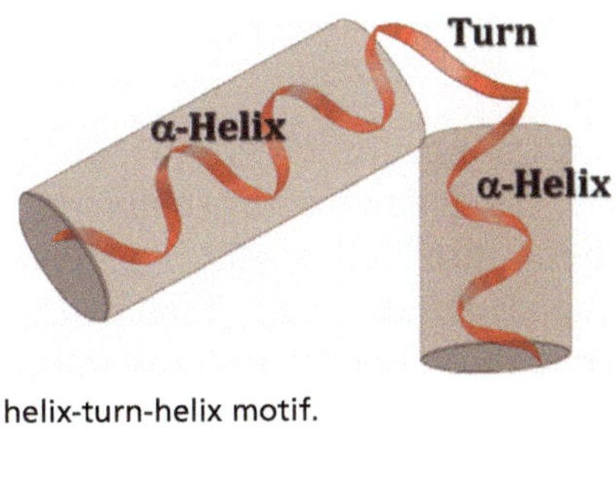

FIG. V.25 Cartoon depiction of the helix-turn-helix motif.

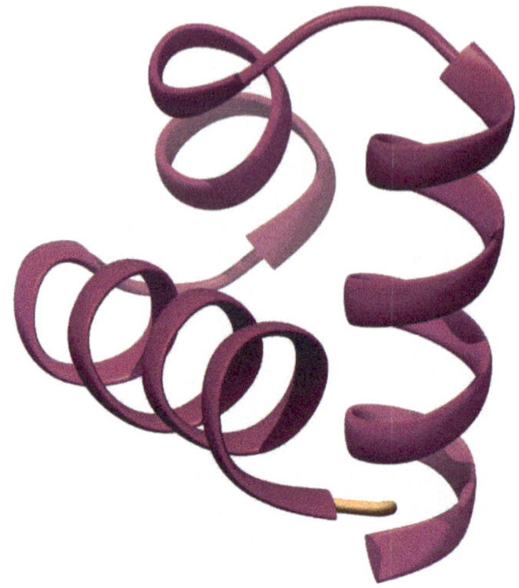

FIG. V.26 A representation of three-helix bundle class of helix-turn-helix containing proteins. *The image was generated from chain: A of pdb1k78 (pax5).*

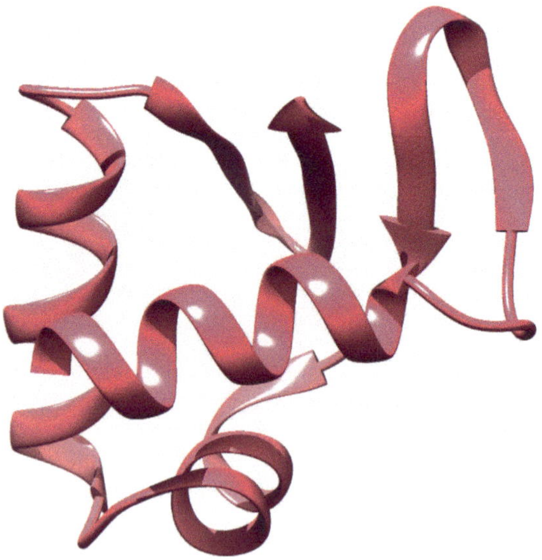

FIG. V.27 The structure of a winged helix-turn-helix domain pdb1cgp.

many variants. Among them, an important class of DNA-binding protein is the winged helix-turn-helix class (Fig. V.27). Winged helix-turn-helix motif contains a β-strand hairpin (the wing) against which the helix-turn-helix packs. The β-strand hairpin is an added element. In several cases, the wing also makes additional DNA contacts apart from the core contacts established by the recognition helix. ETS-1, an important transcription regulator in eukaryotes, belongs to this class.

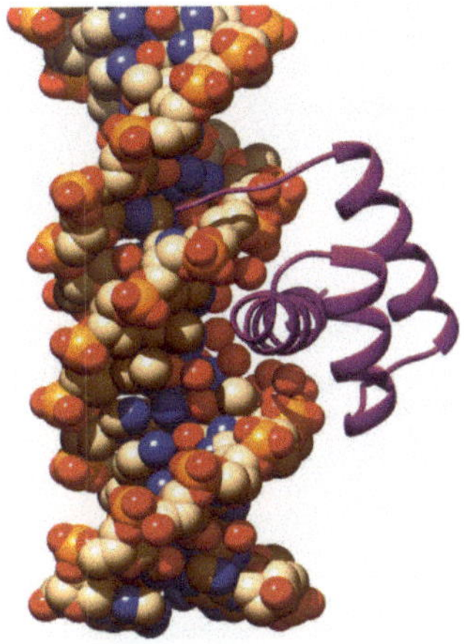

FIG. V.28 The structure of a homeodomain-DNA complex. *The image was created from pdb3hdd.*

Homeodomain containing proteins are another important class of DNA-binding proteins. The homeodomain is approximately a 60-amino acid-long protein domain related to the helix-turn-helix DNA-binding domain. Like the most common form of the helix-turn-helix tri-helical bundle, it consists of three helices (Fig. V.28). The target DNA site of the homeodomain is usually short.

In zinc-finger proteins, helices are used recurrently to generate linked multiple binding elements. The zinc finger itself is a super-secondary structure consisting of an antiparallel β-sheet and an α-helix, held together by a zinc atom, typically coordinated to a Cys_2-His_2 motif (Fig. V.29). The helix is the most crucial recognition element, and the role of the zinc appears to be providing stability to the module. One finger by itself has a low affinity toward its cognate DNA sequence. Adequate affinity and specificity are generated by

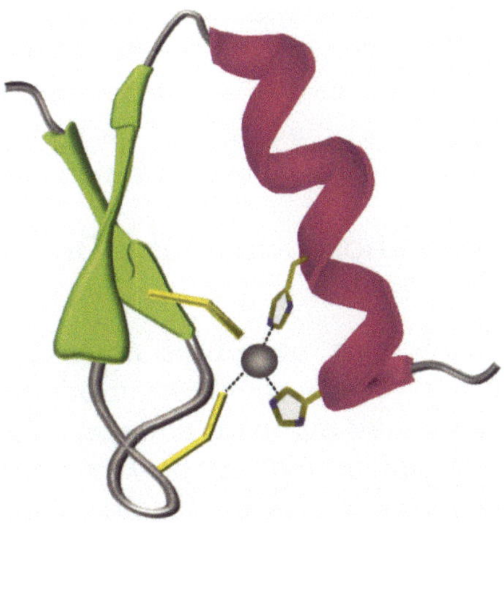

FIG. V.29 The zinc-finger motif.

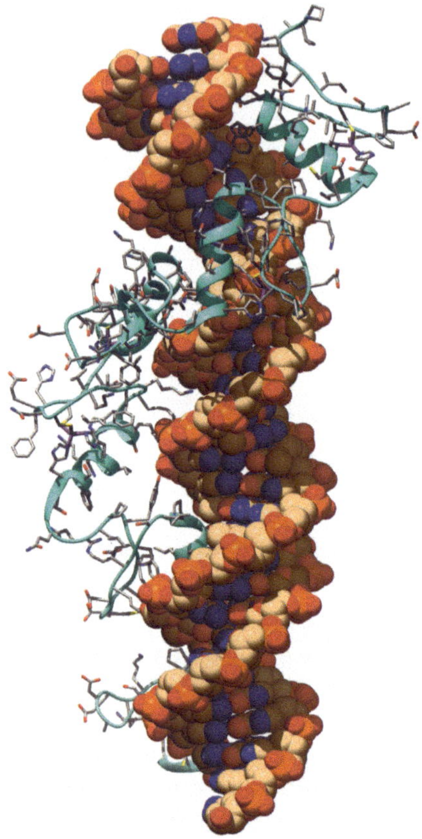

FIG. V.30 Structure of the six-finger TFIIIA-DNA complex. *Created from 1TF6.*

repeating the same mode of interactions with multiple fingers (Fig. V.30). Zinc fingers are nonidentical in general and have a preference for different base sequences. Thus a combination of nonidentical zinc fingers creates a high binding specificity for nonrepeating DNA sequences. Other classes of DNA-binding proteins, such as the helix-loop-helix class of proteins, also use the helix as the principal recognition element.

V.5.4 Recognition of the minor groove of B-DNA by flexible peptide motifs

Short flexible peptide segments have been used as recognition elements in several DNA-binding proteins. As discussed in Section V.4.2, in addition to the recognition helix, an N-terminal arm also plays a crucial role in the homeodomain-containing proteins. The N-terminal arm also plays a similar role in the λ-repressor and its orthologs. The N-terminal arm is not often visible in crystal structures, suggesting its dynamic nature even when bound to its target DNA. However, removal of the arm in the λ-repressor results in loss of binding affinity by more than one order of magnitude. Although a more precise role of this arm in sequence recognition has not been elucidated in the λ-repressor, the role of the N-terminal arm in a structurally related class of protein, the homeodomain, has been elucidated in great detail. We have discussed the role of the N-terminal arm in the homeodomain-containing proteins in Section V.4.2. Many DNA-binding proteins use the α-helix, β-sheet, and flexible elements in different combinations to generate adequate specificity.

V.6 Regulation of gene expression: Chemical principles

Regulation of gene expression is one of the central themes of cellular function. Regulation of gene expression occurs at many levels. Important points of regulation are transcription, translation, and protein degradation. Regulation at these steps plays vital roles and its relative importance varies from gene to gene. Two processes play a dominant regulatory role at the transcriptional level: epigenetic modifications and regulation of transcription initiation by transcription factors. We will discuss the former mode of regulation in a different section. In this section, we focus on the process of regulation of gene transcription initiation by transcription factors.

In previous sections, we have discussed the dense packaging of genomes. This packaging prevents transcription machinery and regulatory factors from gaining access to promoter elements. Controlling access to gene regulatory cis elements is a dominant mechanism of regulation of transcription initiation in eukaryotes. It may also have a role in prokaryotes but the mechanisms may differ. Prokaryotes mainly employ other mechanisms for regulation of transcription initiation. We will describe these mechanisms in detail.

V.6.1 Regulation of transcription initiation in prokaryotes

In prokaryotes, the genomic DNA is condensed to a far lesser degree than the eukaryotes. Here, the access control is operationalized by steric occlusion of the polymerase by regulatory proteins. We may formulate this mode of access control as a three-component equilibrium between the DNA, RNA polymerase, and the regulatory protein often called the repressor in prokaryotic systems. The RNA polymerase and the regulatory protein compete for binding to the promoter. The simplest way this competition occurs is through the overlapping binding sites of the polymerase and the regulatory protein on the DNA. However, this need not be the only way. One can visualize that a regulator protein can bind to a nonoverlapping site and distort the DNA in such a manner that it prevents the polymerase from binding. The competitive mode of access control is thought to operate in the lac operon that is a paradigm for prokaryotic gene regulation. In this model, the lac operator (lac repressor-binding site) binds the lac repressor and prevents polymerase binding to the promoter (Fig. V.31). As long as there is a sufficient concentration of the lac repressor, the polymerase is unable to bind and initiate transcription. When the signal is received

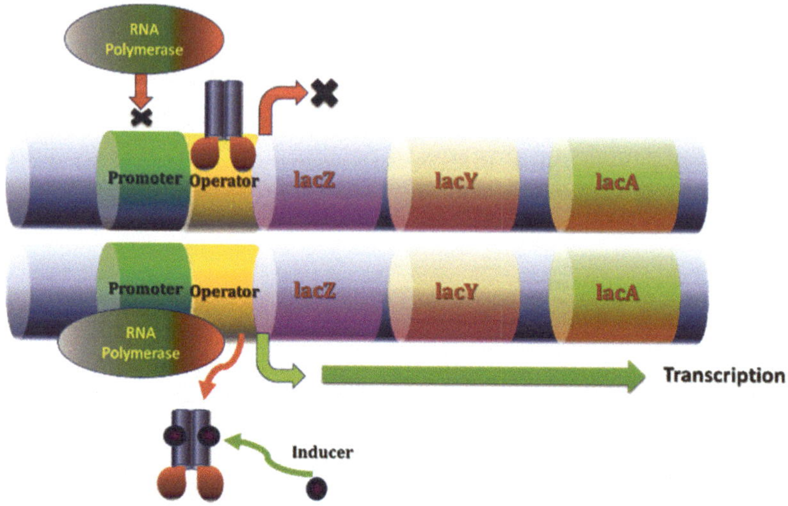

FIG. V.31 A model of regulation of lac operon in *E. coli*. The *upper part* of the figure represents the repressed state, whereas the *bottom part* represents the derepressed active state.

to activate, a small molecule inducer binds to the lac repressor causing the active concentration of the repressor to drop, thus, dissociating it from the operator (the inducer-bound form has much weaker DNA-binding affinity), leaving the polymerase free to bind. Access control in other gene regulatory systems in the prokaryotes may operate through different mechanisms. For example, in *E. coli* GalR and Ara operons, the repressor binds cooperatively to two different sites accompanied by protein-protein interaction, thus, creating a DNA loop (Fig. V.32). The looping of DNA is a crucial element in preventing the RNA polymerase from initiation of transcription, thus, preserving the repressed state. Although DNA looping also occurs in the lac operon, it does not play as dominant a role as in Gal and Ara operons. We will discuss other possible modes of gene regulation in the following section.

In prokaryotes, access control may not be the only mechanism for regulation of gene expression. Apart from the repression, prokaryotic transcription factors also activate gene expression from weaker promoters. The activating transcription factor binds to a promoter-proximal site and augments the rate at which RNA polymerase initiates transcription. Two representative examples are the activation of the lac promoter in the lac operon by cyclic AMP (cAMP) receptor protein (CRP) and the activation of the P_{RM} promoter in bacteriophage λ by the lambda repressor. A model, called differential contact, offers a thermodynamic and kinetic explanation of the activation process. DNA looping and establishment of contact between the transcription factor and polymerase play a crucial role in this model.

The transcription initiation process in the prokaryotes has been analyzed in detail, and the following kinetic equation can minimally describe it:

$$R + P \underset{}{\overset{K_B}{\rightleftharpoons}} RP_c \overset{k_f}{\rightarrow} RP_o,$$

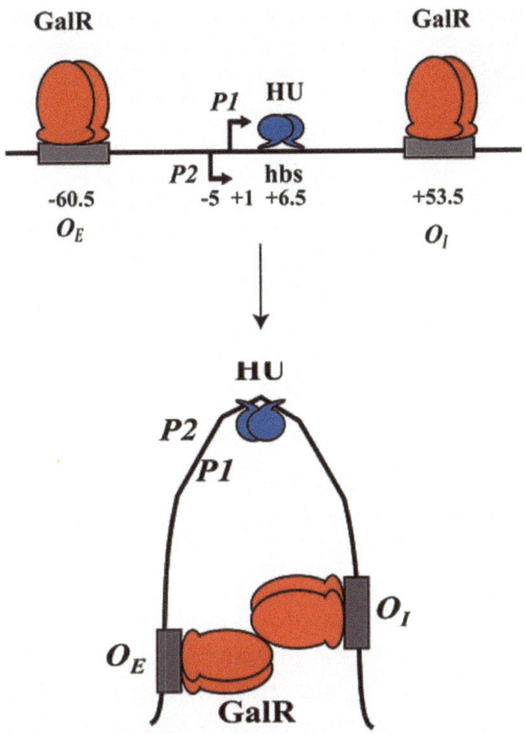

FIG. V.32 A model of looping the Gal operon. *Reproduced from Qian, et al., 2016. Front. Mol. Biosci. https://doi.org/10.3389/fmolb.2016.00074 under CC-BY license.*

where R is RNA polymerase, P is the promoter, RP_c is the closed complex in which the two strands of DNA duplex remain paired, and RP_o is the open complex in which the two strands have separated to create a multibase pair bubble, a necessary intermediate for transcription initiation. K_B is a composite constant and not a true equilibrium constant. It is formally similar to the K_m in Michelis-Menten equation, k_f is the forward rate constant similar to the catalytic constant k_{cat} in the Michelis-Menten equation.

The activation of transcription initiation by a transcription factor bound at or near the promoter can be understood using a formalism based on the absolute reaction rate theory. In this model, the state $[R + P]$ is akin to the reactants, RP_c is an intermediate, and RP_o is the product. Because $R + P$ going to RP_c is a binding event, it is considered to be a barrier-less reaction lacking a transition state. An activation barrier separates RP_c and RP_o, and the polymerase-promoter complex at the top of the barrier is the transition state in quasiequilibrium with the RP_c. Like a chemical reaction, the relative positions of atoms in the complex change as it progresses along the reaction path. However, unlike a chemical reaction, covalent bonds are not broken or formed. The change is purely conformational and alterations of noncovalent interactions.

Let us apply this model to the activation of the lac promoter by cAMP-CRP. In the absence of cAMP-CRP, the rate of transcription initiation is minimal because RNA

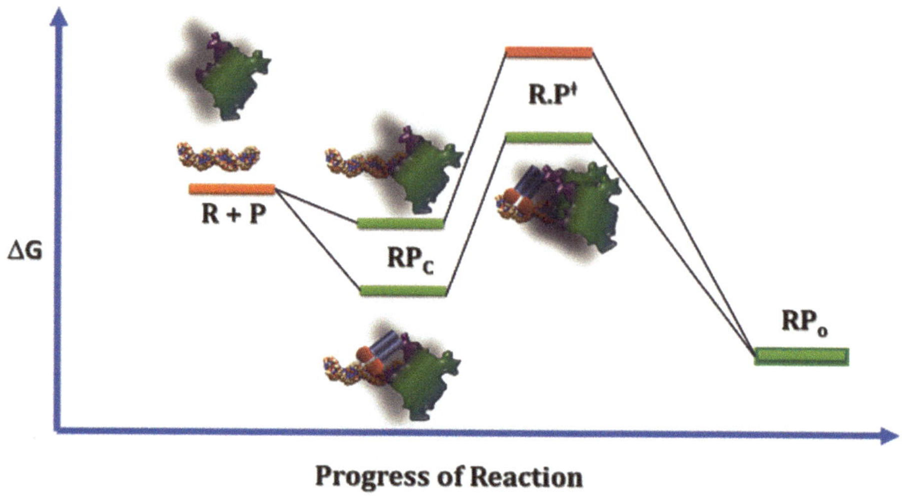

FIG. V.33 Enhancement of recruitment of RNA polymerase by differential contact between polymerase and activator in the closed complex and the transition state.

polymerase has a weak affinity for the lac promoter. Once the cAMP-CRP complex binds to the lac promoter region, it increases the polymerase occupancy by contacting the polymerase. In the differential contact model, the polymerase-cAMP-CRP contact is established at the closed-complex stage and persists throughout the transition state. The establishment of contact leads to the stabilization of the bound RNA polymerase-cAMP-CRP complex leading to enhanced occupancy without significant alteration of the activation energy of isomerization. Augmented occupancy causes the upregulation of transcription initiation (Fig. V.33).

In some promoters, the rate of isomerization step is the slowest and the rate determining. In such promoters, the activation may occur by lowering of the activation energy. The model proposes that new contacts are established during the transition state only, thus, differentially stabilizing the transition state with respect to the ground state(s) (Fig. V.34). This differential stabilization causes a lowering of the activation energy barrier and acceleration of the rate of isomerization from RP_o to RP_c. Because the isomerization is the rate-limiting step in the formation of the open complex, the overall rate of transcription initiation will be enhanced. Activation of the P_{RM} promoter of bacteriophage λ by λ-repressor proceeds through this mechanism.

V.6.2 Regulation by access control: Eukaryotes

In eukaryotes, the genomic DNA is highly packaged and condensed. At the first level, by wrapping the DNA, nucleosomes create barriers for the transcription machinery to access the promoter region. At higher levels, further packaging occurs. The transcription machinery gains access to its target sequence only upon the relaxation of wrapping and packaging (Fig. V.35). An excellent way to visualize this from a chemical thermodynamic point of view is to treat the close (Nu_{cl}) and the open (Nu_{op}) states of the chromatin as a two-state equilibrium.

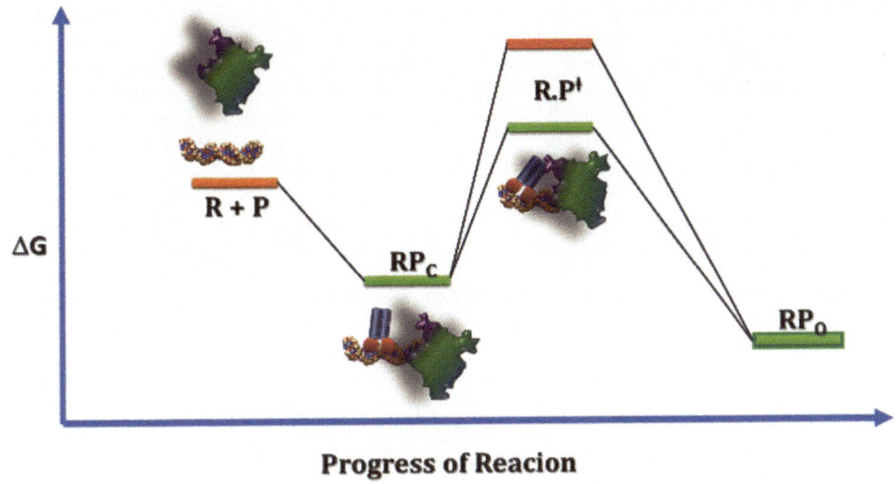

FIG. V.34 Activation of the isomerization reaction by differential contact between polymerase and activator in the transition state.

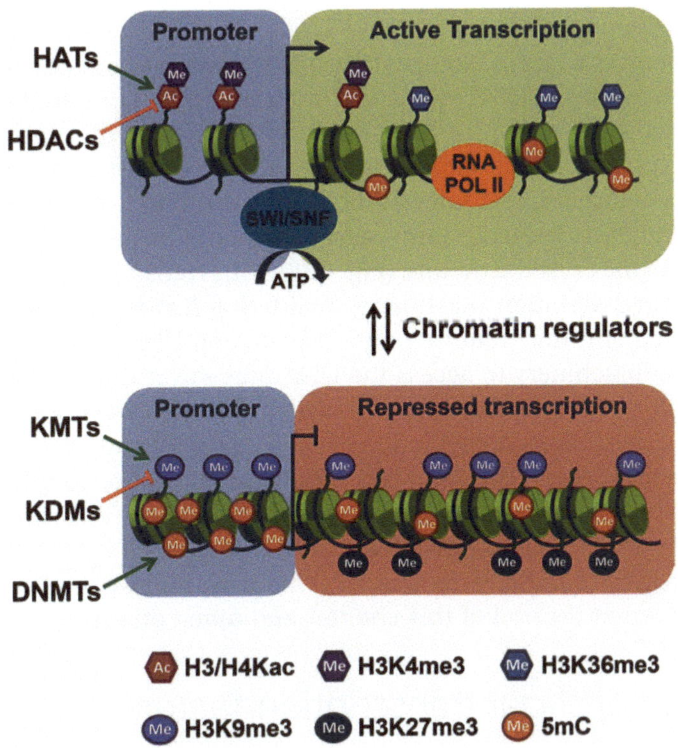

FIG. V.35 Relaxed and repressed states of chromatin and their associated chemical modifications. *Reproduced from Mirabella, A.C., Foster, B.M., Bartke, T., 2016. Chromatin deregulation in disease. Chromosoma 125, 75–93. https://doi.org/10.1007/s00412-015-0530-0. Under the terms of the Creative Commons CC BY license.*

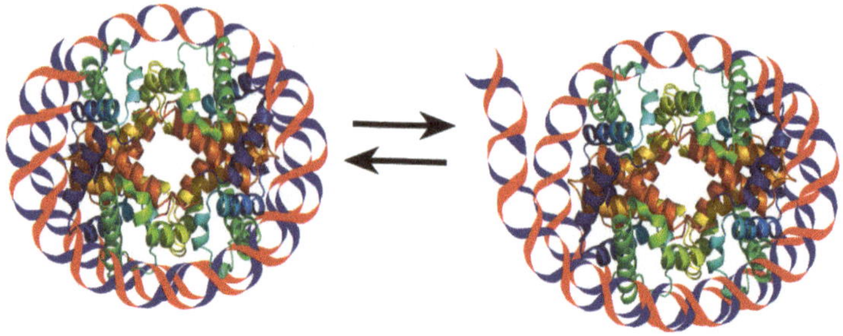

FIG. V.36 Dynamic unwrapping of DNA at the entry and exit point of the nucleosome.

$$\mathbf{Nu_{cl}} \overset{K}{\rightleftharpoons} \mathbf{Nu_{op}} \xrightarrow[k]{\text{RNAP}} \mathbf{RP_o}$$

The transcription machinery can bind only to the open state. We first examine the structures of the closed state and open state of the nucleosome. Nucleosomes are very stable entities requiring a large amount of energy to unravel. $\Delta\Delta G$ of full disassembly could be anywhere between 23 kcal/mol and 34 kcal/mol. Processes requiring such a high amount of free energy are rare in biological systems. However, as was described above, the accessibility of the wrapped DNA is crucial for transcription to initiate. Dynamical features of the nucleosome structure hold the key. As discussed before, proteins, DNA, and their complexes are all dynamic entities. Although the most stable state of the nucleosome is a compact structure as seen in the crystal structure, there are other low-lying states in which the DNA near the entry and exit points of the nucleosome frays from the position seen in the crystal structure (Fig. V.36). This transient opening can give access to proteins of the transcription machinery, facilitating further opening up. This partial unwrapping at each end may require only ~1 kcal/mol. Histone tail modifications that allow transcription machinery to access the DNA may lower this energy, further causing a substantial fraction of the nucleosome to remain frayed. The fraying is believed to be one of the significant mechanisms of regulation of access control and transcription regulation in eukaryotes. We must emphasize that this is only one step in a complex regulatory process. It is by no means a complete description of the regulatory step, and it may not even be the most important one. It is to give the reader a glimpse of an important principle in eukaryotic gene regulation. Interested readers who wish to have a fuller description may see the reviews given at the end of this chapter and many others.

V.7 Designed artificial transcription factors

So far, we have discussed how gene regulatory proteins function. The central importance of gene regulation to biology has prompted many investigators to design artificial transcription factors, some of which are based on totally synthetic molecules. We will take a look at approaches used to develop artificial transcription factors and their applications in gene regulation.

V.7.1 Principles of artificial transcription factor design

In general, transcription factors possess multiple domains. They are almost always modular in design with different functions generally residing in different domains. They perform their tasks by not only occupying a site on the DNA and excluding other proteins from that area but also establishing contact with accessory proteins, often through nonDNA-binding domains. Multiple functions embedded in one protein play crucial roles in gene regulatory capability of transcription factors. A simple modular design is that of the CRP protein of *E. coli*. It has two domains, a DNA-binding domain and a cAMP-binding domain (Fig. V.37). A more complex domain organization is seen in the master eukaryotic transcription factor, p53. The full-length structure of the protein is not yet known at atomic resolutions. However, a model for full-length protein has been described from the extensive biochemical, structural, and crystallographic (domains) data (Fig. V.38). The modularization of different functions is evident from the model.

The modular nature of transcription factors helps to design artificial transcription factors. If the artificial transcription factor has only the DNA-binding function, it may compete for the transcription factor-binding site and counteract the natural one. However, its full utility will come from adding one or more modules to create artificial transcription factors having functions of choice. Most of the examples to date belong to the former category, but a few steps have been taken in the direction of appending additional functions. Both the DNA-binding domain and other functional domains can be peptide-based or nonpeptide synthetic molecules.

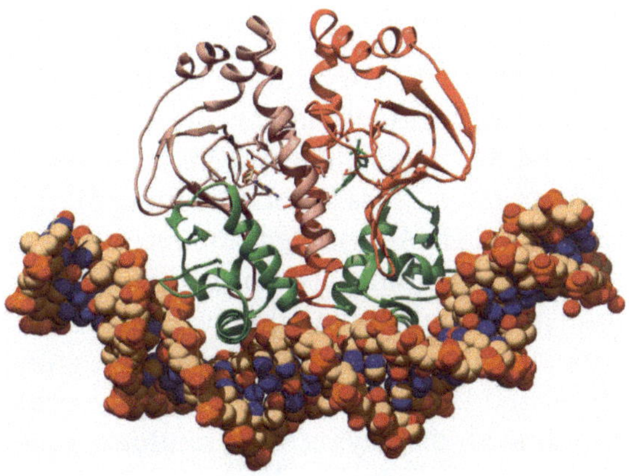

FIG. V.37 Domain structure of the *E. coli* cAMP-cyclic AMP receptor protein complex. The *green-colored* segments are the DNA-binding domains (139–209), the *red-colored* segments are the linkers, and the brown-colored segments are the cAMP-binding domains. *Created with the freeware Chimera from 1o3t.*

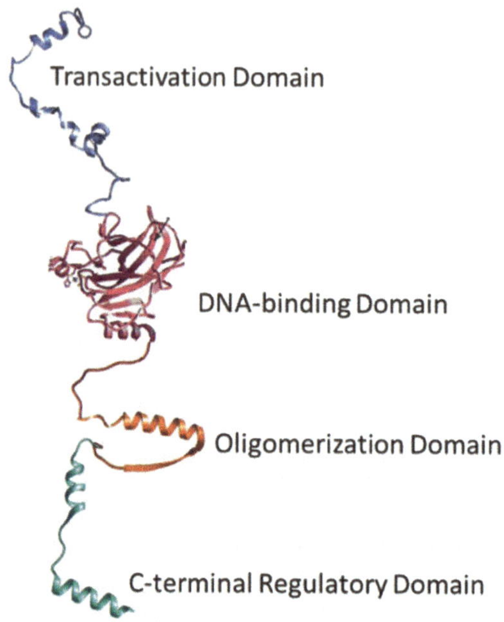

FIG. V.38 Domain structure of p53. *Adapted from Cancers 2018, 10 (9), 288; https://doi.org/10.3390/cancers10090288. License CC-BY.*

V.7.2 Engineered artificial protein transcription factors

An early example of an artificial transcription factor was through grafting of the recognition α-helix of a homologous protein. The grafting changed the specificity of binding to that of the transcription factor from which the grafted helix originated. This established alteration of a recognition helix as a method for creating new DNA-binding specificities.

Attempts to create a truly modular design began with zinc-finger proteins. When bound to their target DNA, the repeating units of zinc fingers follow the path of the major groove of the DNA helix (Fig. V.39). The curvature of the DNA helix requires two helices to be oriented at an angle. This specific orientation results from the conformation of the connecting sequence between the two fingers. The fingers, however, are structurally and DNA interaction-wise nearly autonomous. This autonomy makes it possible to use the fingers as building blocks to create novel DNA-binding proteins using a mix-and-match strategy. In practice, however, many complications arise because of the promiscuity of the multi-finger artificial DNA-binding protein. The method was successful in targeting variants of the cognate sites of existing zinc-finger proteins, whereas designing artificial transcription factors for completely unrelated DNA sequences is far more challenging.

Phage display is a technique for displaying short random sequences of peptides on the phage surface. After the display, the peptides having desired binding properties can be selected. Such an approach led to the isolation of zinc fingers having a significant binding affinity for preselected target sequences. The most successful of these strategies was to use

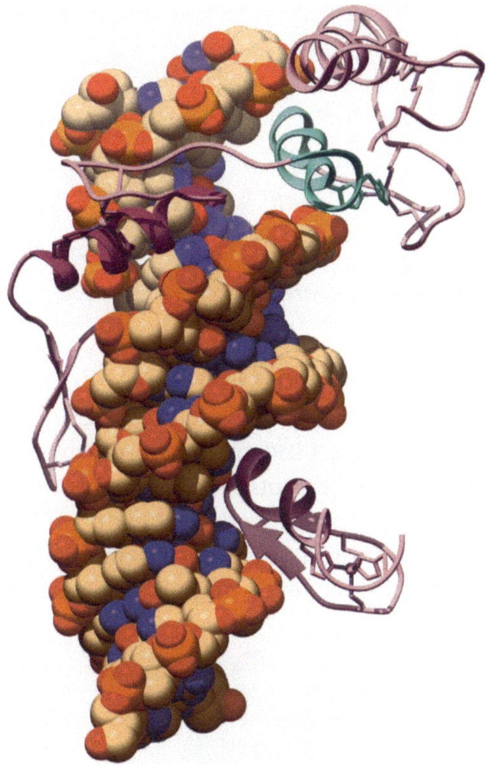

FIG. V.39 Structure of a Zinc-finger protein (Gli) complexed with DNA. Two of the DNA recognition helices are colored differently than the rest of the protein. *Created from 2Gli.*

a two-finger unit to select through phage display and then combining these two-finger units in four- or six-finger proteins. Such a strategy has been quite successful and applied to solve practical problems.

V.7.3 Polyamide-based transcription factors

B-DNA, in its free form, is relatively straight and rod like, particularly at shorter length scales. Given a limited number of naturally occurring base pairs, we may predict that patterns of polar atoms displayed in the grooves are also limited. Thus it may be challenging to develop ligands that can be selectively targeted to specific DNA sequences in a large genome. Nature has evolved complex and nuanced mechanisms to develop adequate specificity using proteins. These lessons are not straightforward to apply to small-molecule ligands. Pyrrole and imidazole-based polyamides were the first major systematic attempt to develop small-sized or medium-sized molecules that bind specifically to a DNA sequence. The polyamides consist of pyrrole or imidazole rings connected by intervening amide bonds (Fig. V.40). The design is based on the natural antibiotics netropsin and distamycin that bind to the minor groove of the B-DNA.

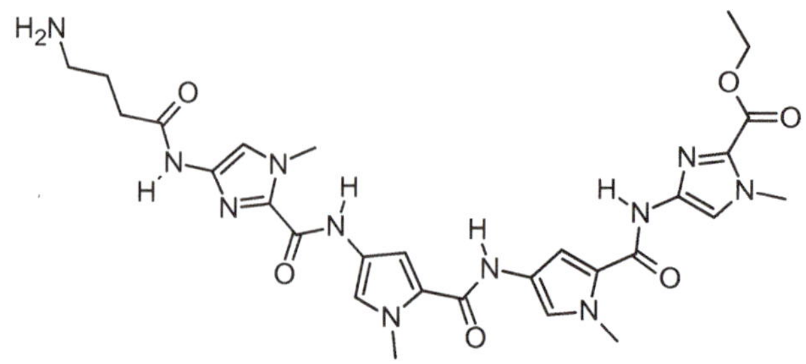

FIG. V.40 Structure of a short imidazole-pyrrole-pyrrole-imidazole polyamide.

Netropsin consists of two *N*-methylated pyrrole rings connected by an amide bond. Distamycin contains three *N*-methylated pyrrole rings connected by two amide bonds. These naturally occurring antibiotics bind to the minor groove of the DNA and prefer the AT-rich sequence (Fig. V.41). For targeting longer and more varied DNA sequences, additional building elements were introduced. Inclusion of *N*-methyl imidazole moiety in the polyamides made the recognition of GC base pairs possible. Further enhancement of the specificity required recognition of even longer DNA sequences. Because the grooves

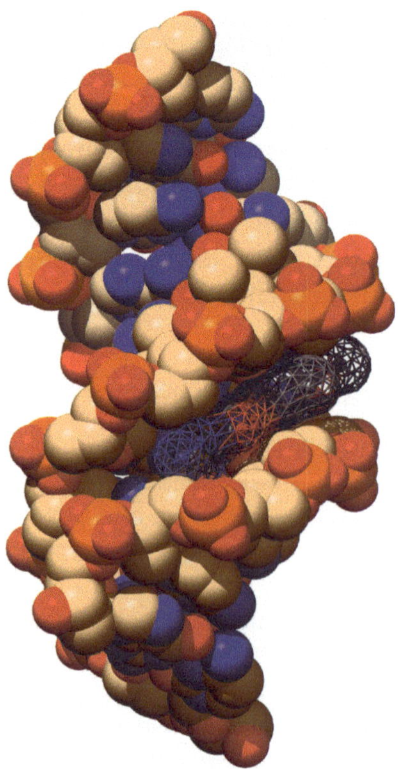

FIG. V.41 Netropsin bound to the minor groove of an oligonucleotide duplex. *Created from PDB121D.*

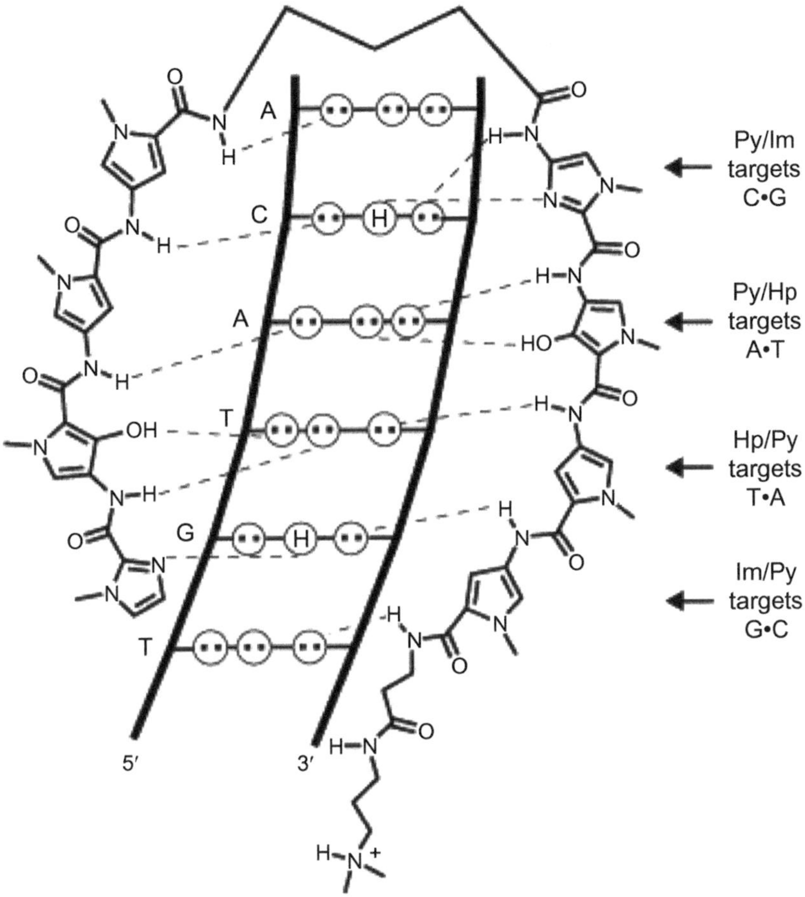

FIG. V.42 Recognition scheme of an oligonucleotide duplex by a polyamide. *Adapted from Dervan, et al., 2003. Curr. Opin. Struct. Biol. 13 (3) 284–299. Elsevier.*

in the B-DNA curve continuously, recognition of long sequences requires flexible elements in the design for tracking the groove. More flexible β-alanine was added to the repertoire for allowing bending of the polyamide. Fig. V.42 shows the recognition of base sequence by a polyamide.

Polyamides have now found significant applications in chemical biology research. In one notable application, a polyamide was synthesized that targets a cluster of GAA tri-nucleotide repeats. In Friedrich's Ataxia, an autosomal recessive genetic disease, the presence of a large number of GAA tri-nucleotide repeats in the Frataxin gene causes severely reduced transcription of the gene. The polyamide was also attached to a molecule, JQ1, that binds a bromodomain protein, BRD4. The small molecule ligand recruits BRD4 to the frataxin gene that in turn recruits P-TEFb (positive transcription elongation factor b) that promotes elongation of stalled RNA pol II. The polyamide was able to promote elongation in cell lines, patient-derived cells, and xenografted mice. In another interesting application, a bromodomain inhibitor as a p300 or CBP (CREB-binding Protein) recruiter was targeted to specific genes by conjugating it to polyamides that bind to sequences proximal to these genes. The application of this molecule to human cells caused a high degree of hyperacetylation of proximal histones, leading to gene activation.

V.7.4 Peptide-based artificial transcription factors

Peptides are natural substitutes for artificial transcription factors due to their protein-like nature. However, we must consider two fundamental issues. First, for a vast majority of transcription factors, a single recognition element is not sufficient to impart enough affinity or specificity in the backdrop of a genome. It has led to many solutions during evolution. As discussed previously, many prokaryotic transcription factors are dimeric. Because of their dimeric nature, two identical recognition elements bind to inverted repeat sequences, leading to augmented affinity and specificity. Eukaryotes have addressed the specificity problem using multiple DNA-bound transcription factors interacting with each other. Second, as pointed out in a previous section, many extant transcription factors use the helix as a major recognition element. Thus most of the peptide-based synthetic transcription factors that mimic a natural one require a designed helix as the major recognition element. When a recognition helix is stripped off from other parts of the protein to which it is embedded naturally, the helical peptide becomes disordered. This disordered conformation of the free peptide implies that upon binding to DNA, additional entropy must be sacrificed to bring it to an ordered conformation. Loss of this additional entropy would significantly reduce the binding affinity.

In yeast, the transcription factor GCN4 uses a basic helical region to interact with its target DNA and a coiled-coil domain to dimerize, thus, linking two recognition elements (Fig. V.43). A strategy of chemical cross-linking of the two recognition elements of GCN4 was used to create an artificial transcription factor. In this strategy, the replacement of the dimer-forming coiled-coil domain of GCN4 by a disulfide bond brought the two recognition elements together (Fig. V.44). This dimeric peptide showed high affinity and significant sequence specificity of DNA binding at 4°C. Instead of combining two recognition elements from a natural transcription factor, a synthetic recognition element may also be used. In one study, a distamycin analog was conjugated to one of the DNA-interacting regions of GCN4 to create a hybrid molecule. This synthetic transcription factor bound the target site with considerably higher affinity than the individual recognition elements by themselves (Figs. V.45 and V.46).

As we have mentioned before, isolated peptides without additional stabilization are not suitable as artificial transcription factors. Additional stabilization and restoration of helical conformation have been accomplished through various means, the most common being the chemical modifications of the peptide (Fig. V.47). The most well known of the helix-stabilizing modifications is the stapling. In an α-helix, one turn of the helix brings the ith amino acid spatially close to i + 3 and i + 4th amino acids on the same face. Covalent linkage between these amino acids forces the stretch of amino acids into a helical

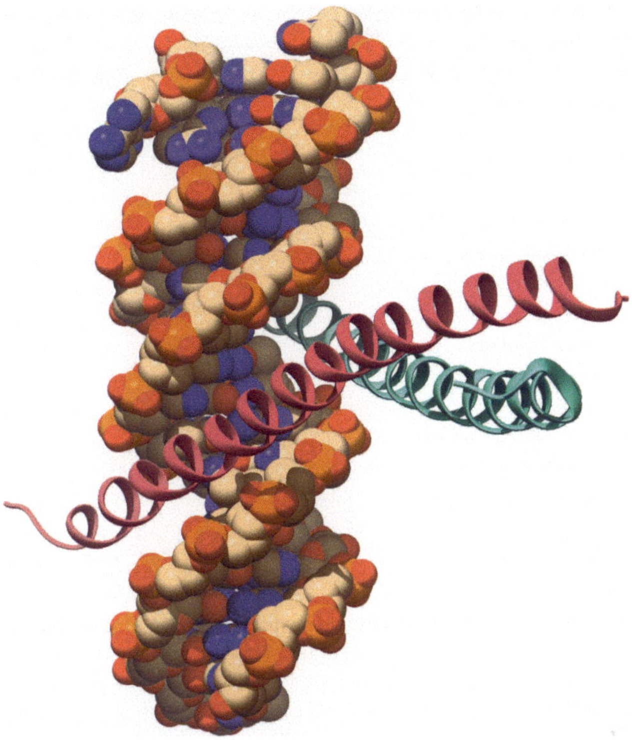

FIG. V.43 The structure of GCN4 bZIP domain in complex with its target DNA. *Created from Pdb1YSA.*

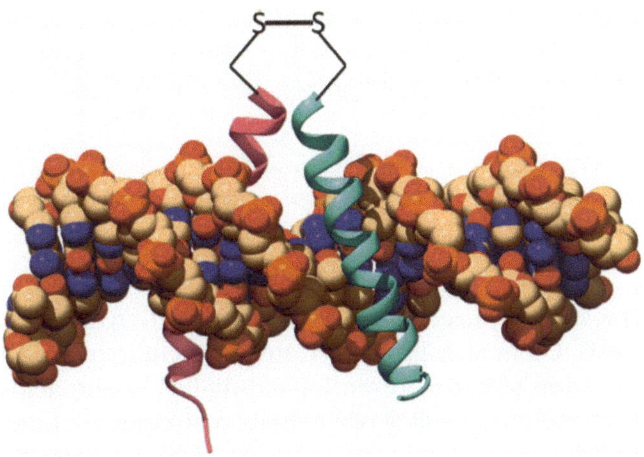

FIG. V.44 The structure of GCN4 bZIP domain peptide cross-linked by disulfide in complex with its target DNA. *Created from 1YSA.*

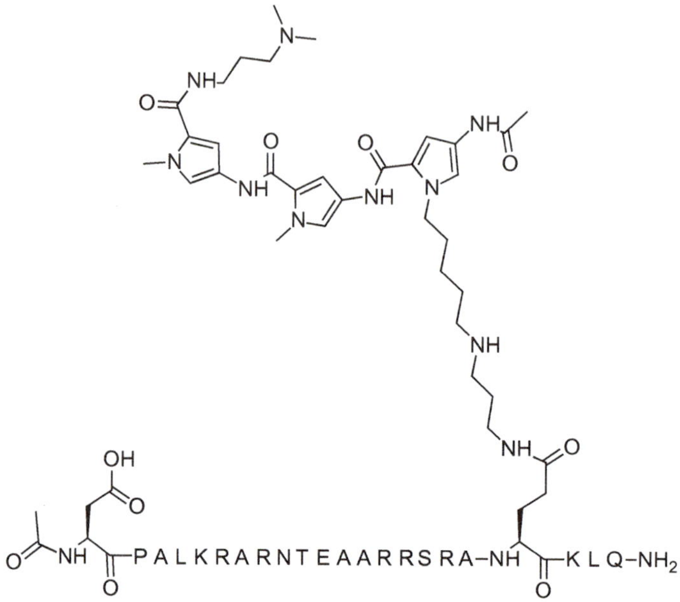

FIG. V.45 Chemical structure of a GCN4-polyamide hybrid synthetic transcription factor.

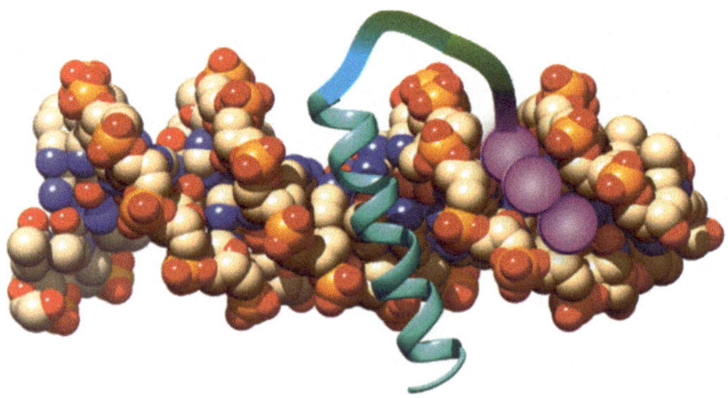

FIG. V.46 The proposed interaction patterns of the peptide-polyamide hybrid synthetic transcription factor, designed based on the GCN4, with the target DNA.

conformation. The most widely used chemistry for the creation of the covalent linker is the ring-closing metathesis. Other stabilization strategies include hydrogen bond surrogates (Fig. V.48) and substitution of helix-promoting α,α-dialkyl amino acids. These modifications are usually performed on the face of the helix that does not interact with the DNA.

DLX-4 is a homeodomain-containing transcription factor involved in many processes, including the regulation of globin genes. As mentioned before, homeodomain transcription factors combine two recognition elements, a recognition helix and a positively

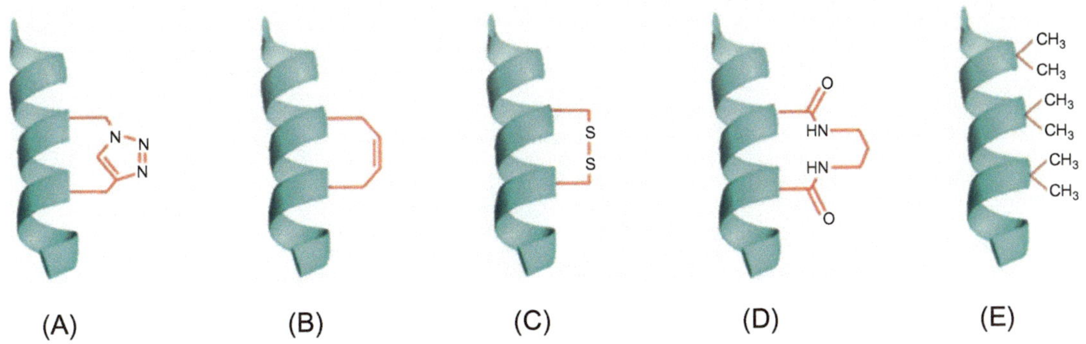

FIG. V.47 Different α-helix stabilization methods for small peptides. For cross-linked peptides, all are between residues i and (i+4). (A) Triazole-stapled helix; (B) hydrocarbon-stapled helix; (C) di-sulfide cross-linked helix; (D) lactam cross-linked helix; and (E) Aib-substituted helix. *Reproduced from Roy, et al., 2018. Biomedicines 6 (4), 118. https://doi.org/10.3390/biomedicines6040118. CC-BY license.*

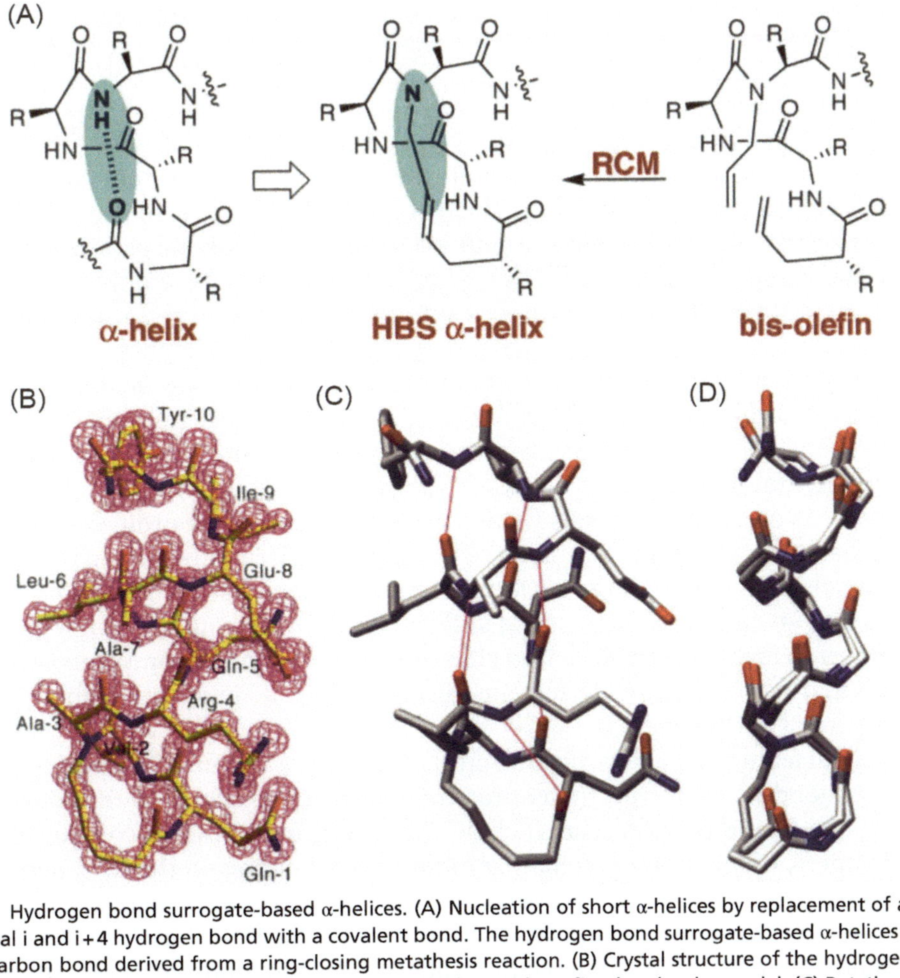

FIG. V.48 Hydrogen bond surrogate-based α-helices. (A) Nucleation of short α-helices by replacement of an N-terminal i and i+4 hydrogen bond with a covalent bond. The hydrogen bond surrogate-based α-helices contain a carbon-carbon bond derived from a ring-closing metathesis reaction. (B) Crystal structure of the hydrogen bond surrogate α-helix with electron density map superimposed onto the refined molecular model. (C) Putative i and i+4 hydrogen bonds *(magenta lines)* in crystal structure-derived molecular model of the hydrogen bond surrogate helix. (D) Overlay of crystal structure and a model of an idealized α-helix. *Reproduced from Henchey, et al., 2008. Curr. Opin. Chem. Biol. 12 (6), 692–697. Elsevier.*

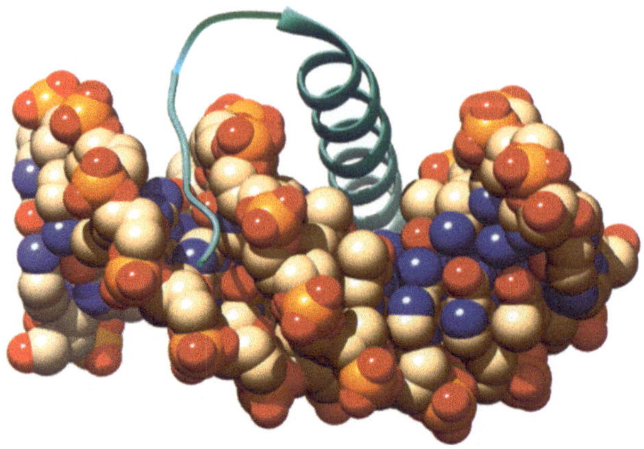

FIG. V.49 A peptide-based synthetic transcription factor based on homeodomains. A designed linker connects the two recognition elements, the N-terminal arm and the recognition helix.

charged N-terminal arm. A synthetic transcription factor, which mimics the homeodomain of DLX-4, was created by combining the recognition helix and the N-terminal arm of DLX-4, by an appropriate linker. The recognition helix was stabilized by the incorporation of α-aminoisobutyric acid as a helix stabilizer. This synthetic peptide-based transcription factor binds to the target DNA with almost DLX-4-like affinity and excellent specificity (Fig. V.49). When transported inside the nucleus of a cell using appropriate cell-penetrating peptide tags, it regulated target genes with considerable selectivity. Using a similar approach, two different elements from two transcription factors, which bind adjacent to each other, were combined to create a chimeric transcription factor. In many cancers, the Ras pathway is constitutively upregulated, promoting cellular growth. The signal from the Ras pathway ends up by activating a pro-oncogenic transcription factor, Elk-1. Among its other activities, Elk-1 binds to the promoter of the proto-oncogene cFOS and upregulates its expression. Serum response factor (SRF), another transcription factor, binds to an adjacent region of the cFOS promoter. The recognition helix from Elk-1 was combined with an unstructured region of SRF that binds to the adjacent minor groove by a suitable linker to create a peptide-based chimeric transcription factor. When applied to a cancer cell that has constitutively activated RAS pathway, it significantly downregulated the expression of cFOS gene.

An essential feature of eukaryotic transcription factors is that upon binding to their target DNA sequence, they recruit many proteins that regulate access to the promoters of targeted genes. These proteins may include chromatin modification complexes, epigenetic machinery, or other gene-regulating proteins. Depending on the recruited protein, the genes are either upregulated or downregulated. In transcription activators, the DNA-binding domain of the transcription factor contributes to the DNA-binding specificity, and an activation domain recruits the activator proteins. This functional architecture was used to design a synthetic transcription factor in which a sequence-specific DNA-binding

FIG. V.50 Structural depiction of polyamide-based synthetic transcription factors. The *green spheres* represent units of the polyamide part, and the *red cylinder* is the activation domain consisting of peptides.

molecule was connected to a peptide known to bind tightly to a histone acetyltransferase (HAT). In one example, such a transcription factor contained a DNA-binding unit of a pyrrole-imidazole containing polyamide and a peptide that binds and recruits HATs. This hybrid molecule activated a target gene in an in vitro transcription assay (Fig. V.50). A similar approach was also successful when applied to a cell. In another instance, a synthetic transcription factor of similar configuration consisted of peptides for both the functionalities.

V.7.5 Combinatorial zinc fingers

Zinc-finger proteins form a class of DNA-binding proteins having a large number of members. The zinc fingers are self-contained modules, and each presents a unique recognition helix that recognizes base pairs through interaction with the major groove. In zinc-finger-containing transcription factors, zinc fingers occur as tandem repeats, and the angle between the recognition helices is such that they can track the major groove. The modular nature of the zinc fingers allows the fingers to be combined in a "mix-and-match" fashion, creating artificial transcription factors of unique DNA-binding specificity. The use of naturally occurring zinc fingers limits one to target these artificial transcription factors to a combination of naturally occurring DNA subsites. Zinc fingers with novel specificity were obtained with the help of phage display, and hence, the capability now exists to create designed polydactyl zinc-finger proteins with novel and unique binding specificity.

V.8 Gene regulatory networks

V.8.1 Principles of multiple transcription factors working in tandem

Transcription factors do not work alone in the cell. They cross-regulate each other and auto-regulate themselves. They work in tandem with other transcription factors to produce a response through the regulation of many genes. The multiple regulatory elements and genes associated with them may come together to act as a unit that defines a gene regulatory network. In gene regulatory networks, a gene product (trans elements) regulates another gene by binding to specific sequences (cis-elements) near that gene and regulate its transcription. Eukaryotic genes, in particular, are regulated by many cis and trans elements, producing complex response output. The theme of regulation of multiple genes through a network of transcription factors is commonly observed in eukaryotes and in simpler forms in prokaryotes.

V.8.2 Different network motifs

A gene regulatory network, such as other biological networks, can be represented by a node-edge graph. The nodes represent genes, and the edges represent the regulation of the genes by the protein product of another gene. A pointed arrow [→] in this graph represents activation. The direction of the arrow points to the gene being activated. The base of the arrow indicates the activator. A dashed-vertical symbol [⊣] serves a similar function but represents inhibition (Fig. V.51).

The lysis-lysogeny decision-making network of the bacteriophage λ is an example of a simpler gene regulatory network (Fig. V.52). Once the bacteriophage integrates into the *E. coli* genome (the lysogenic state), it remains dormant until genotoxic insult damages the DNA. The maintenance of the lysogenic state is performed by the λ repressor. It shuts off all other phage genes but auto-regulates its own to maintain a fixed concentration. When the DNA damage is induced, damage-mediated activation of specific pathways cleaves and lowers the concentration of the repressor, thus, activating the synthesis of another DNA-binding protein, Cro. The Cro protein antagonizes the repressor by binding to the same sites in reverse order of affinity and initiates production of lytic genes and commits the phage to a lytic pathway. This gene regulatory network is an example of how concentrations of transcription regulatory proteins, determined by the rate of transcription, translation, and protein degradation, govern the outcome of a biological process.

Networks such as the one described above have specific motifs. Network motifs are recurrent patterns that occur far more frequently than by chance occurrence. Two of the most abundant and salient network motifs are the auto-regulation and feed-forward loop (Fig. V.53). Auto-regulation can be of two types, negative and positive. In the gene regulatory network that regulates the differentiation of dendritic cells, IRF4, IRF8, TCF4, and other transcription factors show auto-regulation motifs (Fig. V.54). The

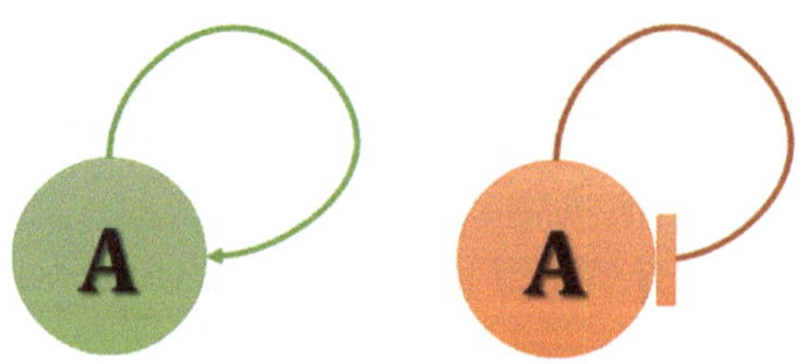

FIG. V.51 Examples of positive auto-regulatory loop *(left)* and negative auto-regulatory loop *(right)*.

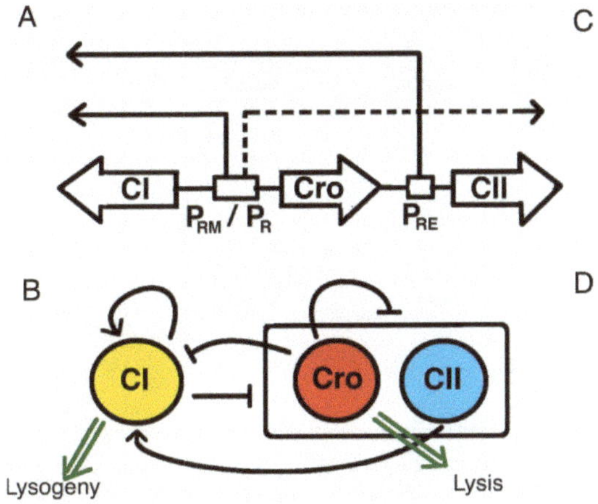

FIG. V.52 The decision-making circuit of the bacteriophage λ. (A) The role of the P_{rm}/P_r and P_{re} promoters involved in lambda phage decision. *Dashed lines* denote transcriptional events that require no activation, whereas *solid lines* denote transcriptional events that require activation. (B) Schematic of the core genetic network involved in lysis-lysogeny decision. CI gene promotes itself and represses the other genes. Cro represses everything, whereas CII promotes CI. (C) Rate of lysogeny as observed from experimental observations in Zeng et al. as a function of viral concentration for different MOI.(Multiplicity of Infection) (D) Rate of lysogeny as a function of MOI for different volumes. Bacterial volumes and viral concentrations are expressed in arbitrary units using the normalized cell lengths following Zeng et al. [24]. *Adapted from Robb, M.L., Shahrezaei, V., 2014. Stochastic cellular fate decision making by multiple infecting lambda phage. PLoS ONE 9 (8), e103636. https://doi.org/10.1371/journal.pone.0103636. CC-BY-4.0 license.*

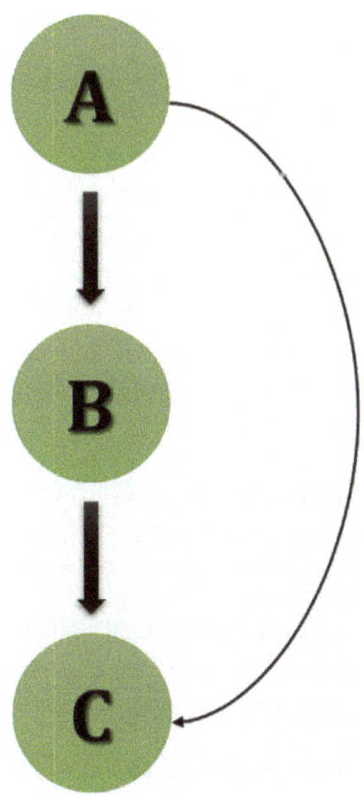

FIG. V.53 Example of a feed-forward loop.

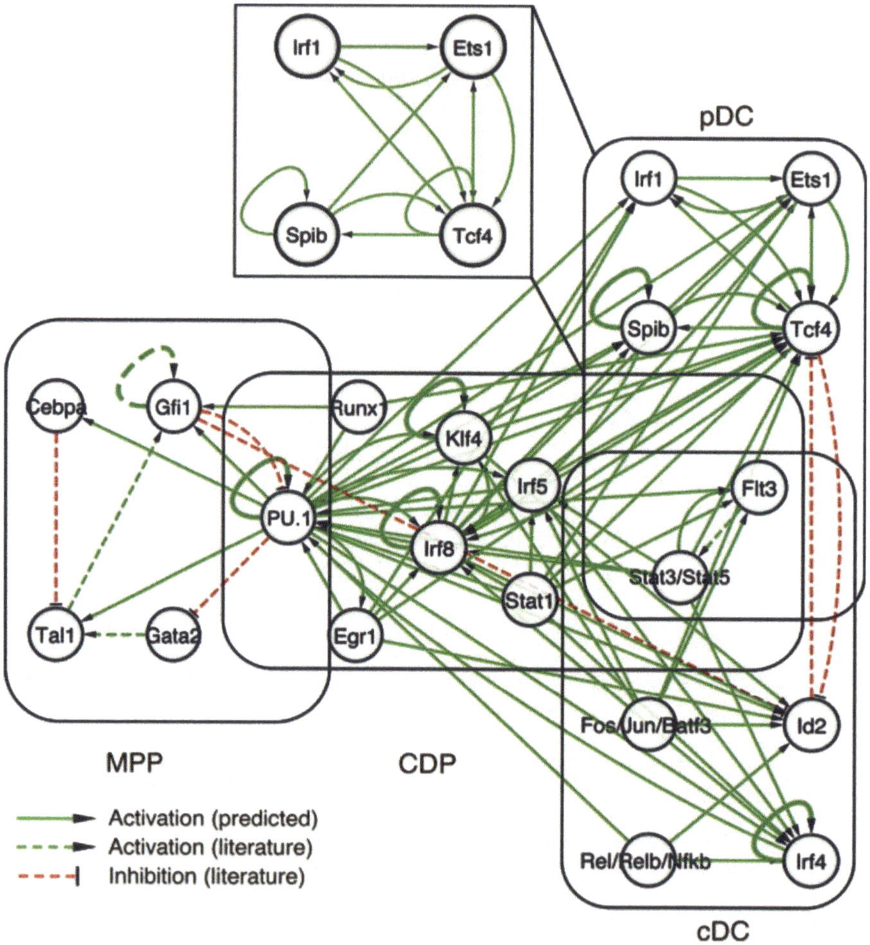

FIG. V.54 The regulatory circuitry of DC development. The network illustration depicts the organization of the integrated DC regulatory circuitry. Nodes represent key regulatory factors involved in DC development. A directed edge from factor a to factor b indicates an active function *(green)* or inhibition function s*(red)* of factor a to factor b. *Dotted edges* represent additional interactions obtained from the literature (6, 39–45). The pDC sub-network of Irf1, Ets1, Spib, and Tcf4 is shown *(gray box)*. *Lin, et al., 2015. Nucleic Acids Res. 43 (20), 9680–9693,*

auto-regulation motif imparts certain characteristic behaviors in the input-response output; for example, negative auto-regulation, that is, the protein product of a gene repressing the transcription from its parent gene speeds up the response time and dampens it more quickly to a lower steady-state value than the same gene without the auto-regulatory loop. It also serves another critical function; it reduces stochastic fluctuations of the steady-state values of the transcription factor. Feed-forward loops are also abundant in the transcription network. Feed-forward loops consist of at least three nodes. In the picture shown, the network consists of three genes (A, B, and C). A activates B that in turn activates C. However, A also activates C directly, setting up a feed-forward loop.

V.8.3 Bi-stability and multistability of gene regulatory networks

Semiquantitative or quantitative input-response behavior of the gene regulatory network can be derived from many different kinds of models, starting from Boolean network-based models that can take only binary values to a series of nonlinear differential equations describing finer properties. However, more complex models necessarily present increasing difficulties in analyzing network behaviors. A series of linear differential equations of the type offers a good compromise:

$$\frac{dX}{dt} = \frac{k_1 X^n}{K_n + X^n} - \gamma X,$$

where X is the concentration of the transcription factor, k_1 is the synthesis rate constant, γ is the degradation rate constant, n is the Hill coefficient, and K is the dissociation constant. For each transcription factor, one such equation describes its behavior. The system of linear differential equations can then be solved numerically to obtain the input-response behavior of the whole network.

Certain classes of networks possess properties that are essential for their biological functions. For gene regulatory networks that operate at timescales of minutes to hours, the existence of steady states is vital for expressing a particular pattern of gene expression. Multiple steady states have been predicted to exist in certain types of networks. An essential criterion for the existence of multiple steady states is the presence of a positive feed-forward loop. Analysis of transcription regulatory networks in both eukaryotes and prokaryotes suggests that many organisms are enriched in such motifs. A steady-state of a gene regulatory network is characterized by a fixed concentration of one or more transcription factors. Fixed concentrations of transcription factors may result in a gene expression pattern that confers a particular character to the cell. In different steady states, transcription factor concentrations may change, producing a different gene expression pattern. Such steady-state-governed gene expression patterns may underpin processes such as lineage commitment.

Chapter summary

This chapter focuses on how different phenomena influence the regulation of gene expression and offer chemical and structural explanations of gene regulation. This chapter also discusses how to construct artificial transcription factors having novel specificity and functions.

- B-DNA, the dominant conformation of genomic DNA, contains sequence-dependent conformational heterogeneity. We have discussed in detail how this heterogeneity is reflected in many structural parameters of individual base pairs.
- DNA and protein are both polyelectrolytes. Counterions and water molecules surround both DNA and proteins in solution. A fraction of the water of hydration is tightly associated with macromolecules. DNA-protein interactions are broadly classified into

two groups: sequence-specific interactions and nonsequence-specific interactions. Interaction between a protein and the DNA leads to the release of counterions and water to the bulk solvent. The release of water and ions is entropically favorable and a significant component of the binding free energy.

- Sequence-specific and nonsequence-specific DNA-protein complexes differ in their structures. Structure elucidation of the sequence-specific complex and the nonsequence-specific complex of lac repressor indicates that in nonspecific complexes, the protein is still very mobile, and the interaction with the DNA is primarily through the backbone. In sequence-specific complexes, the protein often interacts with base atoms in the grooves, primarily the major groove. DNA-bound proteins also interact with each other and may add extra stabilization to the protein-DNA complex. This extra stabilization may be a good source of specificity for multiprotein regulatory complexes that form near eukaryotic genes.
- α-helices are the most abundant elements involved in sequence-specific recognition of DNA. Different protein architectures and motifs used in DNA recognition usually involve α-helix. In some DNA-protein interactions, other peptide segments combine with helices. One of the nonhelical elements is a small flexible peptide that senses the sequence-dependent narrowing of the minor groove in homeodomain-containing transcription factors.
- In prokaryotes, gene regulation by transcription factors occurs through interaction, either through contact or through competition, with the RNA polymerase. In eukaryotes, the DNA is packaged into nucleosomes and higher structures, blocking its access to the transcription machinery. Access control is the primary step, where eukaryotic transcription is regulated.
- Artificial transcription factors have been created by mimicking many of the naturally occurring DNA-protein interactions and by totally synthetic molecules. Polyamides based on naturally occurring DNA-binding antibiotics, netropsin and distamycin, have proved to be effective. Peptide-based transcription factors mimicking naturally occurring transcription factors have also proved to be effective.
- Transcription factors do not function in isolation in a cell. They cooperate with other transcription factors and regulatory proteins to form a gene regulatory network. Many different motifs are present. Motifs may give rise to unique properties having functional implications.

Further reading

Adhya, S., 1989. Multipartite genetic control elements: communication by DNA loop. Annu. Rev. Genet. 23, 227–250.

Affolter, M., Slattery, M., Mann, R.S., 2008. A lexicon for homeodomain-DNA recognition. Cell 133, 1133–1135.

Alon, U., 2019. An Introduction to Systems Biology: Design Principles of Biological Circuits. CRC Press.

Banerjee, R., Basu, G., Roy, S., Chène, P., 2002. Aib-based peptide backbone as scaffolds for helical peptide mimics. J. Pept. Res. 60, 88–94.

Becker, N.A., Phillips, W.J., Wallace, J.P., Schwab, T.L., Clark, K.J., Maher, L.J., 2020. Characterizing the stability of an engineered regulatory DNA loop in living *E. coli* cells. Biophys. J. 118, 380a.

Beerli, R.R., Barbas, C.F., 2002. Engineering polydactyl zinc-finger transcription factors. Nat. Biotechnol. 20, 135–141.

Bogdanove, A.J., Bohm, A., Miller, J.C., Morgan, R.D., Stoddard, B.L., 2018. Engineering altered protein–DNA recognition specificity. Nucleic Acids Res. 46, 4845–4871.

Chakraborty, M., Roy, S., 2017. A peptide-based synthetic transcription factor selectively down-regulates the proto-oncogene CFOS in tumour cells and inhibits proliferation. Chem. Commun. 53, 376–379.

Cho, Y., Gorina, S., Jeffrey, P.D., Pavletich, N.P., 1994. Crystal structure of a p53 tumor suppressor-DNA complex: understanding tumorigenic mutations. Science 265, 346–355.

Dans, P.D., Balaceanu, A., Pasi, M., Patelli, A.S., Petkevičiūtė, D., Walther, J., Hospital, A., Bayarri, G., Lavery, R., Maddocks, J.H., 2019. The static and dynamic structural heterogeneities of B-DNA: extending Calladine–Dickerson rules. Nucleic Acids Res. 47, 11090–11102.

DeHaseth, P.L., Lohman, T.M., Record Jr., M.T., 1977. Nonspecific interaction of lac repressor with DNA: an association reaction driven by counterion release. Biochemistry 16, 4783–4790.

Dervan, P.B., Kurmis, A.A., Finn, P.B., 2018. Molecular recognition of DNA by py–im polyamides: from discovery to oncology. In: DNA-Targeting Molecules as Therapeutic Agents.vol. 7, p. 298.

Erwin, G.S., Grieshop, M.P., Ali, A., Qi, J., Lawlor, M., Kumar, D., Ahmad, I., McNally, A., Teider, N., Worringer, K., 2017. Synthetic transcription elongation factors license transcription across repressive chromatin. Science 358, 1617–1622.

Garcia, H.G., Phillips, R., 2011. Quantitative dissection of the simple repression input–output function. Proc. Natl. Acad. Sci. 108, 12173–12178.

Ghosh, B., Boila, L.D., Choudhury, S., Mondal, P., Bhattacharjee, S., Pal, S.K., Sengupta, A., Roy, S., 2018. A potent conformation-constrained synthetic peptide mimic of a homeodomain selectively regulates target genes in cells. ACS Chem. Biol. 13, 2003–2009.

Henchey, L.K., Jochim, A.L., Arora, P.S., 2008. Contemporary strategies for the stabilization of peptides in the α-helical conformation. Curr. Opin. Chem. Biol. 12, 692–697.

Hochschild, A., Ptashne, M., 1986. Cooperative binding of λ repressors to sites separated by integral turns of the DNA helix. Cell 44, 681–687.

Joshi, R., Passner, J.M., Rohs, R., Jain, R., Sosinsky, A., Crickmore, M.A., Jacob, V., Aggarwal, A.K., Honig, B., Mann, R.S., 2007. Functional specificity of a Hox protein mediated by the recognition of minor groove structure. Cell 131, 530–543.

Kalodimos, C.G., Biris, N., Bonvin, A.M., Levandoski, M.M., Guennuegues, M., Boelens, R., Kaptein, R., 2004. Structure and flexibility adaptation in nonspecific and specific protein-DNA complexes. Science 305, 386–389.

Karlebach, G., Shamir, R., 2008. Modelling and analysis of gene regulatory networks. Nat. Rev. Mol. Cell Biol. 9, 770–780.

Lewis, D.E., Adhya, S., 2015. Molecular mechanisms of transcription initiation at gal promoters and their multi-level regulation by GalR, CRP and DNA loop. Biomol. Ther. 5, 2782–2807.

Li, G., Reinberg, D., 2011. Chromatin higher-order structures and gene regulation. Curr. Opin. Genet. Dev. 21, 175–186.

Liepinsh, E., Otting, G., Wüthrich, K., 1992. NMR observation of individual molecules of hydration water bound to DNA duplexes: direct evidence for a spine of hydration water present in aqueous solution. Nucleic Acids Res. 20, 6549–6553.

Luger, K., Mäder, A.W., Richmond, R.K., Sargent, D.F., Richmond, T.J., 1997. Crystal structure of the nucleosome core particle at 2.8Å resolution. Nature 389, 251–260.

Madhumalar, A., Bansal, M., 2005. Sequence preference for BI/BII conformations in DNA: MD and crystal structure data analysis. J. Biomol. Struct. Dyn. 23, 13–27.

Mahalakshmi, R., Balaram, P., 2006. Non-protein amino acids in the design of secondary structure scaffolds. In: Protein Design. Springer, pp. 71–94.

Mazumder, A., Batabyal, S., Mondal, M., Mondol, T., Choudhury, S., Ghosh, R., Chatterjee, T., Bhattacharyya, D., Pal, S.K., Roy, S., 2017. Specific DNA sequences allosterically enhance protein–protein interaction in a transcription factor through modulation of protein dynamics: implications for specificity of gene regulation. Phys. Chem. Chem. Phys. 19, 14781–14792.

McDermott, M.L., Vanselous, H., Corcelli, S.A., Petersen, P.B., 2017. DNA's chiral spine of hydration. ACS Central Sci. 3, 708–714.

Morgunova, E., Taipale, J., 2017. Structural perspective of cooperative transcription factor binding. Curr. Opin. Struct. Biol. 47, 1–8.

Müller-Hill, B., Oehler, S., 1996. The Lac Operon. Walter de Gruyter, New York.

Nair, S.K., Burley, S.K., 2000. Recognizing DNA in the library. Nature 404, 715–717.

Oda, M., Nakamura, H., 2000. Thermodynamic and kinetic analyses for understanding sequence-specific DNA recognition. Genes Cells 5, 319–326.

Olson, W.K., Zhurkin, V.B., 2000. Modeling DNA deformations. Curr. Opin. Struct. Biol. 10, 286–297.

Onufriev, A.V., Schiessel, H., 2019. The nucleosome: from structure to function through physics. Curr. Opin. Struct. Biol. 56, 119–130.

Ptashne, M., 2005. Regulation of transcription: from lambda to eukaryotes. Trends Biochem. Sci. 30, 275–279.

Roeder, R.G., 2019. 50+ years of eukaryotic transcription: an expanding universe of factors and mechanisms. Nat. Struct. Mol. Biol. 26, 783–791.

Rohs, R., Jin, X., West, S.M., Joshi, R., Honig, B., Mann, R.S., 2010. Origins of specificity in protein-DNA recognition. Annu. Rev. Biochem. 79, 233–269.

Rohs, R., West, S.M., Sosinsky, A., Liu, P., Mann, R.S., Honig, B., 2009. The role of DNA shape in protein–DNA recognition. Nature 461, 1248–1253.

Ross, P.D., Subramanian, S., 1981. Thermodynamics of protein association reactions: forces contributing to stability. Biochemistry 20, 3096–3102.

Roy, K., Mazumder, A., Ghosh, P., Naiya, G., Ghosh, B., Roy, S., 2018. A peptide-based synthetic transcription factor selectively activates transcription in a mammalian cell. Chem. Commun. 54, 1611–1614.

Roy, S., Garges, S., Adhya, S., 1998. Activation and repression of transcription by differential contact: two sides of a coin. J. Biol. Chem. 273, 14059–14062.

Schleif, R., 1992. DNA looping. Annu. Rev. Biochem. 61, 199–223.

Schultz, S.C., Shields, G.C., Steitz, T.A., 1991. Crystal structure of a CAP-DNA complex: the DNA is bent by 90 degrees. Science 253, 1001–1007.

Sera, T., 2009. Zinc-finger-based artificial transcription factors and their applications. Adv. Drug Deliv. Rev. 61, 513–526.

Spolar, R.S., Record, M.T., 1994. Coupling of local folding to site-specific binding of proteins to DNA. Science 263, 777–784.

Struhl, K., 1989. Helix-turn-helix, zinc-finger, and leucine-zipper motifs for eukaryotic transcriptional regulatory proteins. Trends Biochem. Sci. 14, 137–140.

Taniguchi, J., Feng, Y., Pandian, G.N., Hashiya, F., Hidaka, T., Hashiya, K., Park, S., Bando, T., Ito, S., Sugiyama, H., 2018. Biomimetic artificial epigenetic code for targeted acetylation of histones. J. Am. Chem. Soc. 140, 7108–7115.

Tidow, H., Melero, R., Mylonas, E., Freund, S.M., Grossmann, J.G., Carazo, J.M., Svergun, D.I., Valle, M., Fersht, A.R., 2007. Quaternary structures of tumor suppressor p53 and a specific p53–DNA complex. Proc. Natl. Acad. Sci. 104, 12324–12329.

Vázquez, M.E., Caamaño, A.M., Martínez-Costas, J., Castedo, L., Mascareñas, J.L., 2001. Design and synthesis of a peptide that binds specific DNA sequences through simultaneous interaction in the major and in the minor groove. Angew. Chem. Int. Ed. 40, 4723–4725.

Vázquez, M.E., Caamaño, A.M., Mascarenas, J., 2003. From transcription factors to designed sequence-specific DNA-binding peptides. Chem. Soc. Rev. 32, 338–349.

Venters, B.J., Pugh, B.F., 2009. How eukaryotic genes are transcribed. Crit. Rev. Biochem. Mol. Biol. 44, 117–141.

Walensky, L.D., Bird, G.H., 2014. Hydrocarbon-stapled peptides: principles, practice, and progress: miniperspective. J. Med. Chem. 57, 6275–6288.

Weirauch, M.T., Hughes, T., 2011. A catalogue of eukaryotic transcription factor types, their evolutionary origin, and species distribution. In: A Handbook of Transcription Factors. Springer, pp. 25–73.

Wharton, R.P., Ptashne, M., 1985. Changing the binding specificity of a represser by redesigning an α-helix. Nature 316, 601–605.

Wolfe, S.A., Nekludova, L., Pabo, C.O., 2000. DNA recognition by Cys2His2 zinc finger proteins. Annu. Rev. Biophys. Biomol. Struct. 29, 183–212.

Wunderlich, Z., Mirny, L.A., 2009. Different gene regulation strategies revealed by analysis of binding motifs. Trends Genet. 25, 434–440.

VI

Elucidation of genome sequence and functional architecture by chemical biology approach

VI.1 Genome technology

Ability to determine the sequence of nucleobases in the genome revolutionized the study of genomes and much of Biology. Genomes are dynamic entities. They are organized in three-dimensions and are changing in response to signals. Genome sequencing has become a vital tool in Biological Sciences and a critical tool for studying genome organization and its dynamic alteration. In Chapter II, we have discussed parallel sequencing methods briefly. In this chapter, we look at how sequencing methods have evolved and how it is applied today to study genome structure and function.

VI.1.1 Rapid evolution in the sequencing methods

The code of life is written in the form of a DNA sequence. Information required for the functioning of all forms of life is packaged in a chemical code of four different nucleotides, A, T, G, and C. Thus sequencing of the genome of different species is revolutionizing our understanding of life. Since the discovery of the di-deoxy DNA sequencing method by Frederick Sanger in 1977, there has been a rapid and dramatic improvement in DNA sequencing technology, increasing the speed and reducing the cost by leaps and bounds (Fig. VI.1).

Sanger sequencing depends on enzymatically extending the complementary strand of a single-stranded DNA. In the Sanger method (which could be termed the "first-generation" approach), random incorporation of di-deoxynucleotide triphosphates (ddNTPs) during the in vitro synthesis of complementary strand terminates the DNA synthesis at different points of the newly synthesized chain, producing a mixture of chains of different lengths. These fragments of different lengths can be separated by gel electrophoresis. Specific termination of DNA synthesis, with either radioactively or fluorescently labeled nucleotides, is used to read the presence of that nucleotide (e.g., A) at different positions of the DNA. By using different colored fluorescent tags for each ddNTPs, the sequence could be deciphered in an automated DNA sequencer. In 2001, this method brought great success in decoding the human genetic blueprint, that is, the initial draft of the human genome, the first success of the Human Genome Project (HGP). However, the success costs ten years of time and 3

Chemical Biology of the Genome. https://doi.org/10.1016/B978-0-12-817644-3.00008-8

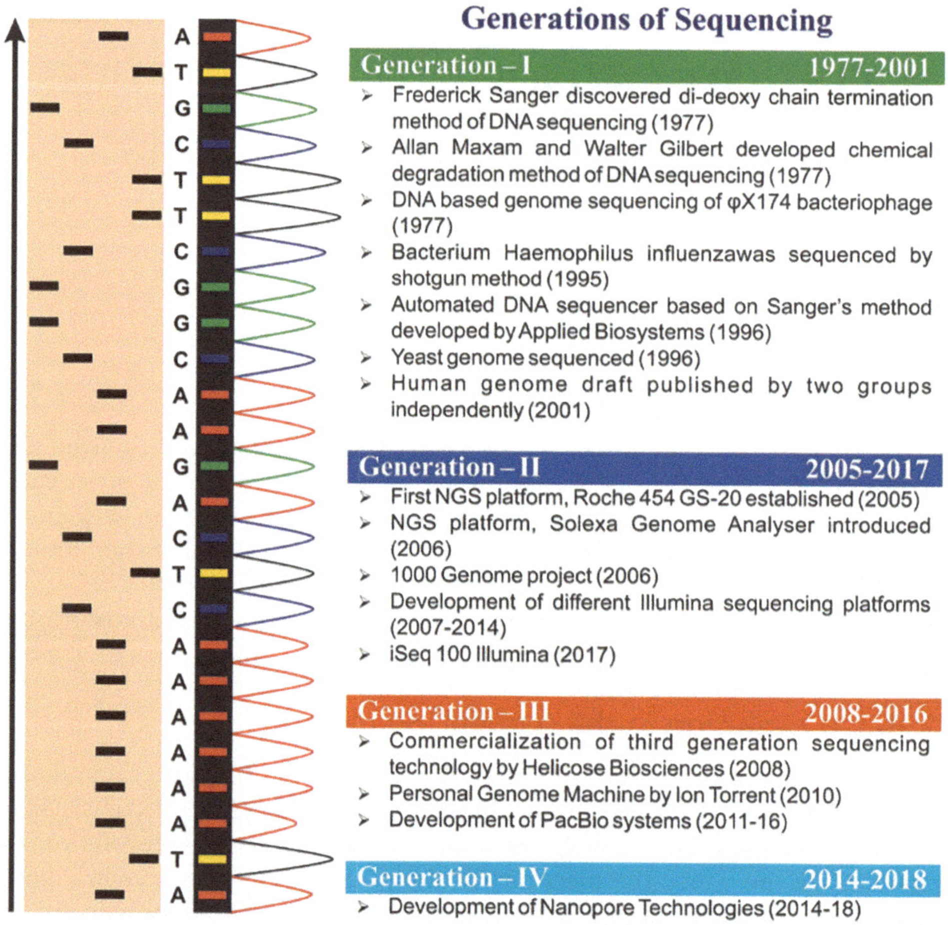

FIG. VI.1 Chronological development of sequencing technologies.

billion USD, as this method could read only 800–1000 bases at a stretch. Thus it was understood at this juncture that an entirely new approach is needed to understand the genome biology of complex organisms.

VI.1.2 Next-generation sequencing

The "next-generation" sequencing (NGS) is a massively parallel sequencing method in which many DNA strands are sequenced at the same time. In one of the earliest developed methodologies, a new kind of chemistry was incorporated other than the di-deoxy sequencing method. In some NGS methods, the "sequencing by synthesis" (SBS) protocol is followed. In this method, DNA is sequenced in short stretches, which allow many pieces of DNA to be sequenced simultaneously. This method has been depicted in Fig. VI.2 (also see "Pyrosequencing technology").

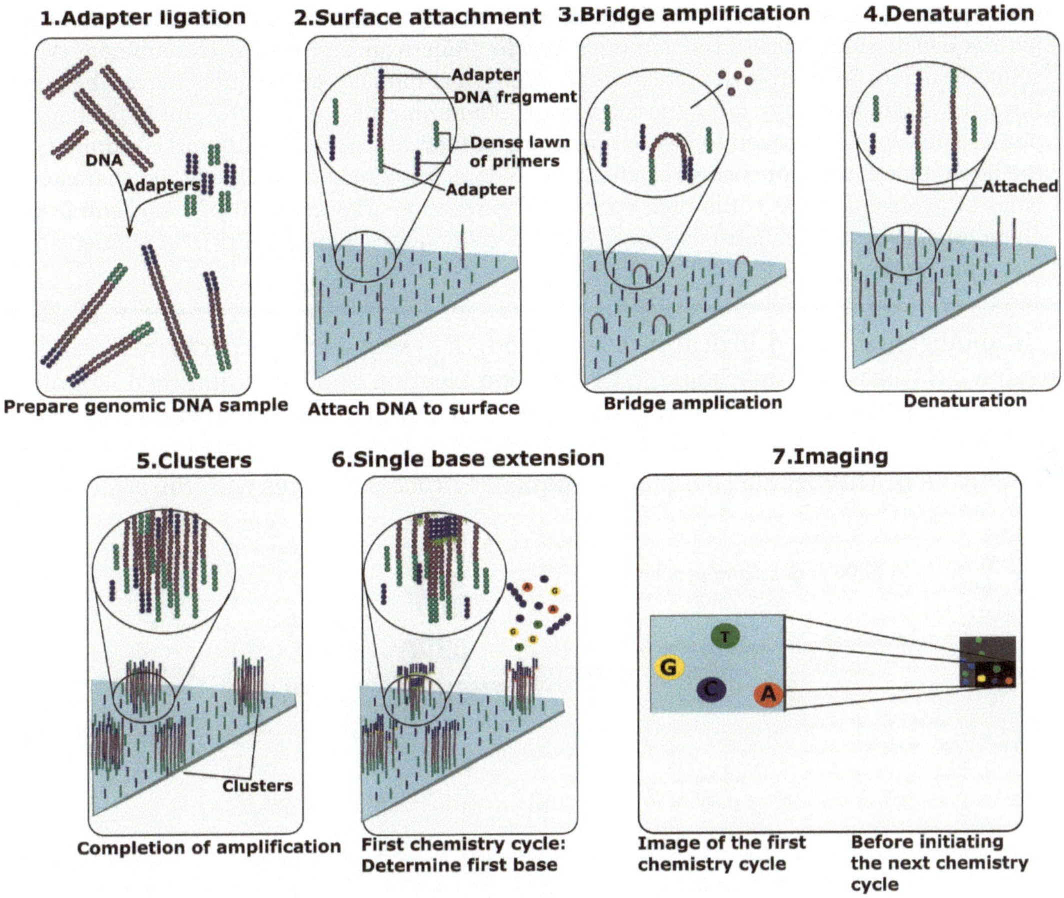

FIG. VI.2 High speed next-generation sequencing methods. DNA to be sequenced is fragmented, adapters are ligated with the DNA fragment and subjected to bridge amplification. Each fragment generates cluster on amplification. The clustered fragments are then denatured, annealed with the primer and sequenced. (1) Randomly fragment genomic DNA and ligate adapters to both ends of the fragments. (2) Bind single-stranded fragments randomly to the inside surface of the flow cell channels. (3) Add unlabeled nucleotides and enzymes to initiate solid-phase bridge amplification. (4) Denature the double-stranded molecules. (5) On completion, seven million dense clusters of double-stranded DNA are generated in each channel of the flow cell. (6) To initiate the first sequencing cycle, add all four labeled reversible terminators, primers and DNA polymerase enzyme to the flow cell. (7) After laser excitation, capture the image of the emitted fluorescence from each cluster on the flow cell. Record the identity of the first base for each cluster, before initiation the next cycle, the blocked 3′ terminus and each fluorophore from each incorporated base are removed. *Source: Mardis, E.R., 2008. Trends Genet. 24 (3), 133–41. Elsevier.*

Pyrosequencing technology

The core methodology of pyrosequencing is the sequencing by synthesis (SBS) approach invented in 1993. During the incorporation of a nucleotide in the complementary strand, a pyrophosphate molecule is released. In this method, the released pyrophosphate (PPi) during

DNA synthesis by DNA polymerase is detected, often using chemiluminescent methods. For pyrosequencing, luciferase from *Photinus pyralis* (American firefly) and recombinant ATP sulfurylase from *Saccharomyces cerevisiae* are used. Operationally, pyrosequencing can be performed two different ways: (a) In the solid phase, after immobilizing the DNA; (b) In the liquid phase, a nucleotide-degrading enzyme, apyrase, is needed. Since no solid support is required in the liquid phase, the sequencing reaction can be done in a single tube. One of the characteristics of pyrosequencing is the high-speed of the reaction. The addition of a nucleotide to chemiluminescent detection takes only 3–4 s. It tremendously speeds up DNA sequencing (see Chapter II).

In another high-speed, high accuracy method, DNA sequencing is performed using the enzyme DNA ligase. In this Sequencing by Oligo Ligation and Detection (SOLiD), oligonucleotide adapter-linked DNA fragments are attached to a 1-mm magnetic bead decorated with complementary oligos. The attached DNA fragments are amplified by emulsion PCR (Fig. VI.3). This ligation-dependent sequence process initiates with the annealing of

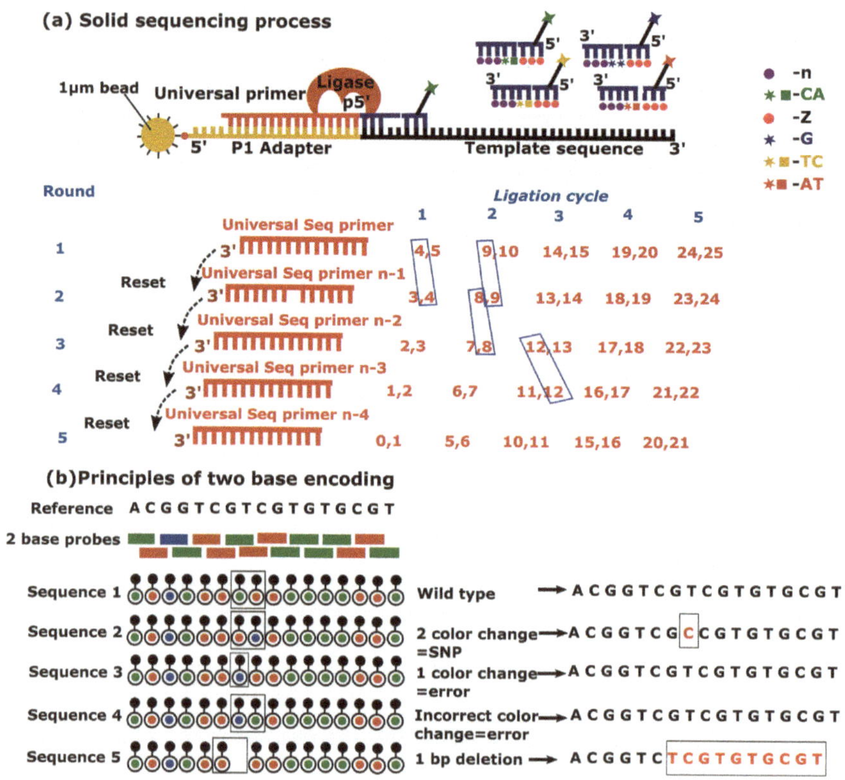

FIG. VI.3 Sequencing by Oligo Ligation and Detection (SOLiD). In this method, first (A) the universal primer anneal with the P1 adapter, which then is ligated with the specific 8mer labeled primer which get recognized in each cycle of sequencing. (B) Two-base probe helps to detect the real polymorphisms. *Source: Mardis, E.R., 2008. Trends Genet. 24 (3), 133–41. Elsevier.*

the universal primer with the P1 adapter. A library of semi-degenerate oligonucleotide octamers is added in the next step, each carrying a dinucleotide combination and six degenerate bases. Each oligo contains a different fluorophore for identification. The oligonucleotide that matches the template binds next to the universal primer, and the ligase seals the phosphate backbone. Three bases from the 5′-end were then cleaved off. The cleaved fluorescent base of the primer identifies the dinucleotide. The hybridization and ligation go on to read the sequence in a regular interval. After washing away the newly made complementary strand, the whole process is then repeated with another universal primer in a different phase. This sequencing method is one of the most cost-effective and accurate, but the read length is relatively short.

VI.1.3 Nanopore sequencing

The most recent development in sequencing technology is nanopore sequencing (Fig. VI.4). In this method, the nanopore made up of protein or solid-state material is placed in an electrolyte solution. The membrane containing the nanopore splits the chamber into two-parts. On applying a voltage, an electric field is created. DNA molecules in this chamber reach near the capture region of the pore by Brownian motion. The double-stranded DNA molecule separates and only one strand passes through the pore due to different forces in the chamber: electrophoretic, electroosmotic, and sometimes thermophoretic forces. As the bases of the DNA pass through the pore, the ion current that normally passes through the pore is disrupted. The disruption pattern is a complex

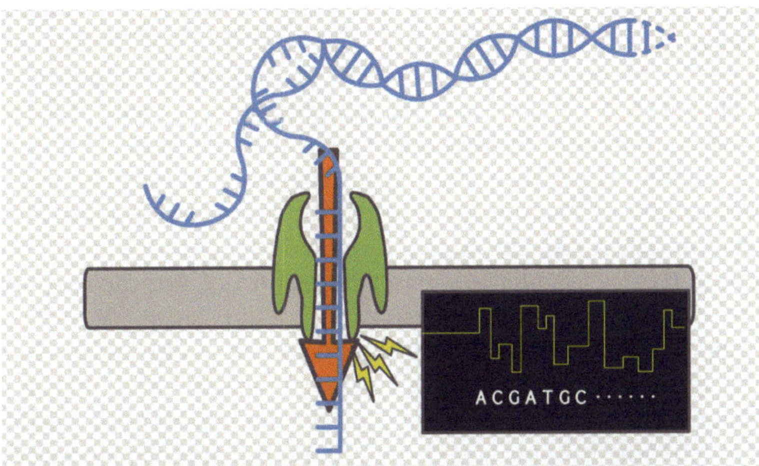

FIG. VI.4 Nanopore sequencing technology. The sequencing chamber is filled with ionic solutions and is separated by a membrane containing a CsgG nanopore. The DNA *(blue)* to be sequenced is driven through the pore electrophoretically in a controlled manner due to the presence of motor protein *(green)*. The double-stranded DNA molecule is unwound, and only one strand passes through the pore. The translocation of the DNA single-strand generates a current signal resulting in a "squiggle plot" for the sequencing. *Source: Wikipedia contributors, 2021, Page ID: 997595357. In Wikipedia, The Free Encyclopedia.*

function of the DNA bases passing through the pore. This pattern of ion current is then deconvoluted to arrive at the DNA sequence.

A single molecule of DNA or RNA can be sequenced very fast by nanopore sequencing without PCR amplification or chemical labeling. It is also considered low-cost genotyping. This method has become highly useful for identifying viral pathogens, especially during this unprecedented COVID19 pandemic. It could also be useful in food safety monitoring, monitoring of antibiotic resistance, and other areas. However, this improvement of sequencing technologies now has made it feasible to sequence hundreds of genomes. It is now being used in personalized medicine, exploring the variations in one person's DNA.

VI.1.4 Chromatin technologies for elucidating the dynamic genome

A functional genome is much more than a simple linear sequence of DNA. The genome is organized into a highly dynamic nucleo-protein complex, especially in a eukaryotic cell. Although the high-speed, low-cost sequencing available today is by itself significantly facilitating our understanding of genome organization and function, several techniques, such as super-resolution microscopy, and computational tools are being used in combination to elucidate context-dependent genome organization (Fig. VI.5). We have discussed some of these tools in Chapter II.

Unprecedented improvements in imaging techniques have made it possible to visualize the functional genome much better than ever before. Multiplex fluorescence in situ hybridization (FISH) has been used to determine the chromosome structure and organization of TADs, briefly discussed in Chapter I. In this method, tens of genomic regions of TAD are observed simultaneously by FISH. The combined multiplexed RNA FISH further improves the image. This improved imaging, when combined with super-resolution stochastic optical reconstruction microscopy (STORM), the highly dynamic chromatin could be visualized. However, these methods are not sufficient to catch the bulk chromatin motion. With the combination of light microscopy and computer vision (Optical Flow) technology, the motion of bulk chromatin dynamics could be captured at sub-pixel accuracy (nanoscale resolution). Technically these methods are called Dense Flow reconstruction and Correlation (DFCC) for the spatial motion of the chromatin and high-resolution diffusion mapping (Hi-D) for physical characterization of the moving chromatin. The combination of deep-photoactivated localization microscopy (deep-PALM) and HiD could also be highly useful to understand the dynamics of chromatin domains further. Besides imaging, chromatin manipulation by optical tweezers and nuclear manipulation helps to understand the physical nature of the genome (chromatin) organization.

Chromatin interacting proteins and chromatin modifications play a critical role in the functional organization of the genome. The binding of transcription factors on the nucleosome or nucleosomal DNA recruits the chromatin-modifying enzymes and initiates the enhancer assembly and chromatin looping. Non-histone chromatin proteins like MENT, DECK, PC4, and HP1s are directly involved in the chromatin organization. To understand the specific roles of these proteins in the genome organization and function, several techniques, all of which are fundamentally based on the NGS, have been invented. In the

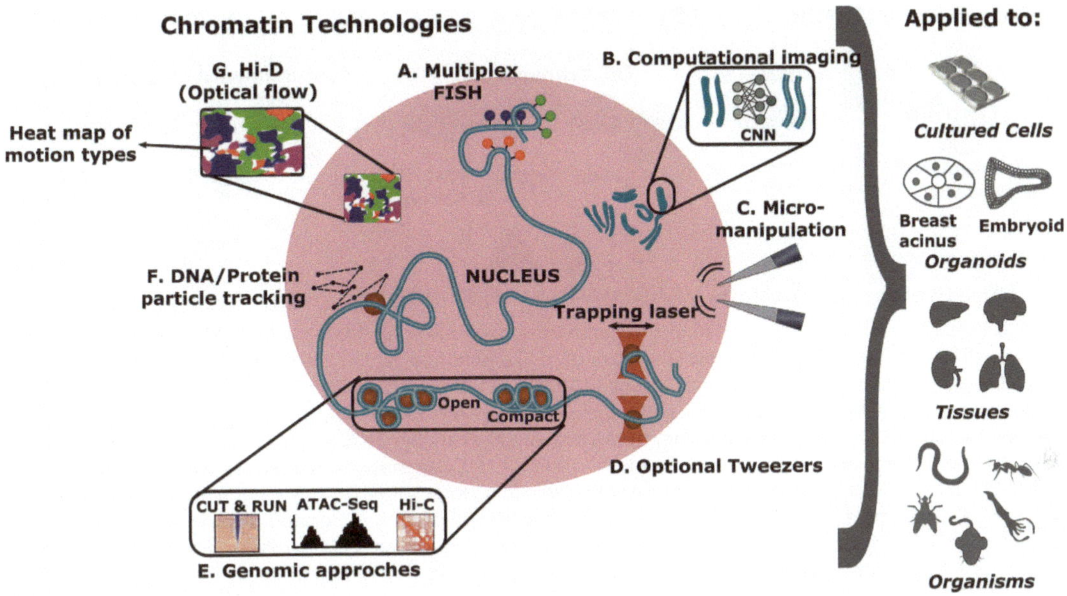

FIG. VI.5 Chromatin technologies to understand the dynamic genome functions. (A) Multiplexed fluorescent in situ hybridization (FISH) imaging, in combination with the super-resolution imaging, could be used to probe the specific TADs in the whole-genome context. (B) Computational imaging adding significant advancement in imaging technology to improve the image quality from low resolution to high resolution. (C) In different environmental conditions, the physiochemical conditions of the nuclei change. The forces acting on the nuclei could be measured by different micromanipulation. (D) Optical tweezers or the single-beam gradient force trap has a great potential to study genome architecture in the living cell. (E) Highly efficient sequencing technology, such as Hi-C, ATAC-Seq, and CUT&RUN, makes it possible to elucidate the functional folding of the genome. (F) The protein particle tracking method is employed to elucidate the chromatin dynamics. (G) The optical flow-based Hi-D technique is used to understand the global chromatin dynamics in a different context. *Source: Agbleke, A.A., et al., 2020. Mol. Cell 79 (6), 881–901. Elsevier.*

three-step CUT&RUN (cleavage under target & release using nuclease) method, first, chromatin-bound protein factors (e.g., transcription factors) or specific histone modification is targeted by a magnetic bead conjugated antibody, upon permeabilizing the cell using digitonin; second, protein A or protein G conjugated nuclease (MNase) is used in which protein A or protein G recognize the immunoglobulin (antibody) and bind to the transcription factor bound antibody. Consequently, MNase cleaves the neighboring DNA; finally, the DNA fragment is subjected to NGS. These sequencing data provide the genome-wide distribution map of a given transcription factor. The same technique could be used for any other chromatin-bound protein or specific histone modification. The data obtained from the CUT&RUN method have much less noise compared with data obtained by chromatin immunoprecipitation (ChIP). It is also a fast and cost-effective method compared with the ChIP-seq analysis (Fig. VI.6).

The genome-wide accessibility of chromatin can be assayed by ATAC-seq (Assay for Transposase-Accessible Chromatin using sequencing). A hyperactive mutant Tn5 Transposase is inserted in the open or accessible region of the chromatin (the process is termed

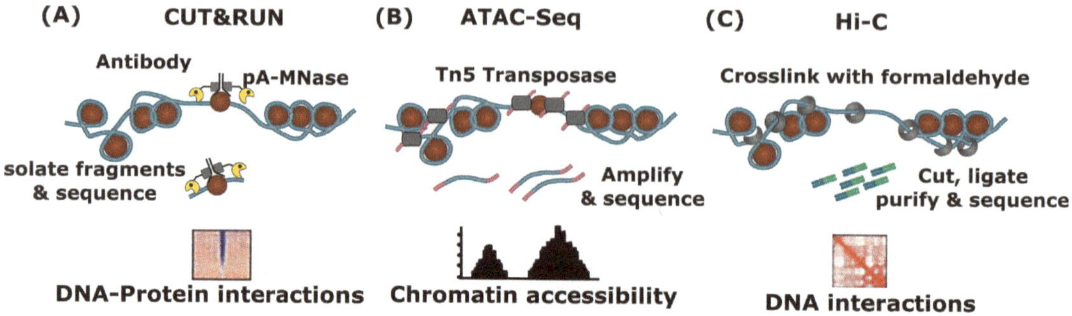

FIG. VI.6 Probing functional organization of chromatin through sequencing. (A) In the CUT&RUN method, chromatin is cut in a site where sequence specific protein factor (e.g., Transcription factor) binds on a DNA using highly specific antibodies against the factor and Protein A-MNase (pA-MNase) fusion protein. Once the protein factor is recognized by the antibodies, pA-MNase binds to the antibodies, as Protein A preferentially binds to the antibodies (Immunoglobulin G). On calcium stimulation, the MNase enzyme cuts the DNA. Sequencing of the cut DNA provides the molecular architecture of the genome involved in the context dependent function. (B) To measure the accessibility of chromatin genome-wide in a given condition (e.g., Knockdown of a chromatin organizing protein), by employing Tn5 transposase, chromatin is cleaved, and sequencing adaptor is integrated at the open site. The adaptor tagged DNA fragment is then sequenced, and the data is used to generate an ATAC-seq peak plot. (C) Hi-C sequencing of chromatin fragments provides the 3D information of the chromatin, as in this method, live folding of chromatin is captured using formaldehyde cross-linking followed by ligation and sequencing. *Source: Agbleke, A.A., et al., 2020. Mol. Cell 79(6), 881–901. Elsevier.*

as "tagmentation") in this method. Unlike the natural transposase, the mutant transposase is hyperactive and cleaves and tags the double-stranded DNA with sequencing adaptors (Fig. VI.6). The purified DNA is then amplified by PCR and sequenced using NGS. The data obtained from ATAC-seq provide information regarding the openly accessible region of the chromatin and the precise positioning of the nucleosome.

The interactions of genomic loci in 3D space in the cell could be analyzed by Chromosome Conformation Capture techniques, termed as 3C. In this method, the number of interactions, which are not close by in linear scale, is quantified. In 3C, interactions between two specific fragments can be quantified. However, in Hi-C, all possible interactions are quantified simultaneously using the deep sequencing method of NGS. In 3C or Hi-C analysis, chromatin loops are cross-linked using formaldehyde, cut by restriction enzymes, ligated by a ligase, and then subjected to sequencing. By using 3C or Hi-C, different kinds of chromatin looping, including enhancer-promoter interactions, could be analyzed.

VI.2 Genome-wide cooperation: GWAS and EpiGWAS

VI.2.1 Genome wide mapping of DNA single strand break (Okazaki-fragment)

Several genome-wide studies of biological phenomena became possible with the invention of fast, highly accurate, and cost-effective DNA sequencing methods. In this section, we shall discuss a few of these studies. DNA strand-breaks occur during many natural and disease processes. It is useful to know where these breaks have occurred in the genomic DNA. DNA single-strand breaks (SSBs) are very common lesions in the genome that often arise spontaneously due to DNA replication. Although the genome-wide picture of DNA double-strand breaks is known in different contexts, the genome-wide map of SSBs was not known until recently. We will illustrate how a precise picture of the genome-wide occurrence of single-strand breaks was obtained using Okazaki fragments that occur during replication, as an example.

A single-strand break generates 3′-OH ends. By using genome-wide ligation of 3′-OH ends followed by sequencing (GLOE-Seq), a comprehensive map of SSBs has been established (Fig. VI.7). Besides the map, these data also provide insight into the maturation of *Okazaki fragments* (Box VI.1). In this method, a biotinylated adaptor is ligated with the 3′-OH end of the DNA strand, followed by fragmentation and capturing of the DNA strand by streptavidin beads. The captured strand is then sequenced by second-strand synthesis to identify the lesion site. These data could accurately map the Okazaki fragments in the genome.

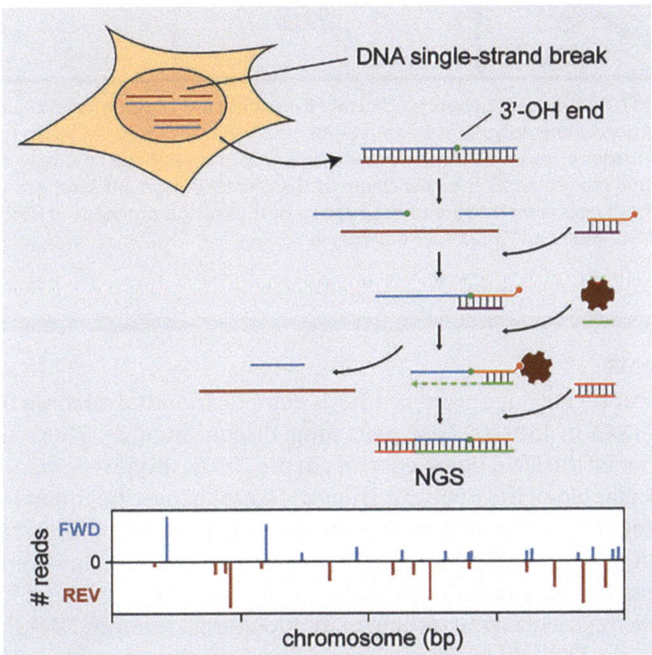

FIG. VI.7 DNA single-strand break mapping. Mapping of single-strand breaks in the genome is important to understand the density of Okazaki fragments (DNA replication), spontaneous DNA break, and other DNA lesions. Capturing of 3′-OH ends followed by sequencing provides this information. *Sriramachandran, A.M. et al., 2020. Mol. Cell 78(5), 975–985.e7. Elsevier.*

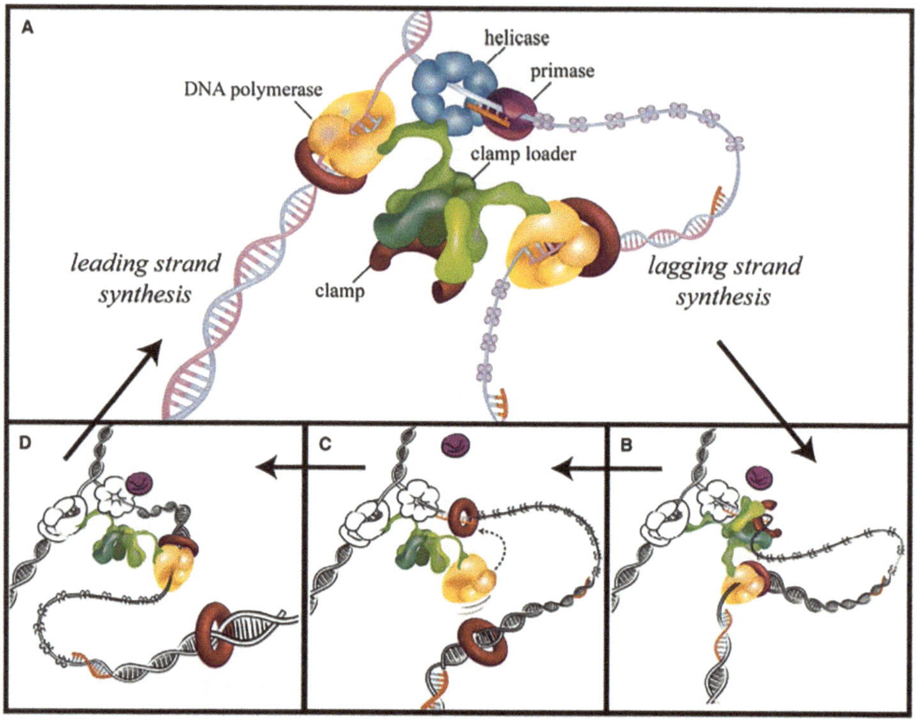

BOX VI.1 Lagging strand DNA synthesis generates Okazaki fragments: (A) Lagging strand synthesis is done in a small fragment where in addition to the polymerase clamp loader, helicase, and primase are required. (B) Primase is displaced by the clamp loader after the RNA primer synthesis, and in the process, the new clamp is loaded. (C) The polymerase is recycled and loaded on to the new clamp as the Okazaki fragment synthesis is completed. (D) New Okazaki fragment synthesis gets completed, and the process of replication proceeds toward completion. *Source: Langston, L.D., et al., 2006. Mol. Cell 23 (2), 155–60. Elsevier.*

Okazaki fragments

During DNA replication lagging strand synthesis complication is overcome by the synthesis of short stretches of DNA of 150–200 base pairs long, discontinuously. These short stretches are then linked together by the DNA ligase enzyme. In the 1960s, this phenomenon was discovered by Japanese molecular biologists Reiji and Tsuneko Okazaki; these fragments are called Okazaki fragments. Since the DNA replication machinery, including the DNA polymerase, can only work in the 5′–3′ direction, at the replication fork, two unwound DNA strands replicate by two different mechanisms. The template strand with 5′–3′ directionality replicates in a discontinuous manner in multiple fragments with the help of multiple small primers. DNA ligase connects the small fragments or the Okazaki fragments into a single continuous strand. Thus the DNA replication process is considered “semi-discontinuous.”

Although the GLOE-Seq technology provides accurate mapping of the SSBs in the genome, the signal-to-noise ratio can be improved further. The spontaneous lesion in

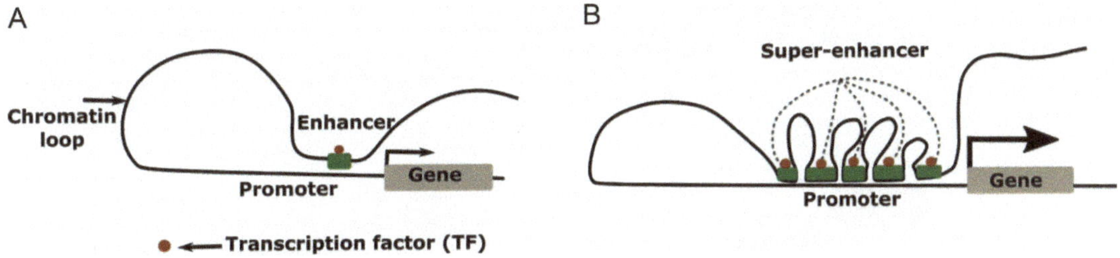

FIG. VI.8 Structure of enhancer and super-enhance. (A) Enhancer where transcription factor binds and to be functional come close to promoter on looping of the chromatin. (B) Closely spaced multiple enhancers termed super-enhancer may cluster together near a promoter to boost the gene expression maximum.

the double-stranded DNA is one of the causes of the noise. However, this method still seems to be highly useful for elucidating the metabolic intermediates of the genome.

VI.2.2 The concept of enhancers and super-enhancers

Enhancers are gene regulatory noncoding cis-element, often situated far away from the gene (promoter). Enhancers could be responsible for activation or repression of the gene, depending on the nature of transcription factors bound to it. The binding of the activator and coactivator to the enhancer leads to transcription activation; on the other hand, if it is occupied by the repressor and corepressor, transcription is repressed. When multiple enhancers cluster together to regulate a single gene, it is called a super-enhancer. A super-enhancer dramatically enhances transcription from its target gene (Fig. VI.8). The predicted number of enhancers in the human genome could be close to one million. However, in a given cell, only a tiny fraction of these enhancers is functional.

Active enhancers are generally located in the open and accessible regions of the chromatin. The histone H3 methylation at lysine 4 (H3K4me1), catalyzed by MLL3/MLL4, and acetylation of H3 at lysine 27 (H3K27AC), catalyzed by p300 are the signature marks of enhancers (Fig. VI.9). Metazoan enhancers function over a long distance, ranging from a few hundred base pairs to several kilobases. Cellular environmental signal induces the transcription factor binding to the enhancer, followed by the recruitment of coactivators or the chromatin-modifying enzymes such as MLL3/MLL4 and p300. With the transcription activator and modified histones, the enhancer gets connected to the promoter by chromatin looping. Mediator complexes and cohesin support and stabilize the interactions of enhancer-bound transcription activator/coactivators and promoter-bound basal transcription machinery.

Genome-wide distribution studies of enhancers using histone modification-specific antibodies revealed the diversity of enhancer and super-enhancers in the genome. Even though most enhancers get activated by the MLL3/MLL4-mediated monomethylation of histone H3, in the absence of MLL3/MLL4 catalytic activity (SET domain mutant), almost 30% enhancers were found to contain H3K4me1 (Fig. VI.10). The enzymes responsible for the methylation of these sites are not known for sure.

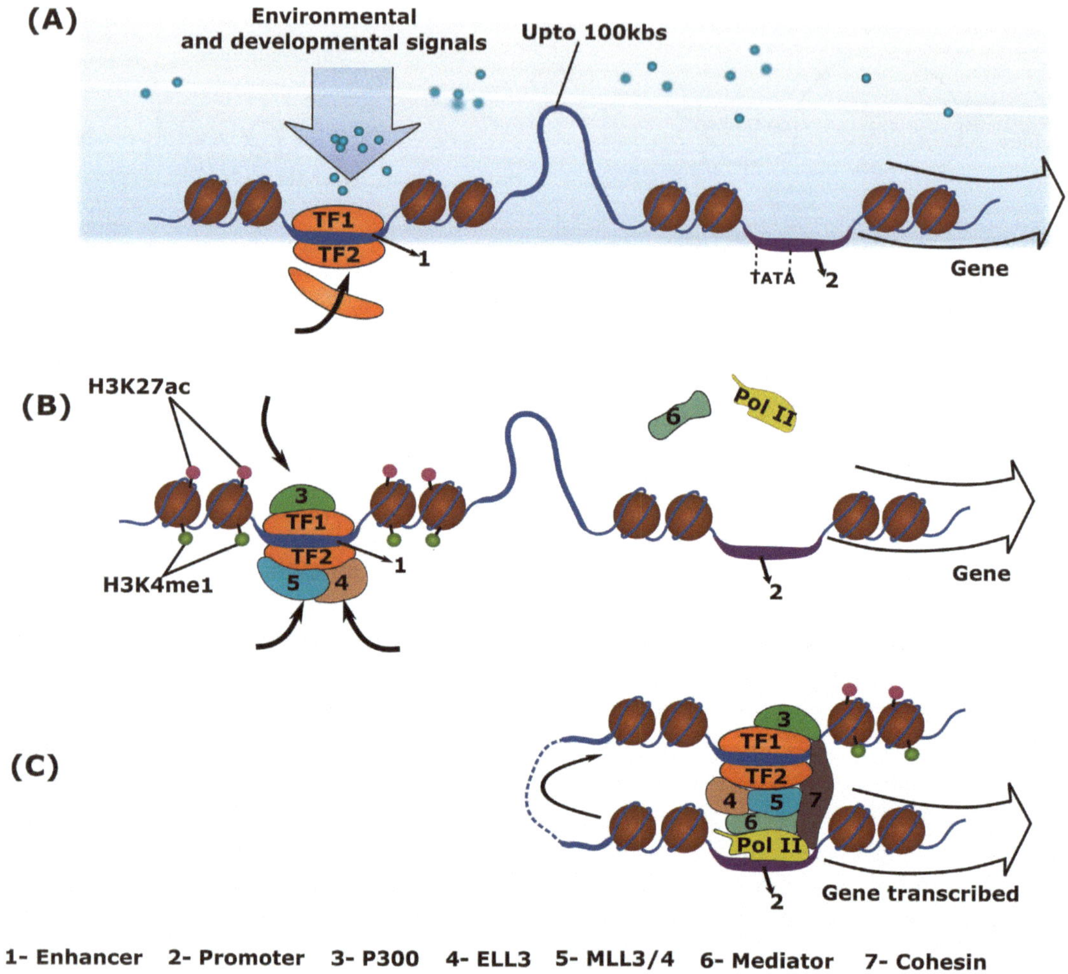

FIG. VI.9 Mechanisms of enhancer function. (A) Environmental signal induces the binding of a transcription factor to the enhancer, (B) which in turn recruits the chromatin modification enzymes: CBP/p300, MLL3/MLL4. The p300 acetylates histone H3 on lysine 27 (H3K27AC) *(pink)*, and MLL3/MLL4 monomethylate histone H3 at lysine 4 (H3K4me1) *(green)* in the enhancer. (C) Enhancer connects the promoter bound RNA polymerase via mediator (complex) on looping out of the chromatin. *Source: Herz, H.M., et al., 2014. Mol. Cell 53 (6), 859–66. Elsevier.*

Alteration of any component of enhancers (*cis* or *trans* elements) of disease-relevant genes may causally be related to the disease manifestation. In genome-wide studies, it was found that single-nucleotide polymorphisms, insertion or deletion at disease-associated enhancers, can modulate the gene expression positively or negatively, affecting the recruitment of transcription factors or cofactors. Disease-specific mutations in enhancers have been discovered for X-linked deafness, Hirschsprung's disease, Crohn's disease, multiple sclerosis, and systemic lupus. Recent evidence suggests that alteration

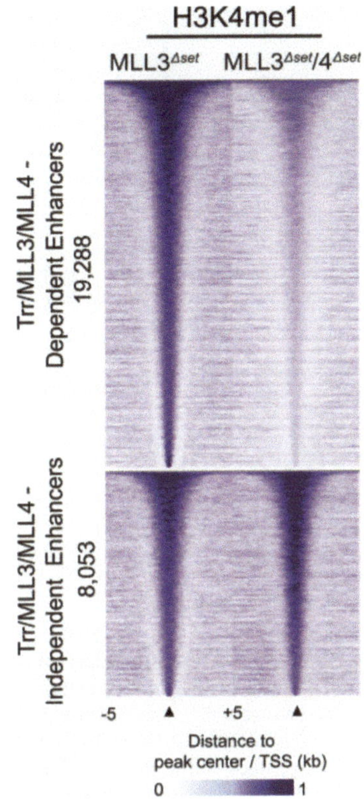

FIG. VI.10 Histone H3K4-monomethylation is critical to establish a functional enhancer. The distribution of H3K4 monomethylation (H3K4me1) at enhancer sites in the human colorectal carcinoma cell line was studied. A significant loss of the H3K4me1 site was observed in the absence of MLL3 and MLL4 (compare column 2 with column 1 in the first panel). There are very few enhancers that are independent of MLL3 and MLL4 (second panel). *Source: Herz, H.M., et al., 2014. Mol. Cell 53 (6), 859–66. Elsevier.*

in the enhancer could be closely linked to the development of neoplastic growth. Enhancer function may directly involve the repression of the tumor suppressor gene and the activation of the oncogene.

Histone H3 monomethylation at K4 (H3K4me1) site is a critical epigenetic signature of the enhancer. The catalytic activity of MLL3 and MLL4 is predominantly responsible for H3K4 monomethylation in the cell. MLL3 and MLL4 genes are mutated in several cancers, including bladder cancer, breast cancer, colon cancer, gastric cancer, liver cancer, medulloblastoma, and non-Hodgkin's lymphoma. Mechanistically, how a mutation in MLL3 and MLL4 may cause cancer could be explained as depicted in Fig. VI.11. The mutation may cause the loss of function of MLL3/MLL4 due to its reduced affinity for the transcription factor or destabilization of the enhancer complex in the presence of mutant proteins, or loss of catalytic activity. However, loss of MLL3/MLL4 function may diminish the function of an enhancer of a tumor suppressor gene and thereby repressing the gene. On the other

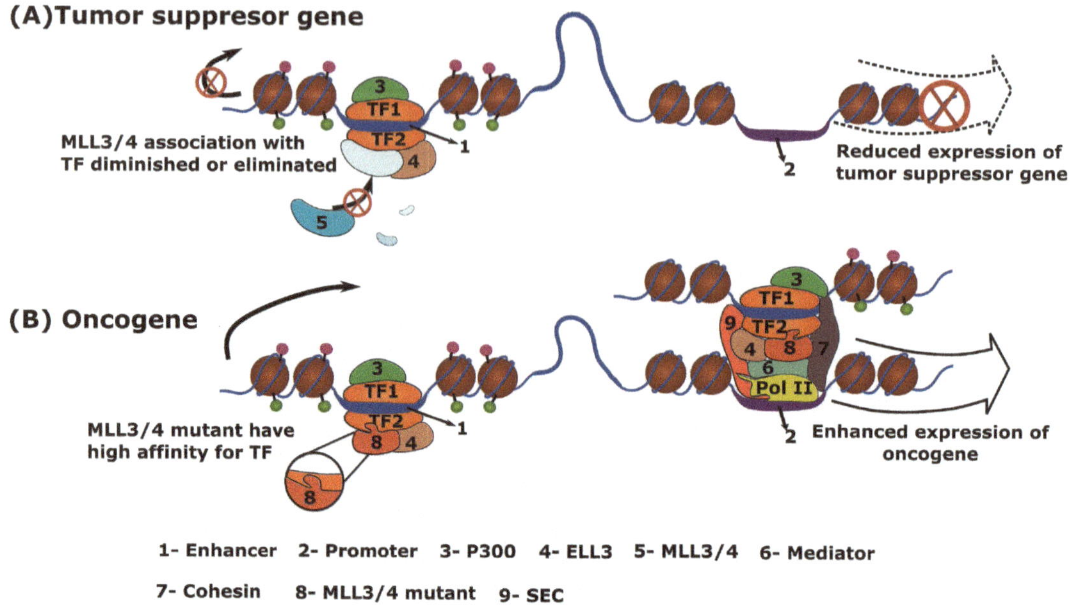

FIG. VI.11 Functional enhancers are critical for the tumor suppressor and oncogene function. MLL3/MLL4 methyltransferases are responsible for the H3K4me1, a signature mark of an enhancer. MLL3/MLL4 mutations may cause enhancer-mediated repression of tumor suppressor genes or activation of oncogenes. (A) When due to mutation, MLL3/MLL4 function is lost due to the lack of functional enhancer, the tumor suppressor gene gets repressed. (B) On the other hand, if the mutation is gain-of-function in nature, the mutations may lead to either increased stability or increased affinity of MLL3/MLL4 to the transcription factor at the enhancer. These mutations thus will cause a better enhancer function leading to overexpression of the oncogene. *Source: Herz, H.M., et al., 2014. Mol. Cell 53 (6), 859–66. Elsevier.*

hand, MLL3/MLL4 gain-of-function mutation may cause the overexpression of an oncogene. The gain-of-function may be achieved through the increased affinity toward a transcription factor or increased stability or hyperactivity due to the mutation.

The complex but seemingly unique orchestration of enhancer functions is being elucidated using the technologies described earlier. The inactive enhancer becomes active once it establishes long-range interactions with the target genes via specific transcription factors and architectural proteins, for example, linage-specific transcription factors and the chromatin looping protein CTCF/cohesion (Fig. VI.12). Emerging evidence shows that enhancers are not only involved in transcription initiation but also in elongation. During transcription elongation, RNA polymerase II may pause for various reasons. Some enhancers may help to release the paused RNA polymerase II through the activation of the elongation factor, the P-TEFb complex.

A given active enhancer could be part of a *transcription factory*, in which it may share the transcription factors and other essential biochemical ingredients essential for RNA synthesis (Fig. VI.12).

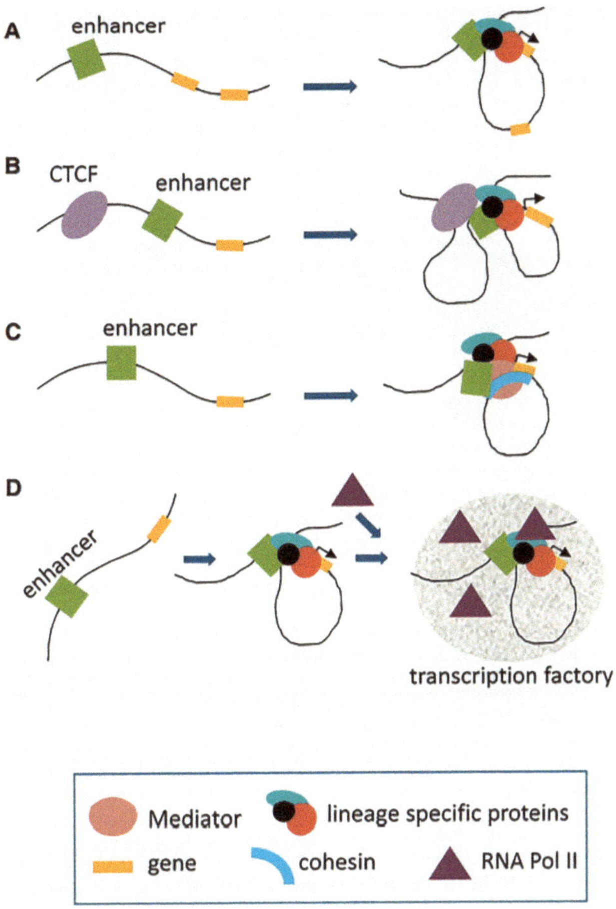

FIG. VI.12 Enhancer assembly creates transcription factories: (A) Enhancer assembly on the promoter with the lineage-specific factors to set the process of differentiation. (B) CTCF could be part of the enhancer complex. (C) Enhancer complex on a single gene promoter. (C) In the presence of RNA Polymerase II enhancer, mediator, and linage specific factors establish the transcription factory and initiate the transcription. *Source: Plank, J.L., et al., 2014. Mol. Cell 55 (1), 5–14. Elsevier.*

Transcription factories

In a eukaryotic cell, transcription occurs at specialized, discrete sites, called transcription factories. In a transcription factory, transcription of 3–4 genes may happen simultaneously. However, each factory may have 4–30 RNA polymerase molecules and a large number of other molecules required for transcription activation and mRNA processing. Since multiple chromatin loops may be present in a factory at a time, the transcription factory also helps in chromatin organization and thereby co-regulates multiple genes (Box VI.2).

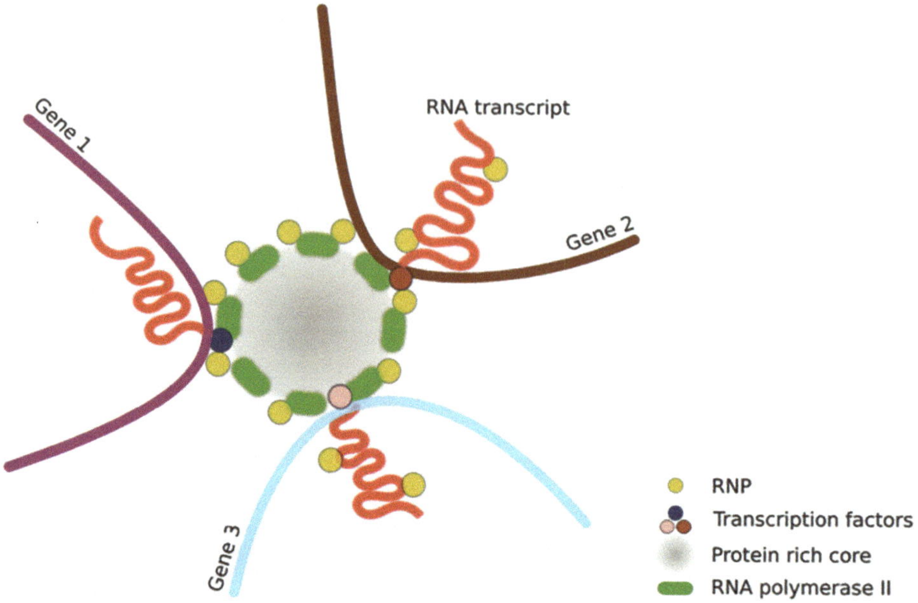

BOX VI.2 Hypothetical structure of a transcription factory. *Source: Rieder, D., Trajanoski, Z., McNally, J.G., 2012. Transcription factories. Front. Genet. 3, 221. doi:10.3389/fgene.2012.00221.*

VI.2.3 Genome-wide distribution and intra-genomic interactions of enhancers

Genome-wide distribution of enhancers and super-enhancers was studied based on ChIP-seq data from the ENCODE project, and available whole-genome sequencing, and transcriptomic data (Fig. VI.13). Several critical observations were made based on this analysis:

1. Enhancers are more abundant, distal from the transcription start sites, and less conserved than super-enhancers.
2. 85% of super-enhancers overlap with only 13% of isolated enhancers.
3. Super-enhancers are transcriptionally more active than isolated enhancers.

It was found that, indeed, super-enhancers are highly enriched with active chromatin marks, like H3K27AC and H3K4me3. Overall, it was observed that super-enhancers are more active and situated in the open chromatin region than the enhancers. A large number of enhancers were found to be present in a poised state. However, enhancer/super-enhancer activity depends upon several protein factors and the sequence of DNA that harbor them. One such important DNA binding factor is MeCP2, which binds to the methyl-CG sequence and contributes to the compaction of the genome, leading to repression of transcription. MeCP2 binding to the DNA directly inhibits the histone modifications involved in enhancer activation (Fig. VI.14). In a MeCP2 knockdown cell, active enhancer marks are highly upregulated, and overexpression of MeCP2 leads to the reverse effect.

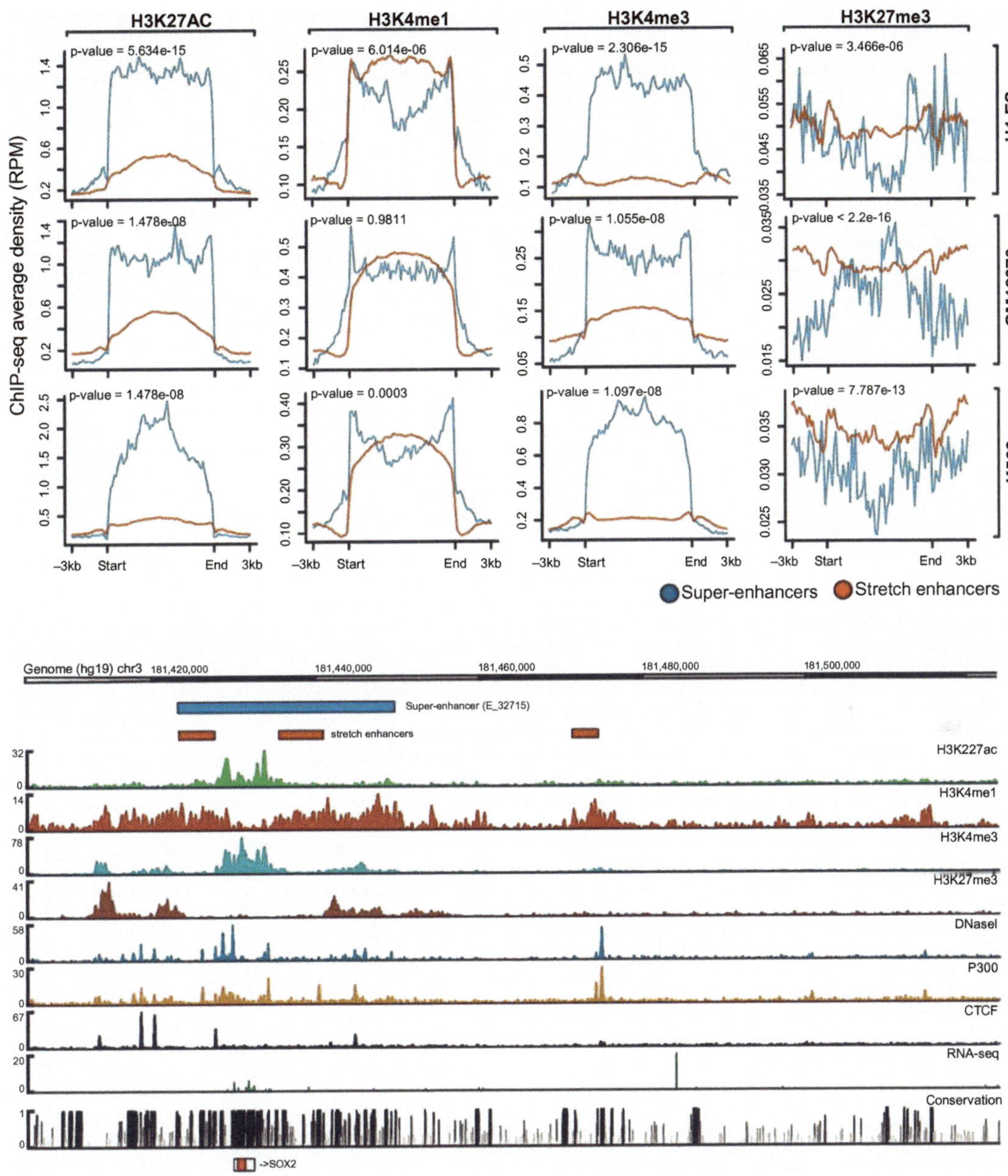

FIG. VI.13 Genome-wide distribution of super-enhancers and enhancers (stretch enhancer): (A) ChIP-seq analysis of H1-ESC, GM12878, and K562 cells using H3K27AC, H3K4me, H3K4me3, and H3K27me3 antibodies at super- and enhancer (stretch enhancers). (B) In-ES cells occupancy of H3K27AC, H3K4me1, H3K4me3, P300, and CTCF; state of chromatin (open or compact, assayed by DNase I); RNA-seq; and conservation at SOX2 gene locus depicted. *Source: Khan, A., et al., 2018. Epigenetics 13 (9), 910–922. Taylor & Francis.*

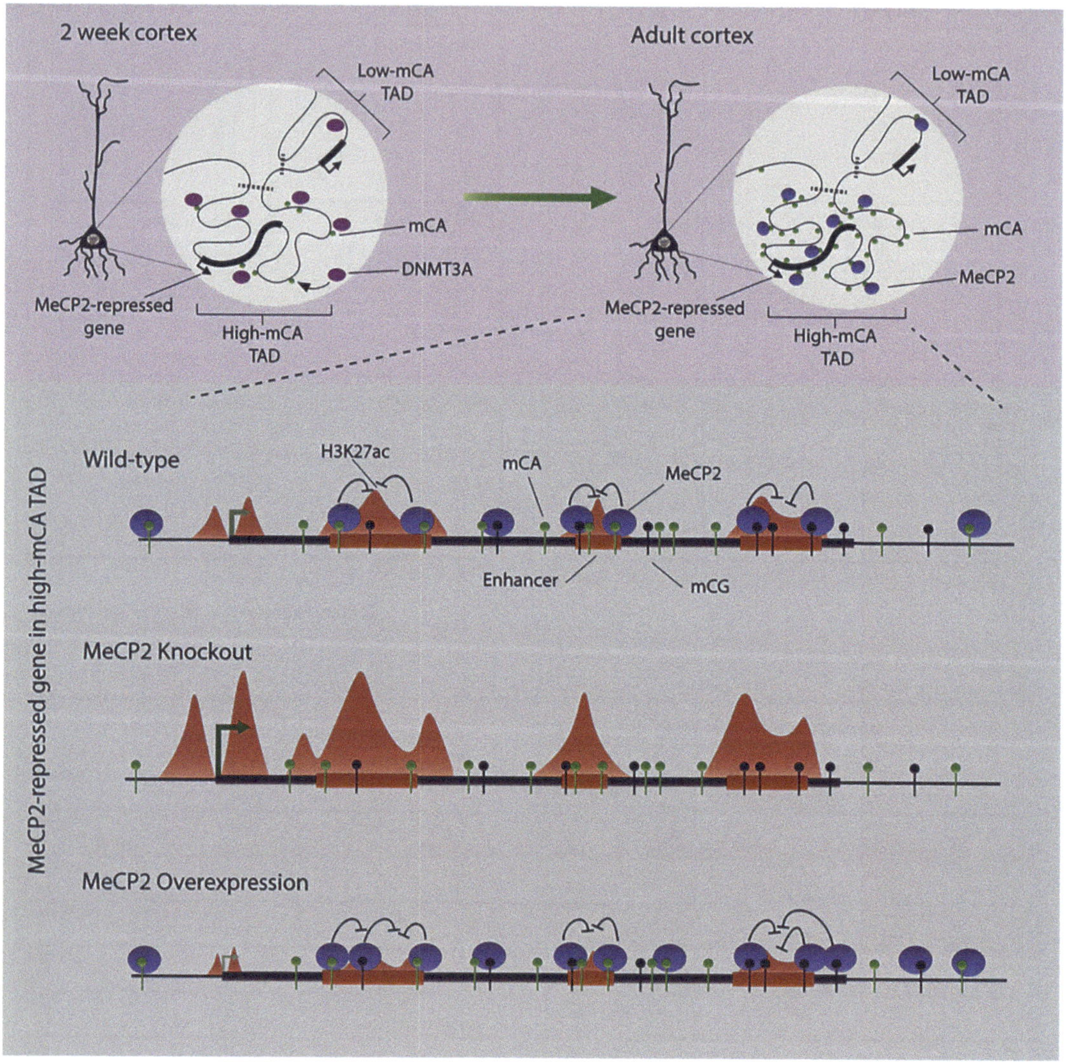

FIG. VI.14 Methyl CpG binding protein MeCP2 distribution determines the TAD configuration. *Source: Clemens, A.W., et al., 2020. Mol. Cell 77 (2), 279–293.e8.*

VI.2.4 GWAS: Regulatory elements, histone modifications, and RNA polymerase II binding sites

To understand the global gene expression activity in a cell, it is essential to know the genome-wide distribution of transcription machinery in the context of chromatin modifications. By using chromatin immunoprecipitation (ChIP) coupled to DNA microarray analysis (ChIP-chip) in human embryonic stem cells, it was observed that RNA polymerase II, H3K4me3, and H3K9, 14Ac modifications, occupy the promoters of most protein-coding genes. Interestingly, it was observed that only a small fraction of these genes, which are also marked by H3K36me3 (a mark for transcription elongation), could produce

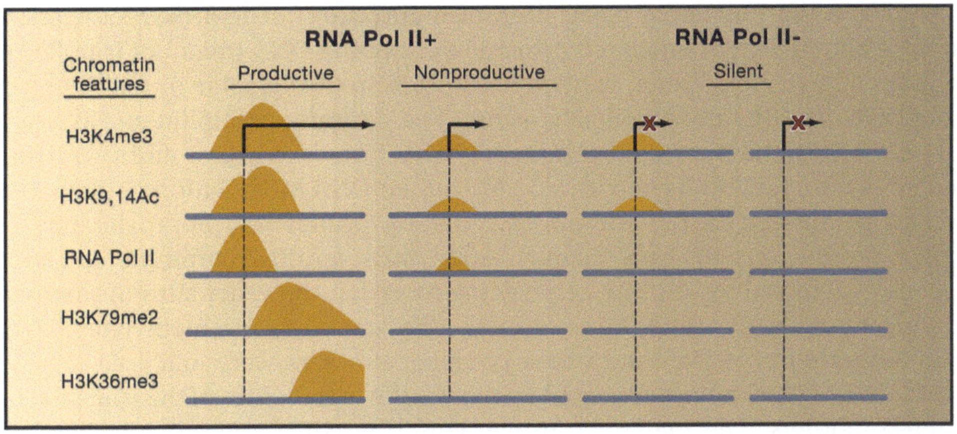

FIG. VI.15 Genome-wide distribution of RNA Polymerase II. Transcriptionally productive RNA polymerase II association with the promoter (at the transcription start site) is found where histone H3K4me3 and H3K9,14Ac modifications around the start site and H3K36me3 and H3K79me2 in the downstream of the promoter are predominant. Nonproductive RNA polymerase II binding is not rare but devoid of H3K36me3 and H3K79me2 in the downstream of the gene. In some classes of these genes, RNA polymerase II may be recruited sporadically. In these gene promoters, H3K4me3 and H3K9,14Ac modifications could be detected around the transcription start site. There could be some genes where chromatin organization is completely refractory to the recruitment of RNA polymerase II. *Source: Lorincz, M.C., et al., 2007. Cell 130 (1), 16–18. Elsevier.*

full-length transcripts. In these transcriptionally active genes, H3K79me2 could also be detected downstream of the promoter of these genes.

The rest of the genes, devoid of H3K36me3, notwithstanding the presence of a suboptimal concentration of RNA polymerase II, could not produce full-length transcripts (Fig. VI.15). In a subgroup of genes, which harbor H3K4me3 and H3K9,14ac without RNA polymerase II were transcriptionally silent. In another subgroup of genes, neither transcriptionally active histone modification nor the RNA polymerase II could be detected. These are probably placed in a compact heterochromatin region of the genome.

VI.2.5 Methylome: The fundamental basis of epi-genome function

Methylation of DNA at the 5-position of cytosine is one of the most important epigenetic modifications involved in epigenetic memory. DNA methylation causes transcriptional repression by different mechanisms. The degree and location of methylation depend on the metabolic state and stress on the organisms. Genome-wide picture of methylation of DNA reveals differential expression of metabolically regulated genes and their biological consequences. To map the genome-wide DNA methylation pattern, the bisulfite sequencing approach is generally adopted. In this method, the treatment of DNA with bisulfite converts the cytosine residue to uracil, but the methylated cytosine does not react with the reagent and remains unchanged. When this bisulfite-modified DNA is subjected to whole-genome sequencing (NGS), the data reveal the genome-wide methylation patterns or the methylome. Hydroxymethylome can also be elucidated by this method.

To elucidate the functional correlation of methylation or hydroxymethylation, the methylated and hydroxymethylated cytosine-containing DNA fragment is pulled down using either mouse monoclonal antibody against 5-methylcytosine or rabbit polyclonal antibody against 5-hydroxymethylcytosine. The immunoprecipitated DNA is then amplified by PCR and subjected to labeling with fluorescent Cy5 and Cy3 dyes. The labeled DNA is then hybridized with a 2.1 M Human DNA Methylation v2 genome tiling array. This commercial array contains 2.1 million probes and approximately 26,210 promoter regions, 27,867 CpG islands, and 730 microRNA promoters. The same DNA is also subjected to bisulfite sequencing to assign the exact sites of methylation.

In a recent study, by using these methods, the genome-wide methylation/hydroxymethylation patterns of lean and obese pregnant mothers were analyzed (Fig. VI.16). The HPLC-ESI-MS technology was used to analyze the metabolites of the placental tissues from these mothers. The metabolic profile could be correlated with the methylome. These data revealed that obesity does alter the methylation and hydroxymethylation pattern. This high-resolution map of the epigenome associated with obesity suggests that environmental factors could be critical to regulate the placental epigenome.

VI.3 The dynamic 3D architecture of the genome

Chromatin organization dramatically changes during several processes, for example, differentiation. As chromatin structure significantly influences gene expression, these dynamic alterations of chromatin organization are essential for faithful differentiation. Embryonic stem (ES) cells can be characterized by the distinct higher-ordered chromatin structure (Fig. VI.17). The pluripotent ES cells are rich in less compact euchromatin, and as the differentiation progresses, chromatin becomes more condensed, and transcriptionally inactive heterochromatin accumulates. In the terminally differentiated state, the heterochromatin regions further increase, and in this state, very few genes remain active. In accordance with this, dynamic modifications of histones also change, favoring the heterochromatinization. Transmission electron-microscopy studies indeed reveal that in retinoic acid-induced differentiating cells, heterochromatin foci are predominant compared with undifferentiated ES cells.

VI.3.1 The fundamental 3D architecture of the genome

In resting human cells, each chromosome is present in a folded state, organized in three-dimensions called chromatin territories (CT). The CT organization has been studied by DNA fluorescence in situ hybridization (FISH), 3C technologies, and electron microscopy (Fig. VI.18). In each CT, chromatin is organized in smaller subdomains known as topologically associating domains (TADs). In a given CT, TADs could be both repressed and active. The active or repressed state of the TAD is determined by the protein factor associated with it. Transcription factors and chromatin looping protein CTCF play crucial roles in determining chromatin folding in the TADs.

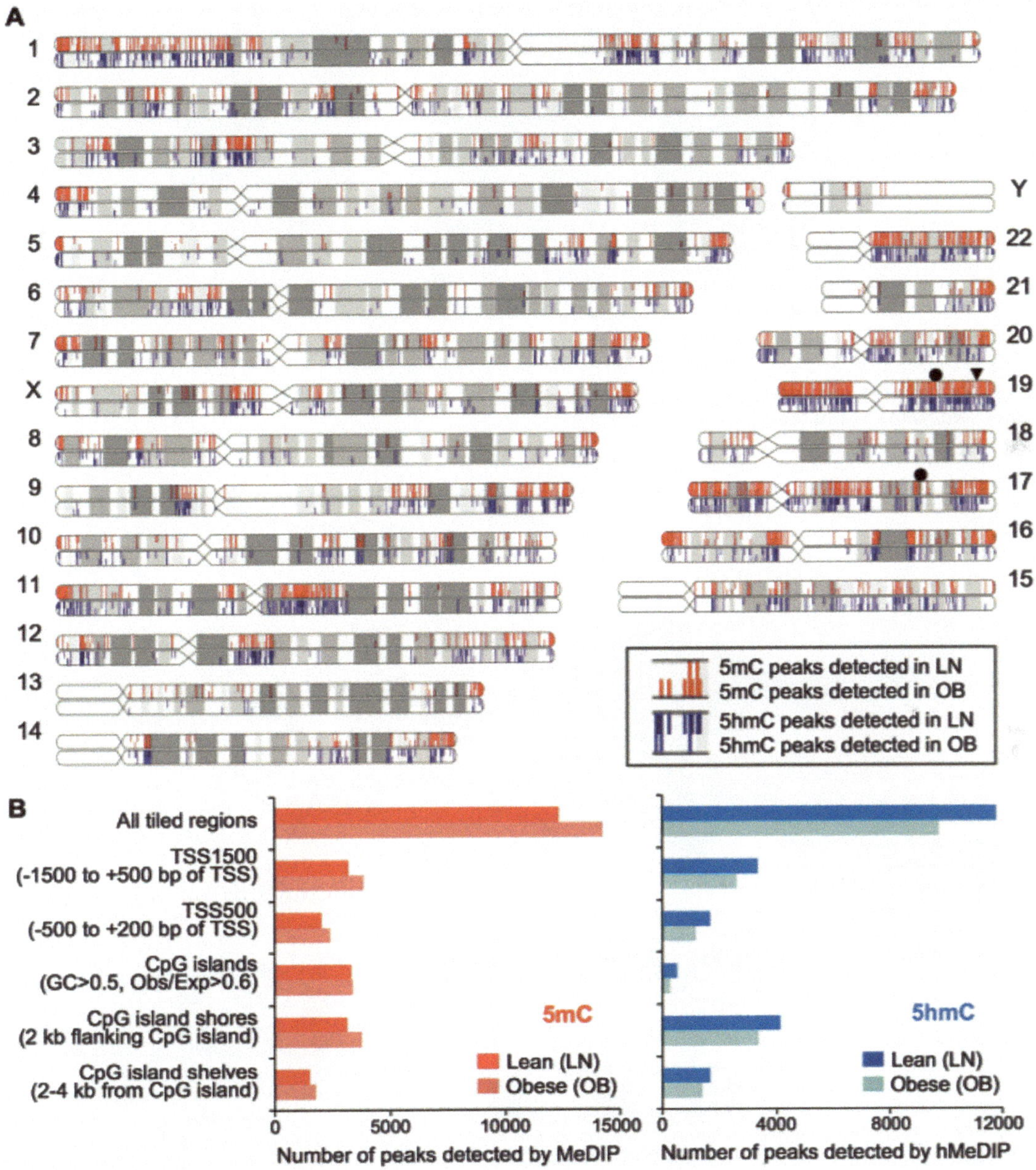

FIG. VI.16 Context-dependent methylome. Genome-wide studies of DNA methylation and hydroxymethylation in placentas of obese and lean mothers showed distinct differences. (A) Chromosome ideograms depict the methylated and hydroxymethylated peak regions in obese (OB) and lean (LN) mothers. *Filled circles* on chromosomes 17q24 and 19q13 indicate the GH-CSH and PSG gene cluster, respectively. On chromosome 19q13, a *filled triangle* indicates the pregnancy-associated miRNA cluster. (B) Methylated peak *(right panel)* and hydroxymethylated *(left panel)* at various parts of the genome are shown. *Source: Mitsuya, K., et al., 2017. PLoS One 12 (10), e0186115.*

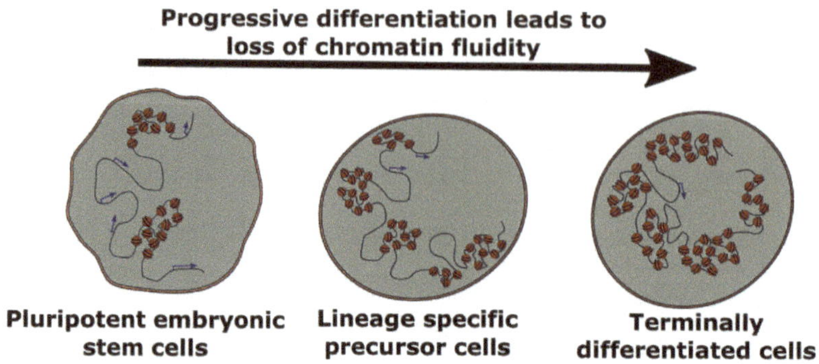

FIG. VI.17 Dynamic genome organization during differentiation. In pluripotent embryonic stem cells, the genome exists in a more open and active form *(indicated by the blue arrow)*. As the cell proceeds toward lineage specificity, DNA becomes more condensed with fewer active genes. In the differentiated cells *(right)*, the genome is highly condensed, with most of the genes in a silent state.

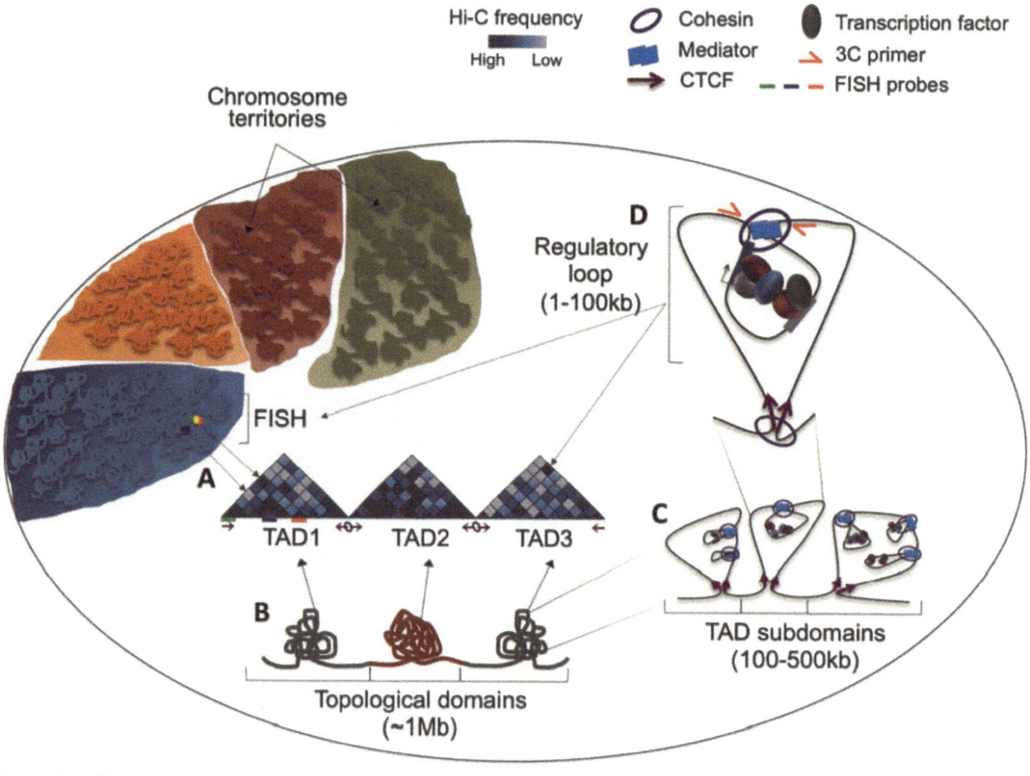

FIG. VI.18 Different layers of chromatin folding lead to the formation of chromatin territories: (A) Gene loci could be detected in the TAD by FISH (here it shows in three loci). Intensities of interactions of loci within the TAD are indicated by *blue color*. *Dark blue* indicates high-frequency interactions. (B) In a given chromatin territory, TADs could be present in the repressed or active state. (C) TADs may be organized in a subdomain. (D) Organization of functional genes in a TAD subdomain. Different protein factors: cohesin, mediator, and CTCF stabilize the subdomain. *Source: Lopes Novo, C., et al., 2016. Brief Funct Genomics 15 (4), 305–314. Oxford University Press.*

Chromosome conformation capture

The three-dimensional organization of chromatin in the cell is analyzed by chromosome conformation capture or 3C. In this method, the number of interactions between genomic loci, separated by long-distance in the linear scale, is determined. Based on the number and complexity of interactions to be addressed, 3C technology could be of different types (as depicted in (Box VI.3). By using 3C technologies, the structural domains of genome organization can be identified.

Chromosome conformation capture technology

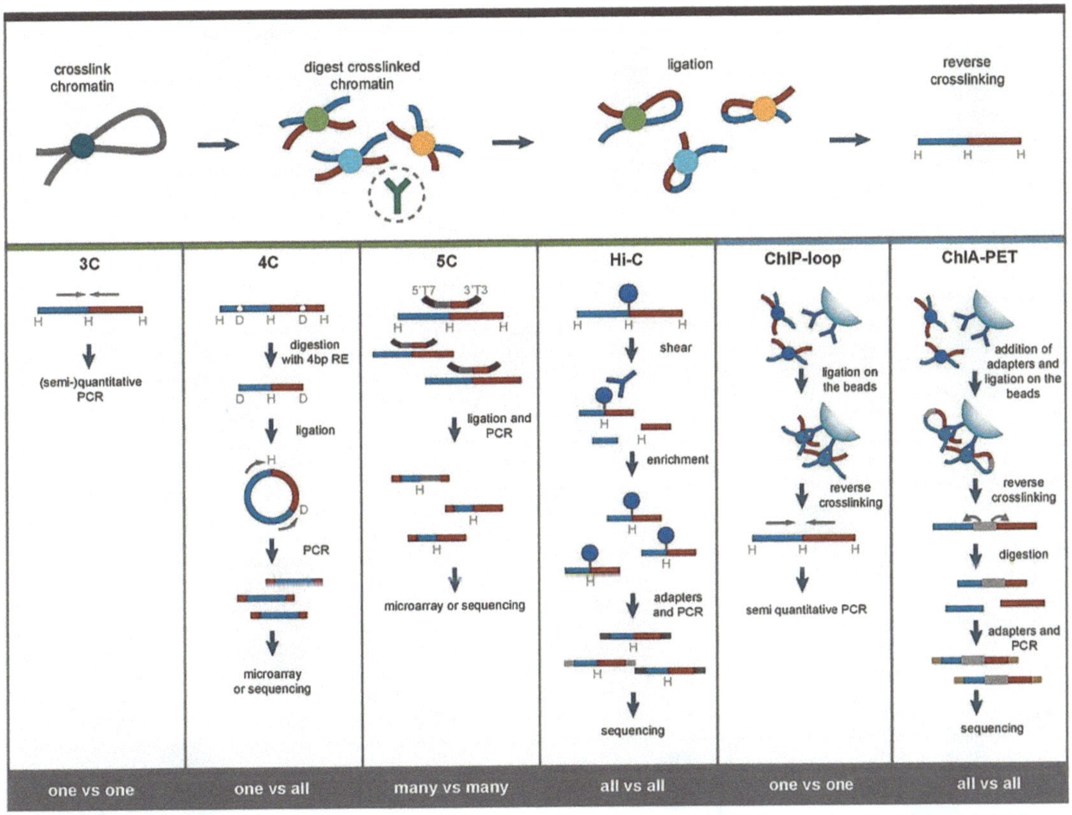

BOX VI.3 In a cell, genome interactions are captured by cross-linking segments close in space with formaldehyde. On cross-linking, the genome is cut with restriction endonuclease (RE). The DNA fragments are then ligated as shown by T4 DNA ligase, amplified by PCR, and identified by microarray or sequencing. *Source: Wikipedia.*

Transcription factors (TFs) sequence-specifically bind to the respective cognate sites on the DNA (including the nucleosomal DNA) and recruits the transcription machinery to activate or repress the transcription. Some TF-binding sites are present near the promoter, but in mammalian cells, most of the TF-binding sites are located on the distant enhancers. The TF-mediated interactions of an enhancer and the promoter via the mediators cause additional chromatin folding. Besides, promoter-enhancer interactions TFs can produce loops by various other mechanisms (Fig. VI.19).

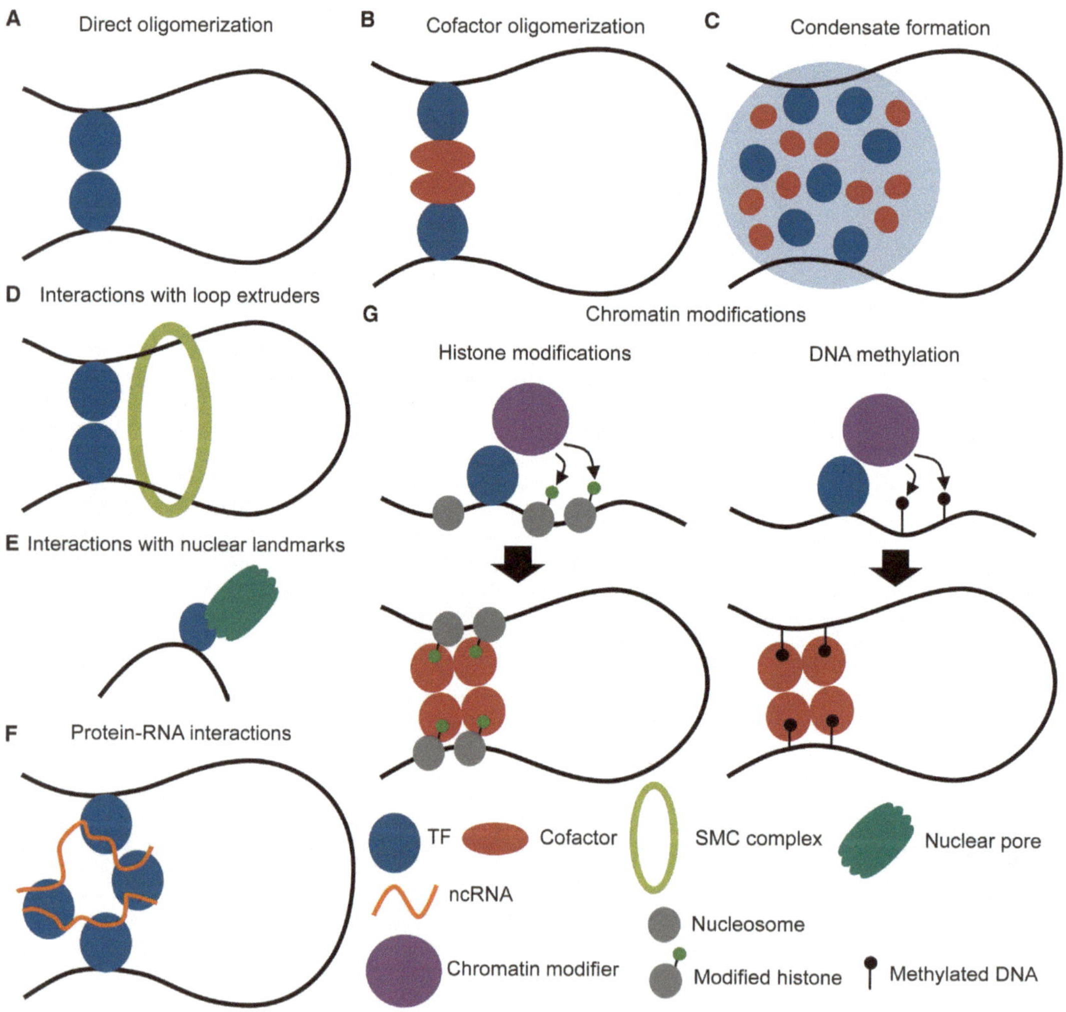

FIG. VI.19 The fundamental role of transcription factors (TFs) in genome folding. (A) Oligomerized TF directly, (B) complex of cofactor and TF together, (C) condensed complex of TF and cofactor, (D) TF and SMC complex, (E) complex of TF and nuclear membrane component such as nuclear pores, (F) TF-RNA complex, (G) TF in conjunction with the modified chromatin (including histone modifications and DNA methylation) fold the chromatin. *Source: Kim, S., et al., 2019. Mol. Cell 76 (2), 306–319. Elsevier.*

VI.3.2 Genome dynamics and development

In the process of development, cellular fate is determined by defined networks of transcription factors. The dynamic chromatin structure acts as a cell type-specific filter in development. Chromatin-bound transcription factors and the epigenetic modifications determine which gene will be transcribed in a particular development stage (Fig. VI.20). Each topologically associating domain (TADs), responsible for a particular development phase, is organized with required histone modifications and DNA methylation to establish an appropriate gene regulatory network to ensure precise developmental steps.

Favorable topological alterations for regulating gene expression may be achieved by reorganizing the enhancer usage (Fig. VI.21). In ground state ES cells, there could be three types of enhancers present: active, inactive, and poised or seed enhancers. These seed enhancers have low levels of H3K27AC and/or H3K4me1. On priming by the required signals, the seed enhancers become active. On movement toward the lineage commitment, the seed enhancers become active. They may cluster and induce to form super-enhancers to boost the transcription of a set of genes required for the cellular identity. Beyond lineage commitment, the clustering of enhancers is also critical for the expression of genes of cellular identity in B cells, myotubes, T-helper lymphocytes, and macrophages.

The availability of whole-genome sequences of different organisms has made it possible to understand the enhancer function both in the context of development and evolution. Emerging information suggests that enhancers are competent targets for

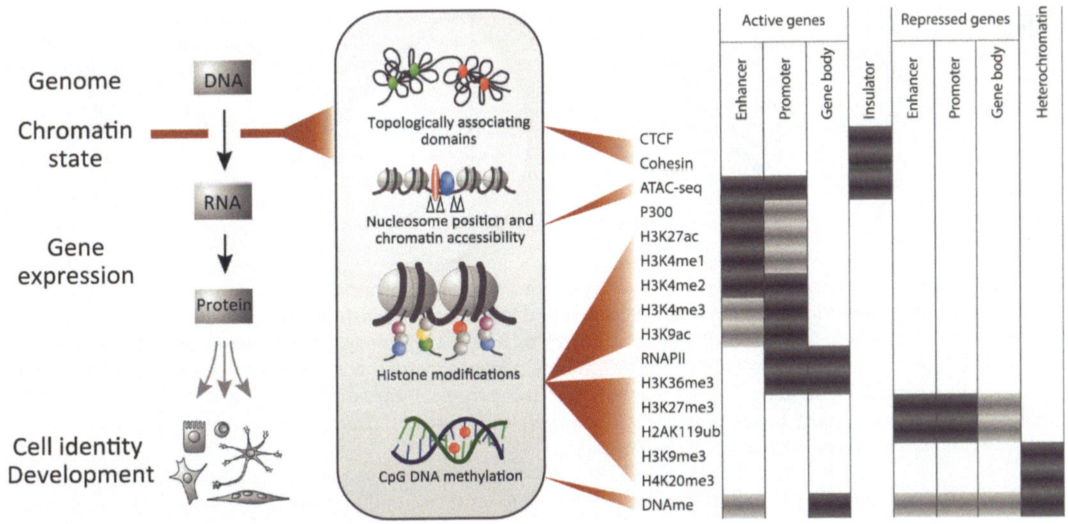

FIG. VI.20 Epigenetic modifications in development. The genomic state of the chromatin plays a vital role in gene expression and, consequently, the cellular identity *(left panel)*. The three-dimensional nucleosomal organization of the genome provides the dynamic nature of the genome through different chromatin modifications *(middle panel)*. *Gray shades* indicate the state of DNA methylation in inactive genes, repressed genes, and heterochromatin. *Lighter gray* indicates that the sequence could be either methylated or unmethylated, depending on the signal. Active promoters generally remained unmethylated. *Source: Perino, M., et al., 2016. Dev. Cell 38 (6), 610–20. Elsevier.*

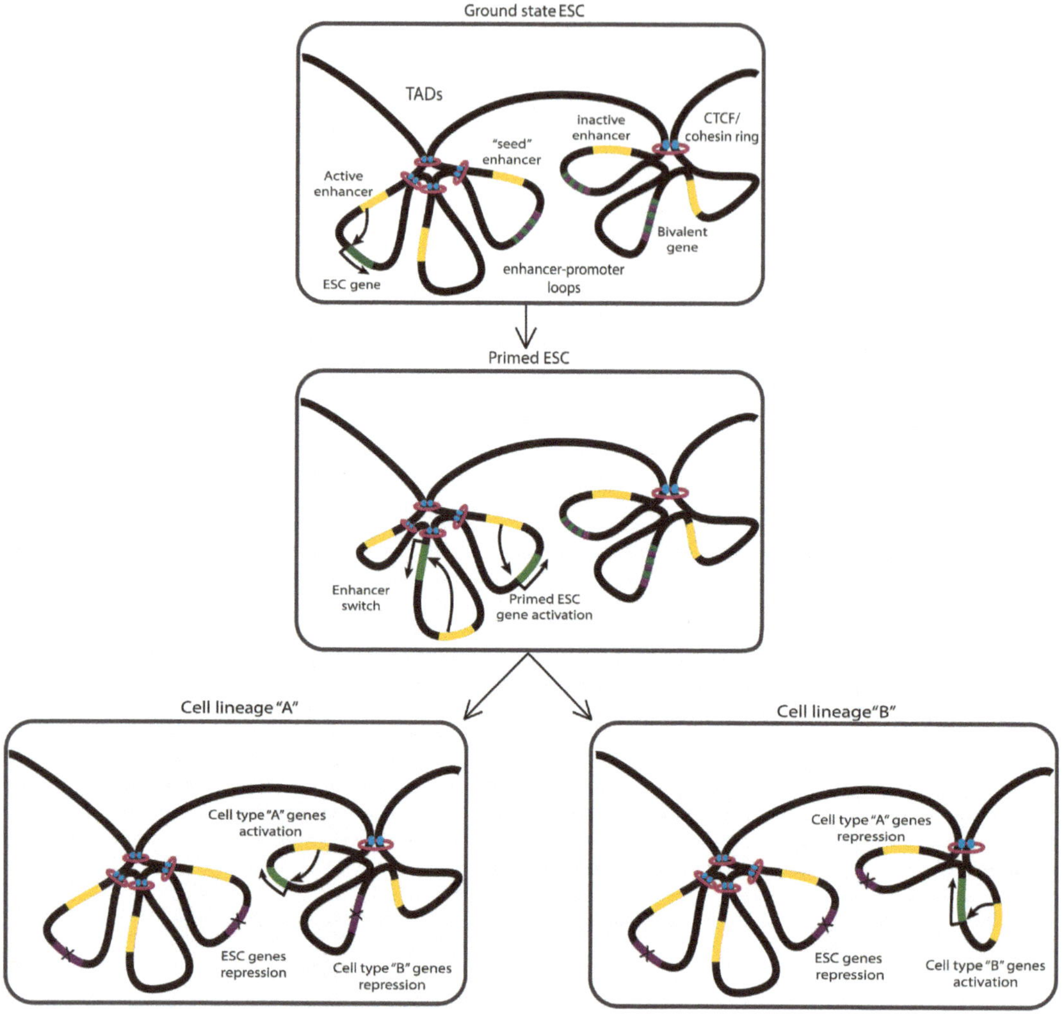

FIG. VI.21 Enhancer organization leads to alteration of chromosome conformation during embryonic stem cell differentiation. As the embryonic stem cells (ESCs) proceed in the path of differentiation from "Ground State" to "Primed State" to different lineages (differentiated state), differential gene expression is achieved by enhancer usage by means of reorganization: Super-enhancers *(yellow)* are placed accordingly. Looping of chromatin, stabilized by CTCF *(blue dots)* and Cohesin *(purple ring)*, is essential in this process. Depending upon the state of differentiation, genes could de active *(green)*, bivalent *(green* and *purple)*, and repressed *(purple gene* with a *black cross)*. *Source: Perino, M., et al., 2016. Dev. Cell 38 (6), 610–20. Elsevier.*

evolutionary changes (Fig. VI.22). The birth of a new enhancer or loss of an enhancer seems to be a common phenomenon in the course of evolution.

Mechanistically, how is this possible? The answer could be in genome duplication. Genomic duplication could be the driving force behind the evolutionary changes. It could create a new gene and a new set of gene regulatory networks and elements, such as the

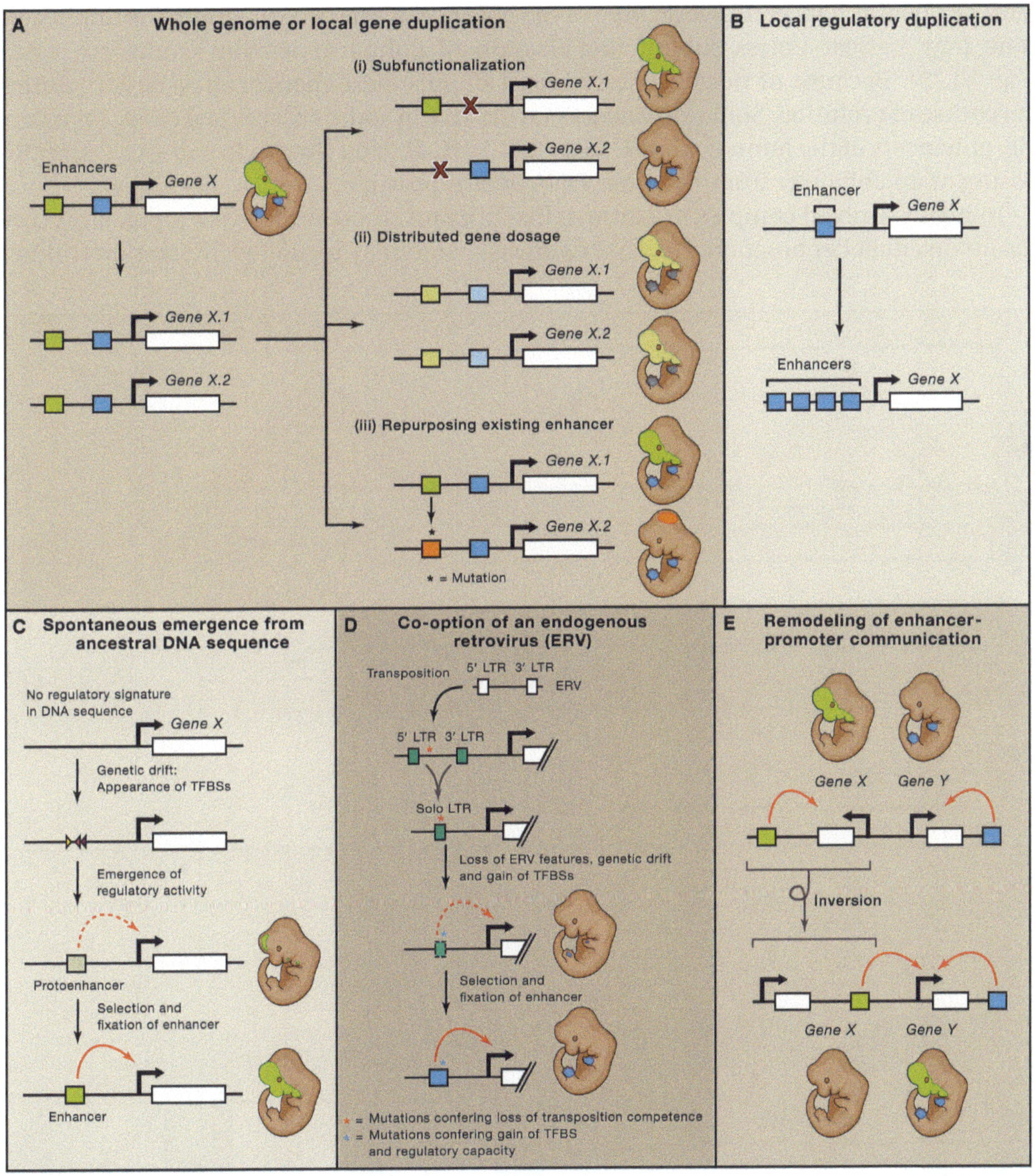

FIG. VI.22 Intragenomic interactions during development: (A) The structure and function of enhancer changes during evolution and development. (B) Gene duplication may increase the number of enhancers. (C) The new enhancer may emerge from ancestral DNA. (D) Active transposable elements or recombination could cause enhancer heterogeneity during the course of development. (E) Gene inversion may lead to a change in the target gene for an enhancer. *Source: Long, H.K., et al., 2016. Cell 167 (5), 1170–1187. Elsevier.*

birth of a new enhancer. However, due to the absence of selective pressure, the duplicated gene may be lost unless subfunctionalization of enhancer activity comes to rescue (Fig. VI.23). Because of negative selection, a duplicated enhancer may be lost during the course of evolution. Some enhancers may survive in some exceptional cases, for example, enhancers of the human apoE/C-I/C-I0/C-IV/C-II gene cluster. In some cases, repurposing of an enhancer from the older DNA is also possible.

In highly evolved complex organisms, like humans, there is always a topological constraint on enhancer function. However, with a substantially regulated interaction within a

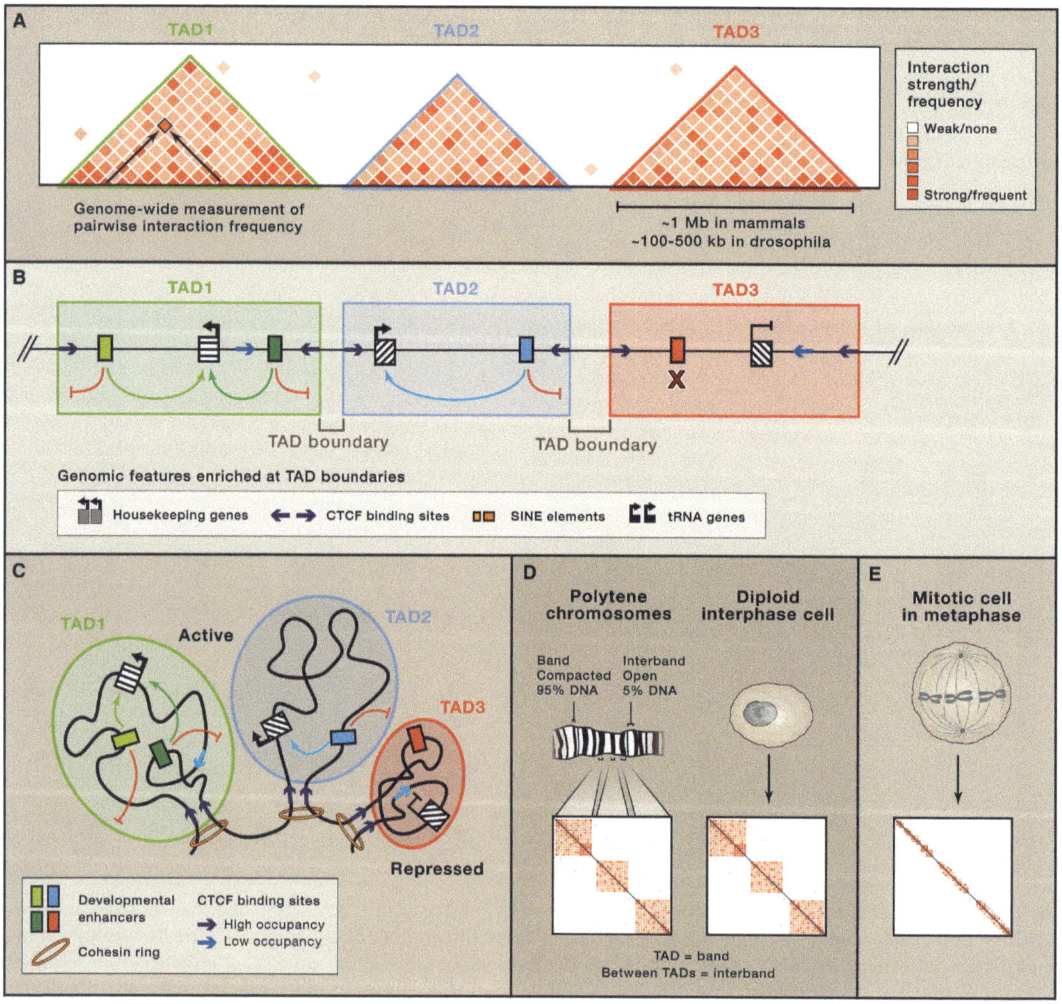

FIG. VI.23 Hi-C analysis to visualize the 3D interactions in the genome. (A) Interactions of TADs in chromosomes are represented in different intensities based on the Hi-C data. (B) Each TAD possesses a boundary to limit its functional interactions and spreading of chromatin modification. (C) Possible folding of chromatin domain keeping TAD territorial heterogeneity. (D) TADs in the polytene chromosome. (E) In mitosis, chromatin gets condensed and forms individual chromosome, thereby the TAD organization is not found. *Source: Long, H.K., et al., 2016. Cell 167 (5), 1170–1187. Elsevier.*

given TAD and restricting the inter-TAD interactions by TAD boundary, enhancer function is controlled. Using Hi-C and 5C, the three-dimensional proximity of DNA sequences in a TAD has been analyzed. The data generated from these analyses revealed that (i) activity of the cis-regulatory element within the same TAD is controlled; (ii) spreading of heterochromatic marks is restricted by TAD boundary. Interestingly, it was found that TAD boundaries are evolutionarily conserved, suggesting that TADs are indeed functional units of the organized genome within which cell-type-specific genomic and epigenomic interactions occur. During mitosis, TAD structures are dissolved to accommodate the highly compact metaphase chromosome conformation. After metaphase, the TAD structures come back and are maintained faithfully by unknown mechanisms.

Chapter summary

In this chapter, we have discussed how the dramatic evolution of DNA sequencing technology has made it possible to gather the whole genome sequence data rapidly and inexpensively. Efficient sequencing facilities could enhance our ability to understand genome biology to a great extent. In brief:

- The automation of the di-deoxy sequencing method discovered by Frederick Sanger helped to decipher the major part of the human genome. A completely new technology, "sequencing by synthesis" (SBS), ushered in the "next-generation" sequencing (NGS) technology currently in large-scale use. The other popular NGS methods are "Oligo Ligation and Detection (SOLiD)" and "Nanopore sequencing" technology.
- Based on NGS, several chromatin technologies have been developed to understand the functional genome organization. Chromatin technologies help to elucidate the signal-dependent folding and unfolding state of the genome in the context of gene expression.
- Modern Genome-wide studies are made possible by advances in rapid and low-cost sequencing technologies. The mapping of single-strand breaks by genome-wide ligation of 3′-OH ends followed by sequencing (GLOE-Seq) portrays the genome's metabolic intermediates.
- Enhancers are gene regulatory noncoding cis-elements, often situated far away from the gene (promoter). The functional organization of enhancers in different physiological and developmental contexts provides the most dynamic picture of the genome. Altered enhancer function may cause several diseases, including cancer.
- Genome-wide distribution of RNA polymerase II (Pol II) in the context of histone modification marks defines the productive association of Pol II. RNA polymerase II binding to promoters, only in the presence of specific histone modification in the transcription start site and downstream catalyze the full-length transcript synthesis.
- Methylation of DNA at cytosine is one of the most important epigenetic modifications involved in epigenetic memory. Genome-wide distribution of methylation (methylome) stably changes with metabolic and environmental cues. Methylome studies are performed by the bisulfite sequencing approach.

- Genome organization dramatically alters during development. As the cells progress toward differentiation, the genome becomes more compact, with very few genes in an active state. Enhancer reorganization is essential for the lineage specificity and cellular identity. In the chromatin territories, the topologically associating domains (TADs) could be functionally heterogeneous. The evolutionarily conserved boundaries between the TADs restrict the spreading of heterochromatin.

Further reading

Agbleke, A.A., et al., 2020. Advances in chromatin and chromosome research: perspectives from multiple fields. Mol. Cell 79, 881 September 17.

Clemens, A.W., Wu, D.Y., Moore, J.R., Christian, D.L., Zhao, G., Gabel, H.W., 2020. MeCP2 represses enhancers through chromosome topology-associated DNA methylation. Mol. Cell 77 (2), 279–293 e8.

Dekker, J., Belmont, A.S., Guttman, M., Leshyk, V.O., Lis, J.T., Lomvardas, S., Mirny, L.A., O'Shea, C.C., Park, P.J., Ren, B., Politz, J.C.R., Shendure, J., Zhong, S., Network, D.N., 2017. The 4D nucleome project. Nature 549 (7671), 219–226.

Gasser, S.M., 2002. Visualizing chromatin dynamics in interphase nuclei. Science 296 (5572), 1412–1416.

Herz, H.M., Hu, D., Shilatifard, A., 2014. Enhancer malfunction in cancer. Mol. Cell 53 (6), 859–866.

Khan, A., Mathelier, A., Zhang, X., 2018. Super-enhancers are transcriptionally more active and cell type-specific than stretch enhancers. Epigenetics 13 (9), 910–922.

Kim, S., Shendure, J., 2019. Mechanisms of interplay between transcription factors and the 3D genome. Mol. Cell 76 (2), 306–319.

Langston, L.D., O'Donnell, M., 2006. DNA replication: keep moving and don't mind the gap. Mol. Cell 23 (2), 155–160.

Long, H.K., Prescott, S.L., Wysocka, J., 2016. Ever-changing landscapes: transcriptional enhancers in development and evolution. Cell 167 (5), 1170–1187.

Lorincz, M.C., Schübeler, D., 2007. RNA polymerase II: just stopping by. Cell 130 (1), 16–18.

Mardis, E.R., 2008. The impact of next-generation sequencing technology on genetics. Trends Genet. 24 (3), 133–141.

Meshorer, E., Misteli, T., 2006. Chromatin in pluripotent embryonic stem cells and differentiation. Nat. Rev. Mol. Cell. Biol. 7 (7), 540–546.

Mitsuya, K., Parker, A.N., Liu, L., Ruan, J., Vissers, M.C.M., Myatt, L., 2017. Alterations in the placental methylome with maternal obesity and evidence for metabolic regulation. PLoS One 12 (10), e0186115.

Perino, M., Veenstra, G.J., 2016. Chromatin control of developmental dynamics and plasticity. Dev. Cell 38 (6), 610–620.

Plank, J.L., Dean, A., 2014. Enhancer function: mechanistic and genome-wide insights come together. Mol. Cell 55 (1), 5–14.

Sriramachandran, A.M., Petrosino, G., Méndez-Lago, M., Schäfer, A.J., Batista-Nascimento, L.S., Zilio, N., Ulrich, H.D., 2020. Genome-wide nucleotide-resolution mapping of DNA replication patterns, single-strand breaks, and lesions by GLOE-Seq. Mol. Cell 78 (5), 975–985 e7.

Strickfaden, H., Tolsma, T.O., Sharma, A., Underhill, D.A., Hansen, J.C., Hendzel, M.J., 2020. Condensed chromatin behaves like a solid on the mesoscale in vitro and in living cells. Cell 183 (7), 1772–1784 e13.

Rieder, D., Trajanoski, Z., McNally, J.G., J.G., 2012. Transcription factories. Front. Genet. 3, 1.

VII

Chemical genomics: Understanding gene regulatory networks in differentiation and disease

VII.1 Functional gene regulatory network

From a simple prokaryotic organism to a complex human being, multiple genes function in a highly orchestrated network behind every biological phenomenon.

In a gene regulatory network (GRN), the input signal induces the expression of a limited set of genes, the primary output. Often, the primary output is the mRNA followed by the synthesis of corresponding proteins. These primary outputs affect the expression of a broader set of genes responsible for altering cellular behaviors. In general, there is a feedback loop in the whole process, where the final gene product also controls the input signal and or steps that follow (Fig. VII.1). This type of gene regulatory motif is exemplified by the bacteriophage λ life cycle, one of the most well-established and earliest discoveries in the field. The virus (phage λ) infects bacteria. Upon infection, it may either incorporate its genome into the bacterial genome and stay in a dormant state, the lysogenic state, or the genome of the virus decodes the information in its genome and produces new viruses, lyses the bacterial cell (lysis state), and infects new cells. The decision to follow a lysogenic or lytic path, the lysis-lysogeny switch (Fig. VII.2), is regulated by a simple but highly well-orchestrated GRN, as described in Chapter V.

Although the same principle of utilizing a GRN for gene regulation is followed in eukaryotic organisms, the process and outcome are much more complex. Behind every observable phenotype, there may be more than one regulatory network involved. The functioning of these networks is invariably dependent upon the direct interactions of genes and the environment and regulated genetically and epigenetically. However, each network influences distinct primary sets of genes. Different primary sets of expressed genes also regulate their target genes differentially. The phenotypic outcomes of such regulation are embryonic development, organogenesis, and disease pathogenesis (Fig. VII.3). Thus the properties of GRNs are not only useful for understanding life in molecular terms but also can be exploited for disease diagnosis and drug designing.

Different types of molecular networks may exist underneath the physiological and pathophysiological phenotypes, including (1) transcriptional regulatory network, (2) protein-protein interaction network, and (3) metabolic network. Furthermore, interactions among these networks fine-tune the phenotype. With the availability of the complete

Chemical Biology of the Genome. https://doi.org/10.1016/B978-0-12-817644-3.00006-4

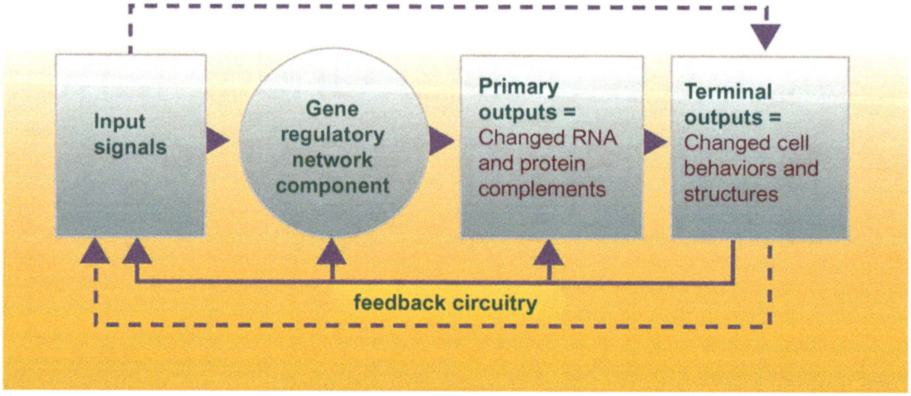

FIG. VII.1 Gene regulatory network circuit. *Source: Wikipedia contributors, 2020. In Wikipedia, The Free Encyclopedia. pp. 987567791.*

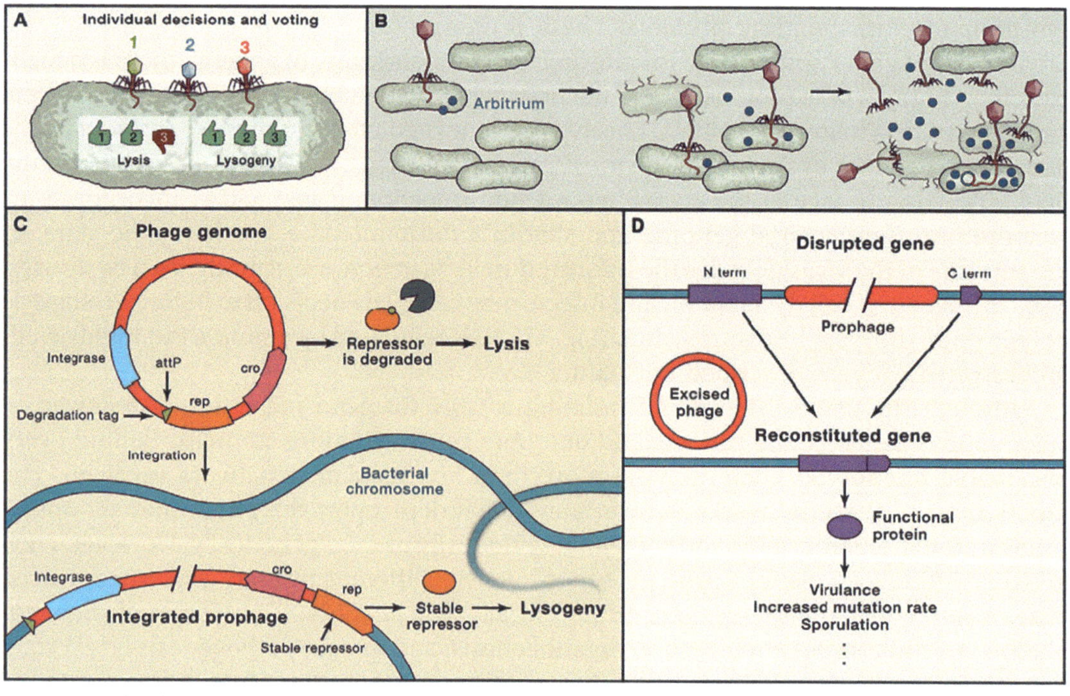

FIG. VII.2 Molecular mechanisms of phage lysis-lysogeny switch. (A) If multiple λ phages infect one *Escherichia coli* bacterial cell, the collective majority decide the lytic or lysogeny cycle. (B) When phage phi3T infects a bacterium, it secretes a six-amino acid peptide, called arbitrium *(blue dot)*; the concentration of the peptide decides the switching. (C) In the case of mycobacteriophages, if prior to integration, the tagged repressor is degraded, the lysis cycle is operated, but following integration, if the tag is separated, the stable repressor is produced, and lysogeny path is established. (D) Integration of the phage genome may disrupt a protein-coding gene. As and when the host needs this protein, the phage genome may be excised out. *Source: Ofir, G, et al., 2018. Cell 172 (6), 1260–1270. Elsevier.*

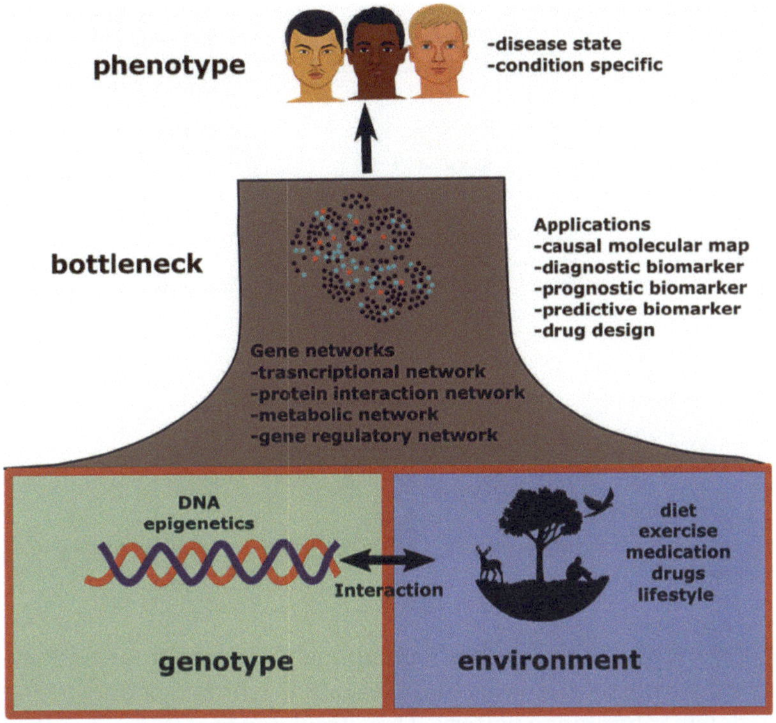

FIG. VII.3 Gene network is behind every phenotype. The expression of genes is regulated by environmental factors, including lifestyle. It is a network that determines (bottleneck) the output of expressed genes. *Modified from Emmert-Streib, F., 2014. Front. Cell Dev. Biol. 2, 38. Frontiers Media SA.*

human genome sequence, we know the sequence of every gene and its location. It has been possible to map mutations causing inherited genetic disorders and elucidate their molecular basis. One important question may be asked at this stage. How many GRNs are there in humans? It is difficult to answer accurately, but considering the number of cell types, we may say that there could be more than 200 GRNs. In this number, the pathological networks are, of course, not counted.

Genes rarely, if ever, operate in a stand-alone fashion. They influence each other, as do their mutant versions (epistasis). However, understanding genetic epistasis in the ever-changing gene network interactions in normal physiological functions and diseases is a Herculean task. We have just started elucidating these influences.

VII.2 GRN in epistasis

The first hint that multiple genes may work together came from the discovery of the phenomenon of epistasis in early research on genetics. The nonallelic interactions in gene function are referred to as Epistasis. This phenomenon was originally discovered by William Bateson in 1907. Bateson and R. C. Punnett found that upon crossing white flower

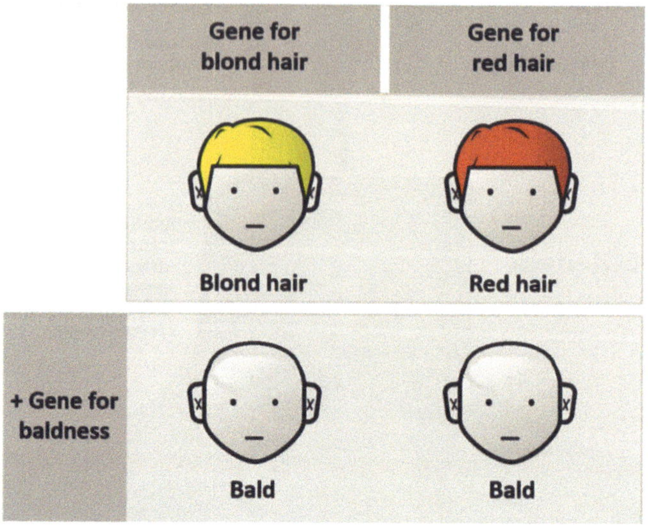

FIG. VII.4 Epistatic regulation of human phenotype. *Source: Wikipedia contributors, 2020. In Wikipedia, The Free Encyclopedia. pp. 986675624.*

peas in F1, all of the flowers were purple, a completely new phenotype. When they crossed the F1, the ratio of purple to white was found to be 9:7, which was unexpected and did not follow the Mendelian inheritance pattern. They predicted that this is due to interactions of at least two genes. We may note here that this prediction was made long before the DNA structure was discovered. Now we know that two genes involved in this phenomenon produce enzymes responsible for anthocyanin biosynthesis. In the homozygous null mutation condition, either of these genes is responsible for the white color.

There is plenty of evidence of epistasis in human life and disease. An interesting example is that of the color of human hair and baldness.

The gene for baldness is epistatic to the gene for hair color (blond or red). The presence of a gene (mutated?) that influences baldness epistatically dominates the gene for hair color. Here, the hair color gene is hypostatic (subordinate). Thus epistasis is a biological phenomenon, which reveals the existence of an underlying gene network in a visible phenotypic term (Fig. VII.4).

Complex diseases like Alzheimer's disease (AD) and diabetes are examples of GRNs involved in disease development. As shown in Fig. VII.5, multiple genes interact negatively or positively in these diseases, and the final output determines the state of disease. Here, we shall discuss AD briefly in the context of GRN.

VII.2.1 Epistatic regulation of gene function in AD

AD is a progressive neurodegenerative disorder associated with a gradual loss of memory and other cognitive functions. It is one of the most common causes of dementia in aged adults. At present, there are approximately 50 million people in the world who are dementia

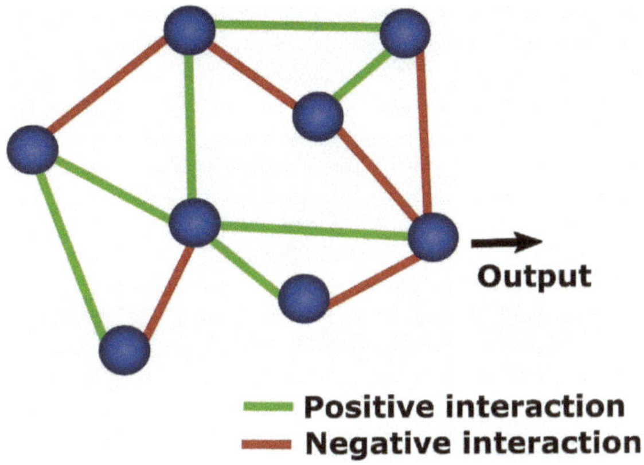

FIG. VII.5 Interactions of multiple genes result in the output that causes disease pathogenesis. The hypothetical genes *(blue circles)* may interact negatively *(red)* or positively *(green)*.

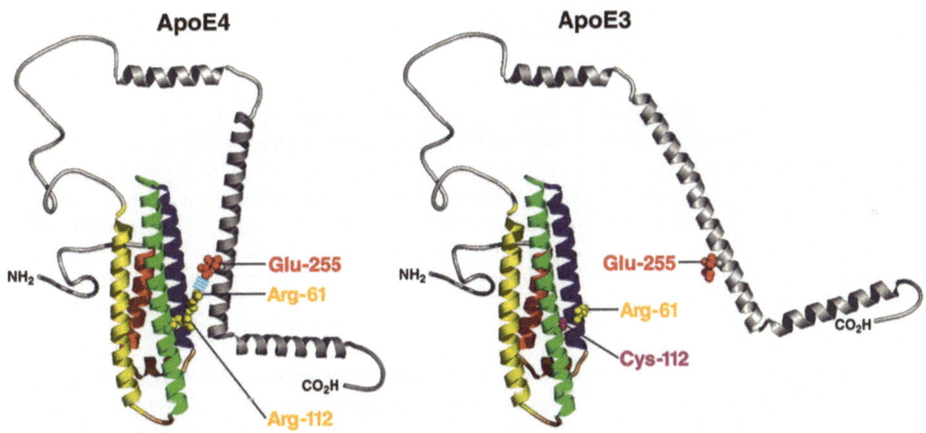

FIG. VII.6 Domain organization of ApoE protein: Critical residues involved in the interactions are indicated. In ApoE4, Arg-112 orients Arg-61 in such a way that it can interact with Glu-255. However, in ApoE3, Arg-61 is buried and not available for interaction with Glu-255. *Source: Najm, R., et al., 2019. Mol. Neurodegener. 14 (1), 24. Springer Nature CC-BY license.*

patients. It is predicted that in another 20 years, this number will be around 75 million. It was found that the apolipoprotein E (ApoE) gene is associated with AD. The polymorphic allele of *APOE* is the key genetic determinant of the disease risk. Any person possessing the ε4 allele is at an increased risk of AD than the commonly present ε3 allele (Fig. VII.6).

This allele-specific risk is due to pathogenesis-specific gene-gene interactions in the development of Alzheimer's. ApoE isoforms differentially regulate the clearance and aggregation of Aβ, one of the fundamental causes of AD. Apparently, there are 27 different epistatic interactions of different groups, namely cholesterol metabolism, beta-amyloid production, inflammation, oxidative stress, etc., which determine the risk of AD (Fig. VII.7).

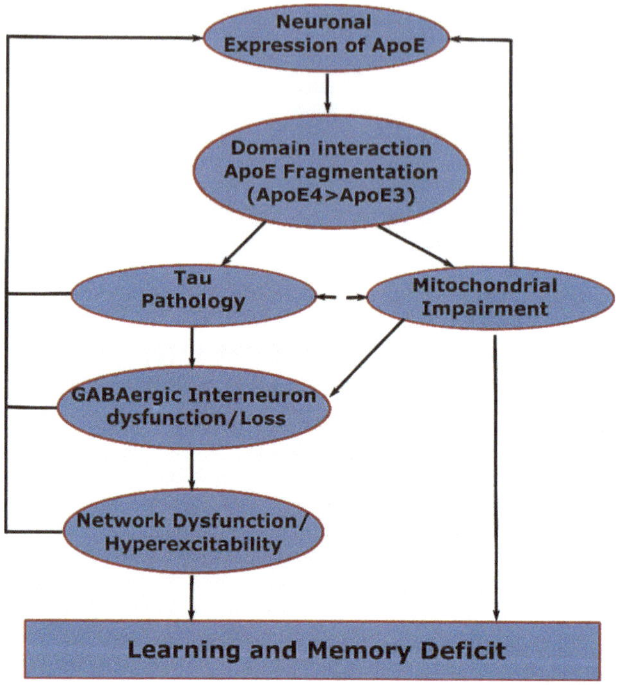

FIG. VII.7 Network signaling behind the pathology of Alzheimer's disease. In neurons, ApoE protein is expressed upon exposure to stress or injury, as well as aging. ApoE helps to repair the neuron. But if a person possesses the risk-enhancing *APOE* allele, there is more fragmentation due to a vulnerable domain conformation, leading to tau pathology. In response to ApoE4 toxicity, some GABAergic interneurons do not function and ultimately degenerate. These events slowly impair learning and memory. *Modified from Najm, R., et al., 2019. Mol. Neurodegener. 14 (1), 24. Springer Nature.*

Besides AD, epistatic gene interactions are also predominantly involved in the pathogenesis of diabetes, cardiovascular disease, hypertension, autism, cleft lip and/or palate, schizophrenia and other neurological disorders, and sporadic breast cancer. However, the details of these interactions and their regulations are yet to be elucidated.

VII.3 Tools to understand GRNs

Tools that are used to analyze GRNs have been discussed briefly in Chapter II. The network perturbation is the first step, in which genes are knocked out or knocked down, using shRNA/siRNA. In the second step, upon perturbation, high-throughput transcriptome profiling (e.g., RNA-seq) is performed. The gene-gene interaction analysis is performed based on these data, which may reveal the specific molecular signatures with functional modules of the network. At the organismal level, GRN analysis helps to understand, in molecular terms, the different biological processes: embryogenesis, hematopoiesis, oncogenesis, and inflammation. It integrates almost all types of interactions: protein-metabolite, protein-protein, protein-RNA, and protein-DNA in the functional context.

VII.3.1 siRNA and shRNA

One of the most widely used techniques to knockdown gene products is through RNA interference (RNAi). We have discussed this pathway in Chapter II. Here, we will discuss specific RNAs used to interfere with gene functions.

VII.3.1.1 siRNA

A 20–27 base pair long noncoding RNA functions in the RNAi pathway, termed the small interfering RNA or siRNA. The naturally occurring siRNAs are phosphorylated at 5′ ends and possess hydroxylated 3′ ends with two overhanging nucleotides. For knockdown in vitro, double-stranded siRNAs are transfected into the cell, which are then processed by the RISC complex (Fig. VII.8), and the resulting single-stranded siRNA targets the specific mRNA and degrades it. siRNA has some similarities to the micro-RNA (miRNA). However, miRNAs may have broader specificity compared to the strict specificity of siRNA (must have 100% complementary to the specific mRNA for effective regulation).

VII.3.1.2 Short hairpin RNA or small hairpin RNA

Short hairpin RNA or small hairpin RNA (shRNA) is an artificial RNA molecule with a hairpin turn having a high affinity toward its target. It is processed by the RNA silencing machinery. The construct for shRNA expression is generally made under promoters that are transcribed by RNA polymerase II. The construct could be packaged in the lentiviral system and transfected to mammalian cells. The recombinant viral genome gets integrated into the chromatin of the cells stably and then transcribed by cellular RNA polymerase II. The transcript similar to primary microRNA is then processed by the RNAi machinery. Besides lentivirus, adeno-associated viruses and adenoviruses may also be used for transfection.

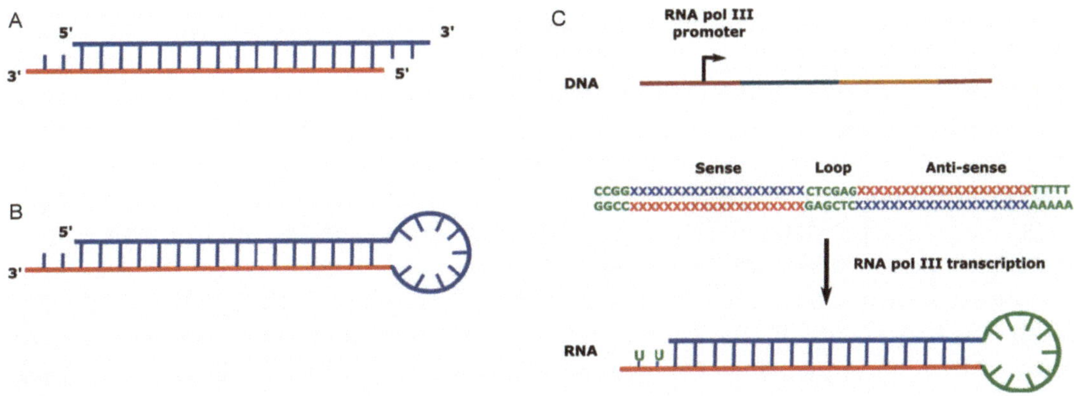

FIG. VII.8 A typical siRNA and shRNA structure. (A) siRNAs possess 2 nt 3′ overhangs. (B) shRNAs has a loop sequence. (C) shRNA in the expression construct, to be packed in the viral vector.

VII.3.2 Inducible expression of shRNA

Stable knocking down of essential genes may be lethal to the cells. Therefore to understand the function of these genes, shRNAs are often expressed in a regulated manner using an inducible system.

We provide below an example of how such inducible shRNAs are delivered inside the cell. In a typical construct, the tetracycline-resistance operon of *E. coli* was used. In this operon, the Tet repressor binds to the Tet operator (similar to the lac operon described in Chapter V). At present, three different configurations of inducible systems are available (Fig. VII.9): (1) Tet operator (TetO) sequence is inserted between a constitutive promoter and the gene of interest (shRNA gene). When the Tet repressor is bound to the TetO, the shRNA gene is repressed (basal state). For inducing expression, cells are treated with Doxycycline (a synthetic antibiotic derived from tetracycline). Doxycycline binds to TetR and prevents its binding to the TetO, allowing transcription of the gene to occur. (2) in another configuration, the TetO sequence is placed before the minimal promoter to make the gene expression transcriptional activator dependent. A chimeric transcription factor (TF) construct consisting of TetR and VP16 (tTA), a eukaryotic transactivator derived from herpes simplex virus type 1, is used to drive the transcription of the desired gene. If doxycycline is available in the cells, it binds to the chimeric protein (at TetR site), as a result of which it can no longer bind to the promoter and activate the transcription. (3) In the Tet-on system, a mutant chimeric protein construct is used, where four amino acids are substituted. This mutant protein needs doxycycline to bind to the promoter. A low dose of antibiotics is essential to express the gene of interest; in this case, the shRNA construct.

Here, we shall discuss one significant gene network of the mammalian system, whose function has been elucidated by the tools mentioned so far: the *Rb pathway.*

VII.4 Gene regulatory pathway of Rb

The retinoblastoma protein (Rb) is a tumor suppressor, often mutated in different cancers (Fig. VII.10). Physiologically, Rb is a key regulator of the cell cycle. The siRNA mediated knockdown of *RB1* gene product dramatically affects the cell cycle (G1/S and G2/M transitions), DNA replication and repair, mitosis, and apoptosis. Mechanistically, Rb, along with the corepressor complex, interacts with the TF E2F and inhibits the E2F-dependent cell cycle, promoting gene expression (Fig. VII.11). When mitotic signal activates Cyclin-CDK4/6, it phosphorylates Rb. The phosphorylated Rb dissociates from E2F, which allows the transcriptional activation by E2F.

Connections of Rb to other regulatory networks were analyzed through knockdown of Rb. The knockdown of Rb was followed by gene expression profiling, either by microarray or by RNA sequencing. This type of analysis revealed that Rb is involved in regulating several important cell cycle regulatory pathways. Knocking down of Rb induces expression of selected genes belonging to important cellular pathways: cell cycle

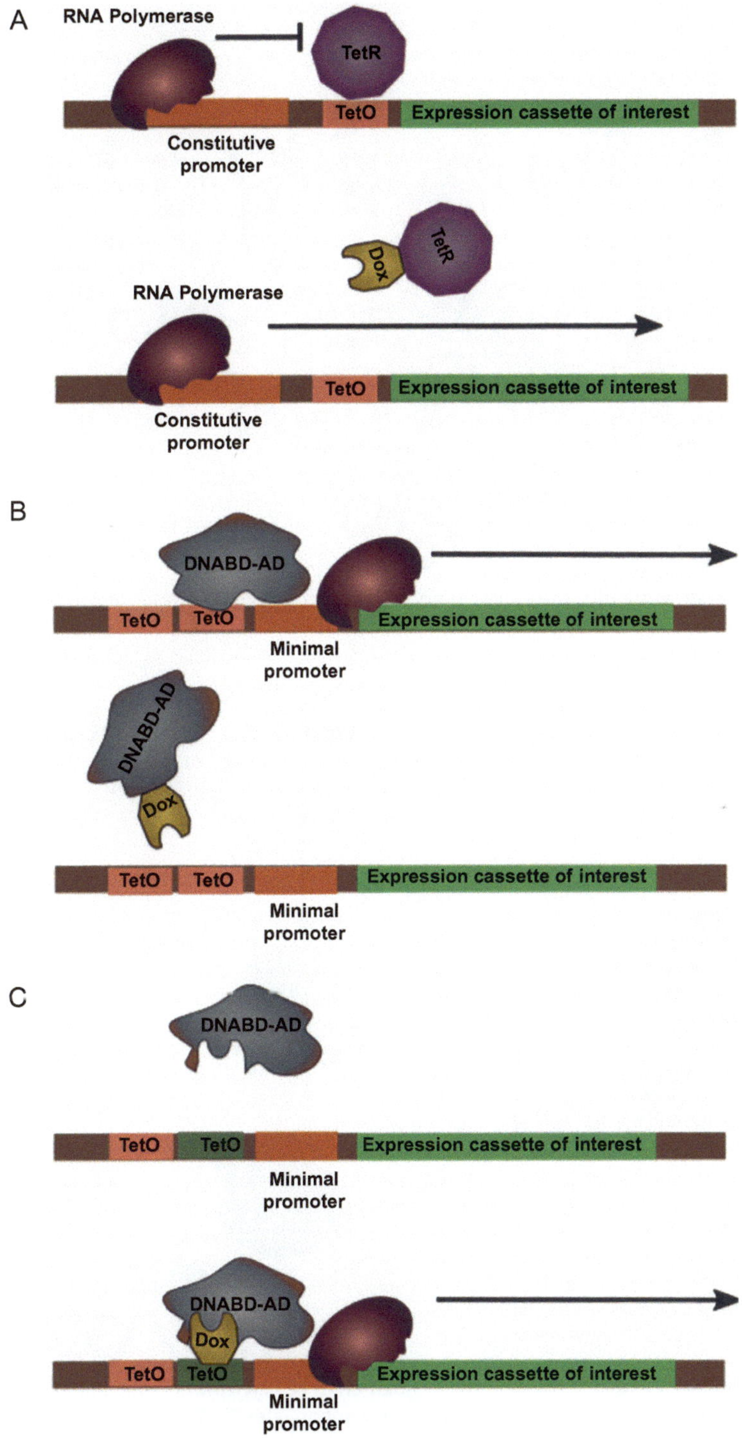

FIG. VII.9 Doxycycline (tetracycline)-regulated gene expression system used to silence specific gene expression by shRNA. (A) TetR-mediated repression could be relieved by Dox. (B) A recombinant transcription factor activates gene expression, which could be inhibited by Dox (Tet-off configuration). (C) A mutant recombinant transcription factor needs DOX to activate transcription (Tet-on configuration). *DNABD*, DNA-binding domain; *AD*, activating domain; *TetO*, tetracycline operator; *Dox*, doxycycline; *TetR*, Tet repressor. *Modified from Kallunki, T., et al., 2019. Cells 8 (8), 796. MDPI.*

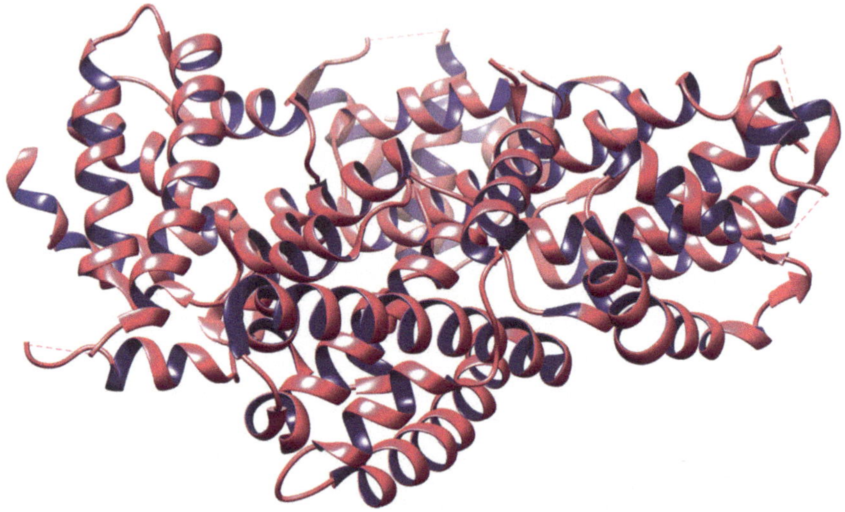

FIG. VII.10 The crystal structure of the inactive retinoblastoma protein (Rb1) phosphorylated at T373 (PDB ID: 4ELJ).

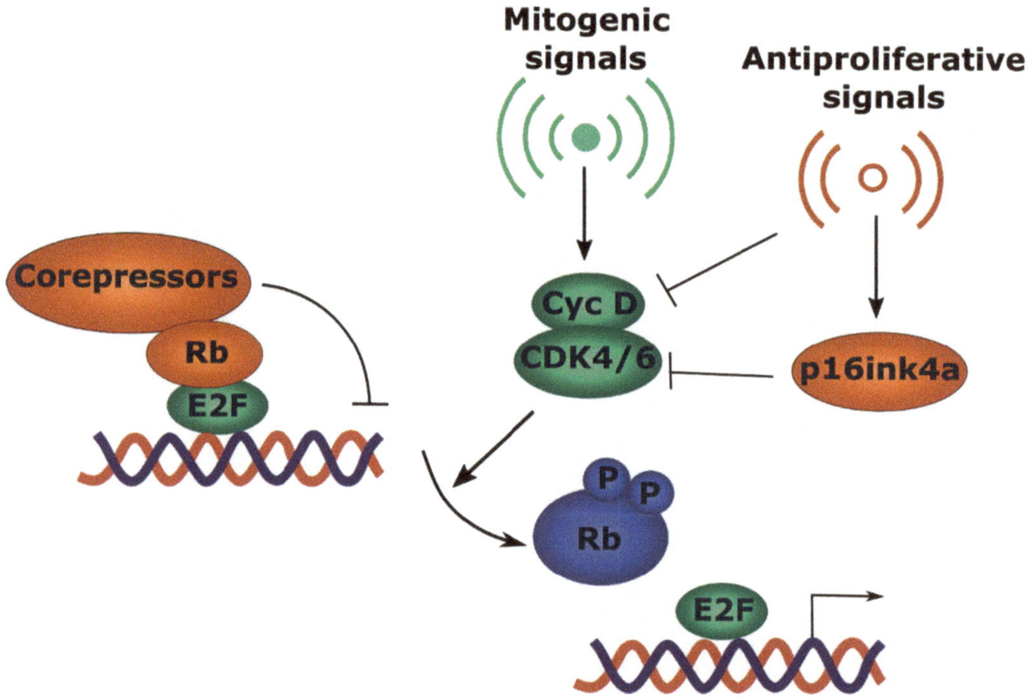

FIG. VII.11 Molecular mechanisms of Rb-mediated regulation of cell cycle: Mitogenic signal activates the cyclin-dependent kinase (CDK4/6 and cyclin D complex), which phosphorylates Rb. The phosphorylated Rb dissociates from E2F, leading to derepression of E2F-regulated genes. Activation of E2F induces the cell cycle progression. We have followed the convention that the protein is abbreviated as Rb and gene as RB1. *Source: Knudsen, E.S., et al., 2019. Trends Cancer 5 (5), 308–324. Elsevier.*

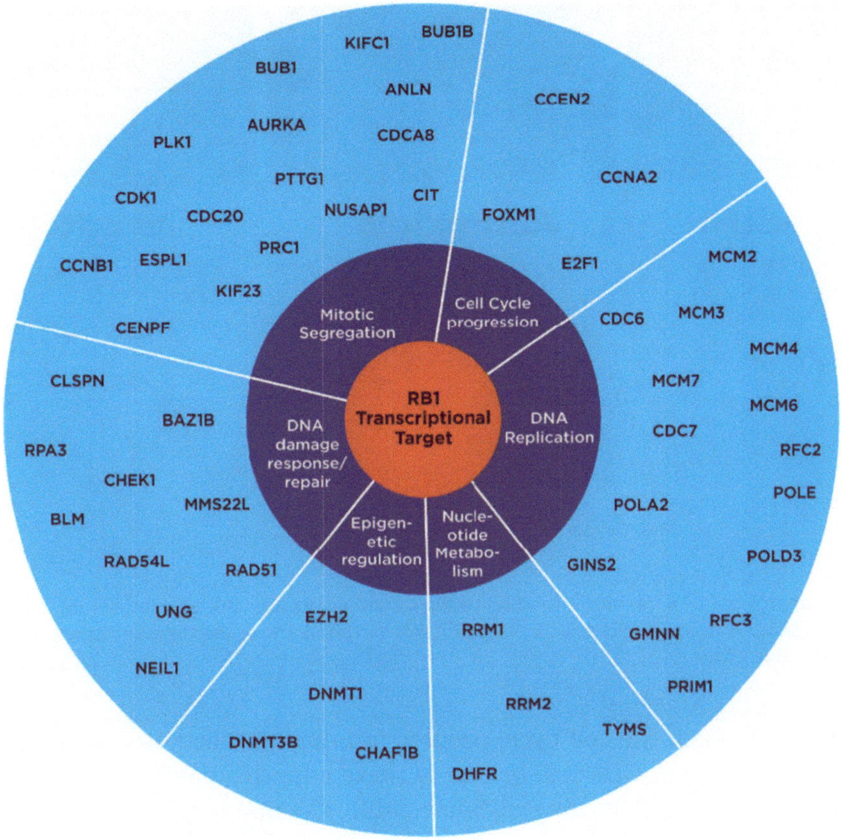

FIG. VII.12 Rb regulates key genes for cell proliferation.

progression, DNA replication, DNA damage response/repair, nucleotide metabolism, and epigenetic regulation (Fig. VII.12). These genes are critical for regulating cellular proliferation.

Mitogenic or oncogenic signals activate the cyclin-dependent kinases (CDK)4/6. When the antiproliferative signals are dysregulated and or *RB1* gene is mutated with loss-of-function, cells step into uncontrolled growth, promoting malignancy. The pharmacological inhibitors of CDK4/6 were indeed effective in stopping tumor growth both in animal models and human patients. These inhibitors were found to be most effective in the estrogen receptor-positive (ER^+) breast cancer patients. Three CDK4/6 inhibitors, palbociclib, ribociclib, and abemaciclib, have received FDA approval and are now being used in the clinics (Fig. VII.13). However, because the pathways regulating the cell cycle have considerable plasticity, resistance to CDK4/6 inhibition may develop rapidly and has been documented. One of the pathways is the loss of tumor-suppressive function of Rb. Another pathway is that the presence of the inhibitor may trigger the activation of an adaptive

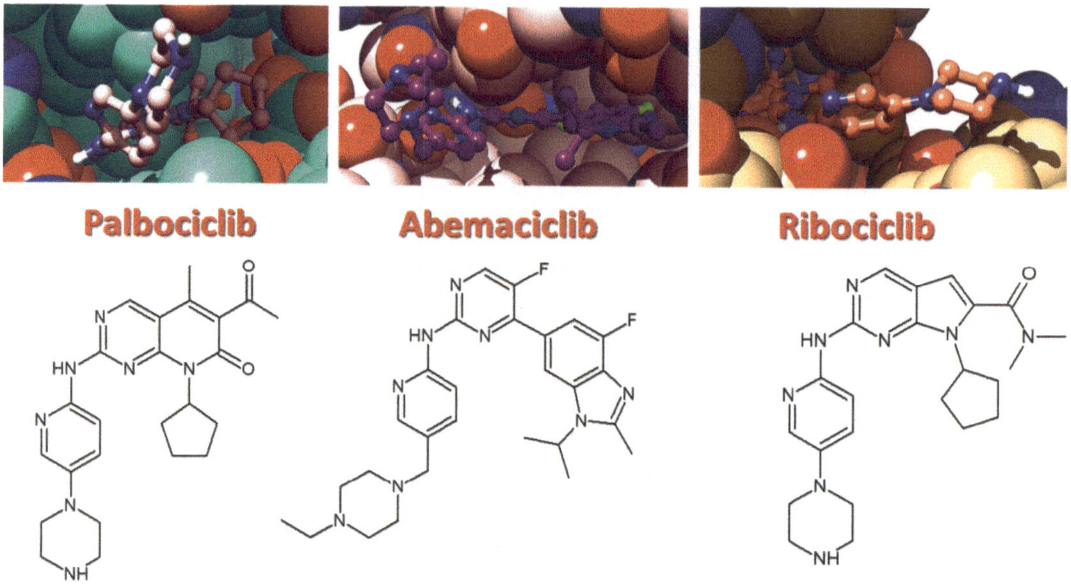

FIG. VII.13 FDA approved CDK6 inhibitors' binding to the target: palbociclib (A), abemaciclib (B), and ribociclib (C), (PDB palbociclib; 5L2T; 5L2S, abemaciclib; 5L2I, ribociclib). *Bottom panel* shows the chemical structures of palbociclib, abemaciclib, and ribociclib.

pathway through which the cell cycle progression may proceed, notwithstanding pharmacological inhibition of CDK4/6. In one of these adaptive pathways, cyclin D1 seems to be upregulated. To combat this resistance, inhibitors of oncogenic signaling, targeting MEK, EGFR, PI3K, and mTOR in combination with a CDK4/6 inhibitor, may be used (Fig. VII.14). The complex relationships of Rb with so many regulatory pathways are still being intensively investigated with an eye to understand drug resistance.

In earlier paragraphs, we have shown the complex role of Rb in the GRNs that control the cell cycle. In reality, the roles of Rb in cellular regulation are even more complex because Rb also functions beyond the regulation of the cell cycle. Quite contrarily, Rb seems to be essential for activating the interferon-response genes. For example, Rb regulates the class II major histocompatibility complex transactivator (CIITA) and NF-kB activity. Treatment with a CDK4/6 inhibitor activates many immunologically relevant genes, including the canonical interferon-responsive genes. The mechanisms of Rb-mediated activation of these genes are not known. One of the possibilities could be through the transcriptional repression of DNA methyltransferase 1 (DNMT1). The DNMT1-catalyzed methylation inhibits the endogenous retroviruses that evoke the interferon response.

Cancers always originate in a specific tissue and bear signatures of its tissue lineage. In some of these cancers, lineage-specific pathways are hijacked for proliferation. For example, breast and prostate cancers often require hormone receptor signaling, a

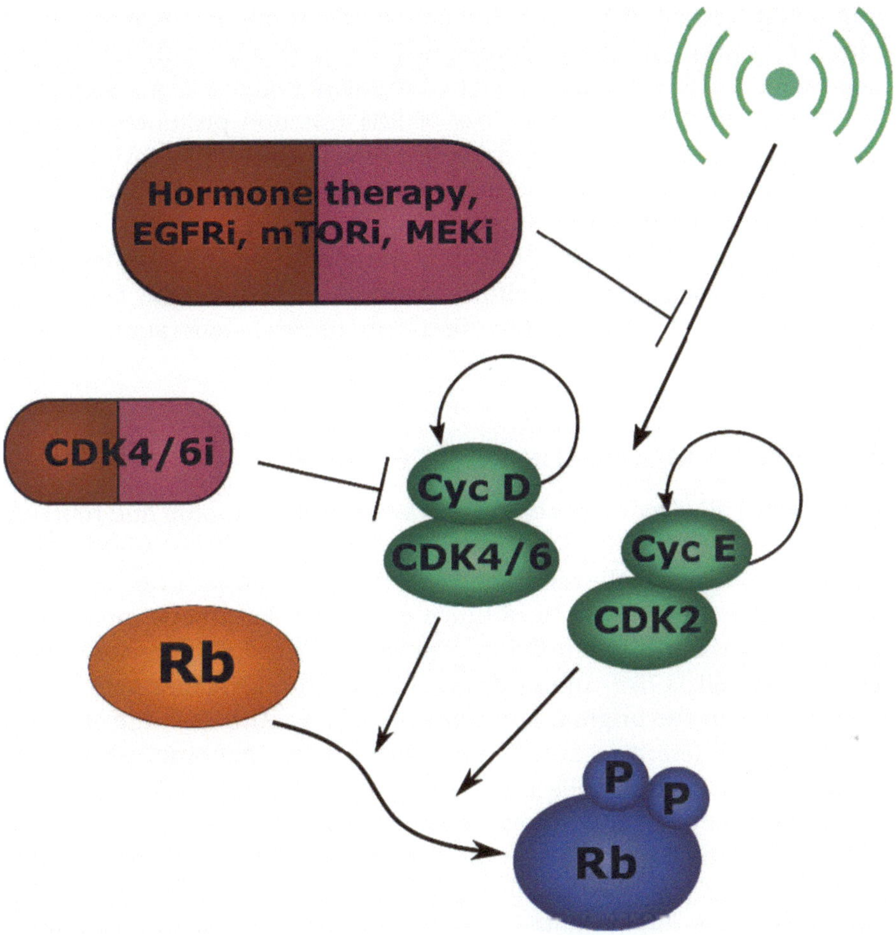

FIG. VII.14 Targeting the Rb pathway for cancer therapy. Cyclin-dependent kinase (CDK)4/6 inhibitor inhibits the phosphorylation of Rb, thereby sustaining the repression of cell proliferation-related genes. However, inhibition of the oncogenic signals that activate the CDK-Cyclin kinase, along with the kinase inhibitor, seems to be a more effective therapeutic strategy. CDK4/6i, CDK4/6 inhibitors (e.g., palbociclib); EGRFi, epidermal growth factor receptor inhibitors (e.g., osimertinib); MEKi, mitogen-activated kinase 2 inhibitors (e.g., trametinib); mTORi, mammalian target of rapamycin inhibitors (e.g., everolimus). *Source: Knudsen, E.S., et al., 2019. Trends Cancer 5 (5), 308–324. Elsevier*

lineage-specific pathway, for growth. Cancer lineage specificity-dependent growth and development is often targeted for therapy. Tissue-specific growth factor signaling is also characteristic of many cancers, and growth factor receptors are often targeted (e.g., Tyrosine Kinase inhibitors for EGFR-dependent cancer). However, cancer cells are capable of switching characteristics of its original lineage signatures. Because of this plasticity, cancer cells often develop resistance to this kind of therapy. Rb can suppress the lineage switching through (1) the repression of the cell cycle, which assist in the stabilization of the epigenome, (2) maintenance of chromatin organization, which is essential for the appropriate gene expression for a particular cell state and chromosomal integrity, or (3) direct inhibition of pluripotency genes (e.g., *SOX2* and *EZH2*) by the Rb-E2F complex. The essential loss of RB1 function promotes reprogramming (lineage switch) through induced pluripotency (see further). The overall loss of RB1 function or suppression of the Rb pathway enhances the lineage plasticity by lifting repression of many of the abovementioned pathways that may also cause the development of drug resistance. The scope of targeting proteins of lineage-specific growth pathways, in combination with CDK inhibitors, seems to be very bright. Elucidation of Rb regulatory network made possible the discovery of new generation of therapeutic targets.

VII.5 Epigenetic landscape and cell fate

Another area where GRNs play a profound role is cell differentiation and reprogramming of cell states.

For a long time, it was believed that cell differentiation is a one-way process, as proposed by C. H. Waddington in his famous epigenetic landscape model. It was proposed that the process of differentiation from an undifferentiated stem cell to a differentiated mature cell is like rolling down the hill into a deeper valley from where it can never go back to the original point of the journey. Although the stem cells (pluripotent cells) and differentiated cells possess identical genome, phenotypically or physiologically, differentiated cells are completely different. This phenotypic difference was attributed to epigenetic programming. However, once the cells are differentiated and placed at the bottom of the hill (i.e., reached the final cell fate state), they cannot go back to the top, the initial pluripotent state. However, recent discoveries revealed that terminally differentiated cells are not totally incapable of changing their states. Differentiated cells could be pushed to travel back up their differentiation path to the original state (pluripotent state) by overexpressing a set of TFs: Oct4, Sox2, Klf4, and cMYC (abbreviated as OSKM). OSKM is also referred to as Yamanaka reprogramming factors. Furthermore, using a specific set of TFs, cells may also be induced to differentiate from one differentiated cell type to another (Fig. VII.15). These discoveries have revolutionized the concept of differentiation and opened up the immense possibility for regenerative therapy.

VII.5.1 Induced pluripotency and Yamanaka factors

Shinya Yamanaka and colleagues discovered that overexpression of four TFs, namely Oct4, Sox2, Klf4, and cMyc, drives the reprogramming of the fibroblast cells to pluripotent stem cells. The role of each of these TFs has been elucidated in the process of reprogramming. Subsequently, the replacement of each of these factors with the other members of their

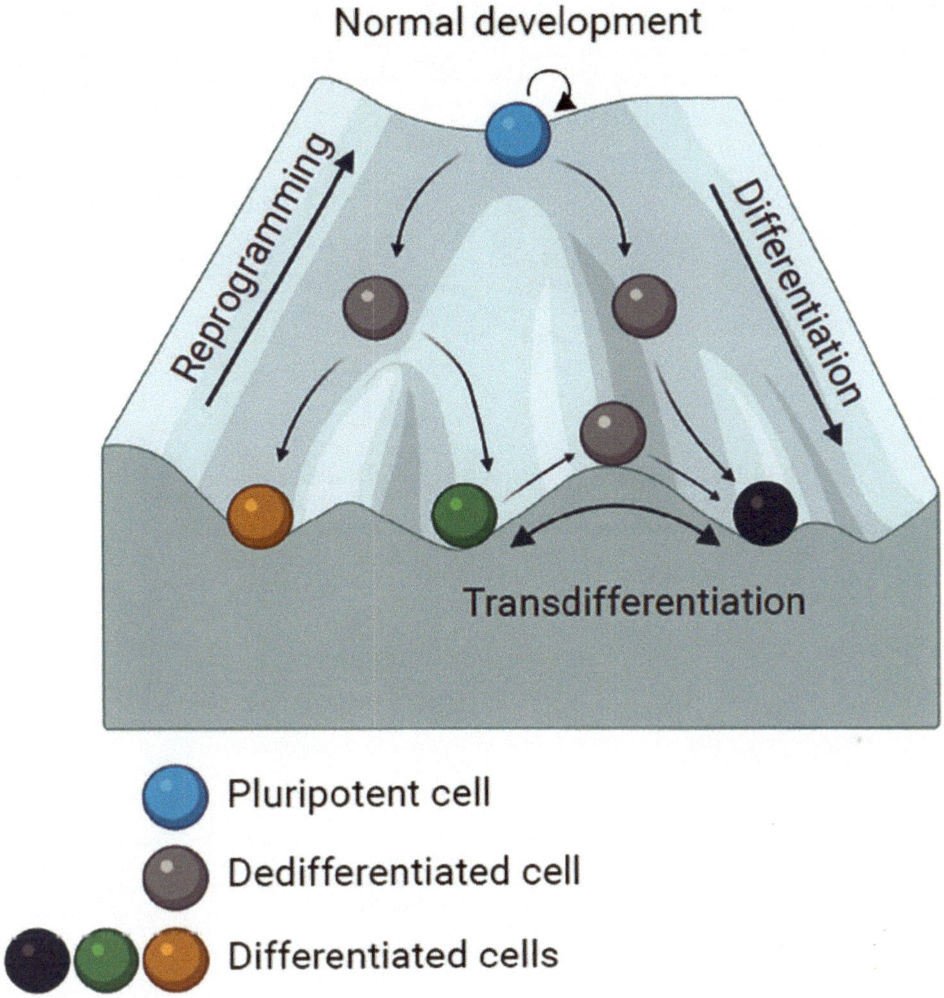

FIG. VII.15 Reprogramming in Waddington epigenetic landscape. *Source: Granados, et al., 2020. Int. J. Mol. Sci. 21 (21), 8274. https://doi.org/10.3390/ijms21218274. CC-BY license.*

family was also successfully shown by multiple groups. From all these studies, it was found that Oct4 is an essential factor in this reprogramming cocktail, which could not be replaced. Very recently, however, it was shown that Sox2, Klf4, and cMyc (SKM) could efficiently induce the fibroblast to induced pluripotent cells (iPSCs), which could produce all iPSC-mice in the tetraploid complementation system. Furthermore, it was found that Oct4 expression induces aberration in the epigenetic landscape, resulting in off-target gene expression. This is a landmark discovery for the application of iPSC technology (Fig. VII.16).

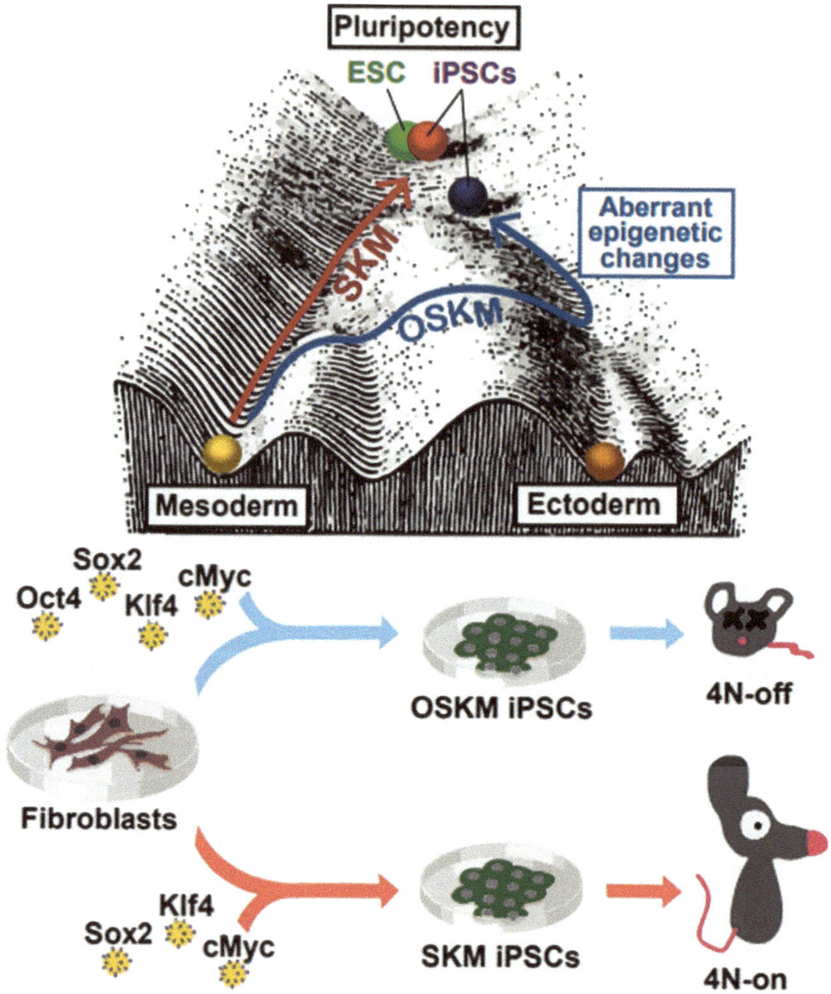

FIG. VII.16 Oct4 transcription factor is not essential in Yamanaka Cocktail (OSKM). *Source: Velychko, S., et al., 2019. Cell Stem Cell 25 (6), 737–53. Elsevier.*

VII.5.2 Cell fate reprogramming by small-molecule compounds

Although different TF-based reprogramming could efficiently induce pluripotency in the differentiated cells and generate lineage-specific cells with efficiency and safety of use for therapeutic purposes or drug molecule screening, this system seems to have innumerable limitations. This problem could be solved by a cocktail of small molecules. For each of the Yamanaka factors, a cocktail of the small-molecule compounds has been discovered. By using these small molecules, both dedifferentiation and

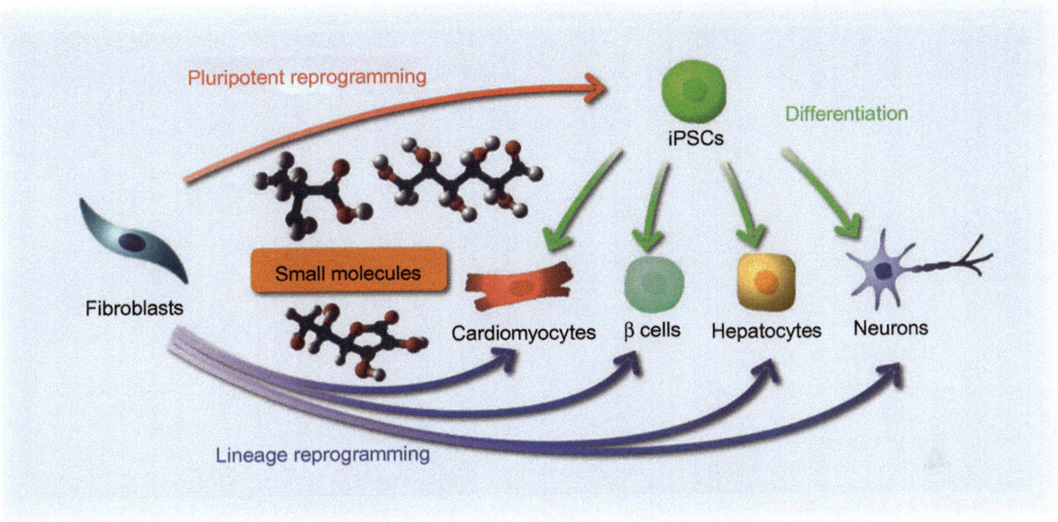

FIG. VII.17 Cell fate reprogramming by small-molecule modulators: The job of Yamanaka factors could be substituted by a cocktail of small molecules to generate pluripotent cells or other functional cell types: cell types, such as cardiomyocytes, hepatocytes, pancreatic β cells, and neurons from fibroblasts cells. *Source: Ma, X., et al., 2017. Protein Cell 8 (5), 328–348. Springer Nature.*

transdifferentiation could be achieved. Different cocktails could be identified for the lineage-specific differentiation as well (Fig. VII.17). A list of these small molecules and their respective cellular targets and reprogramming potential has been shown in Fig. VII.18.

VII.5.3 Cellular reprogramming by small molecule and implications in therapy

As discussed earlier, small-molecule compounds could be used to induce pluripotency and/or lineage-specific cells by using cells from patients' lineage-specific cells that could be transplanted for organ regeneration purposes. These cells could also be used for high-throughput screenings of therapeutically potential molecules (Fig. VII.19). This technology may be useful for biomedical studies and clinical applications for curing degenerative diseases, injuries, and aging.

By employing patient cell-derived stem cells, many therapeutically potential molecules are being screened and processed further for the drug discovery program (Fig. VII.20). Eventually, this drug discovery route may replace several steps of conventional drug discovery, especially by reducing the preclinical animal experiments (Fig. VII.21).

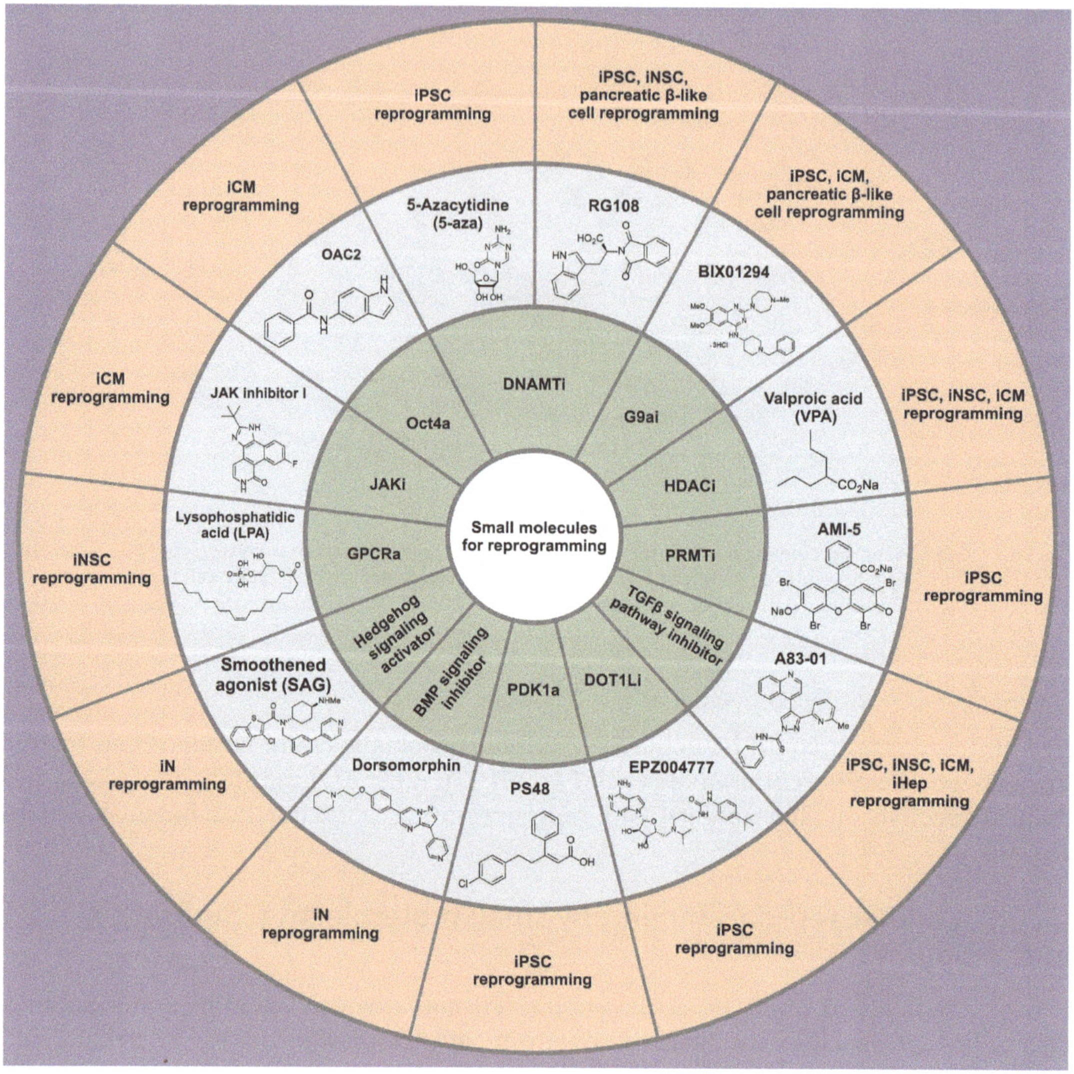

FIG. VII.18 Small-molecule compounds induce cell reprogramming toward either induced pluripotent stem cells (iPSC) or cells of specific lineage: induced cardiomyocytes-like cells (iCM), induced neural stem cells (iNSC), induced neuron (iN), and induced hepatocyte-like cells (iHep). For each reprogramming, however, a cocktail of molecules is used. *Source: Self, based on information from Ma, X., et al., 2017. Protein Cell 8 (5), 328–348. Springer Nature.*

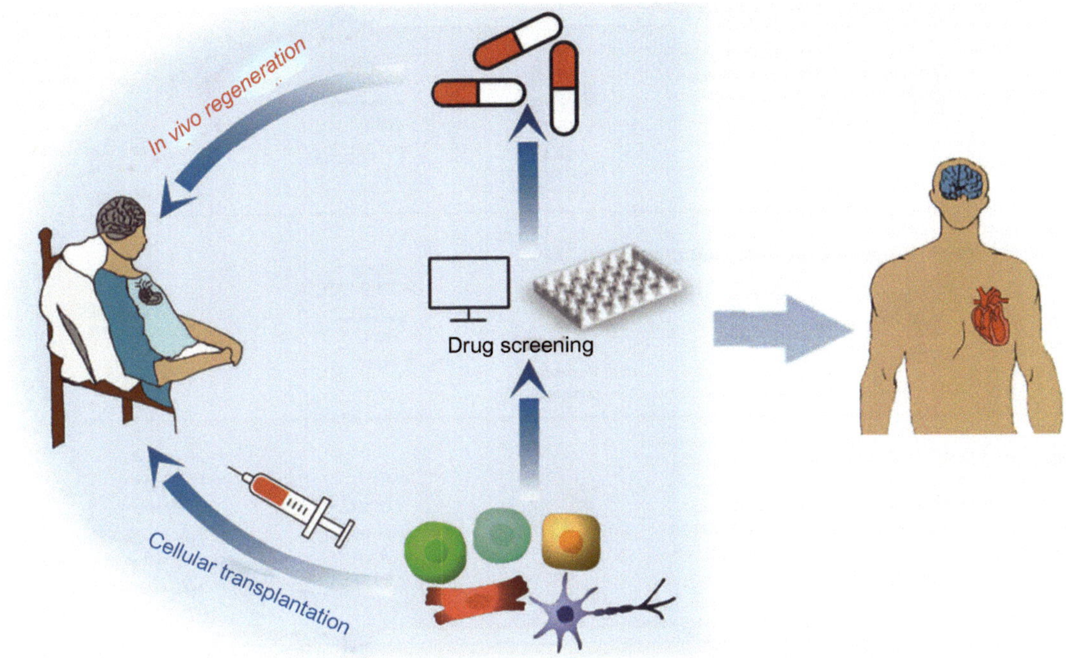

FIG. VII.19 Cellular reprogramming could be useful for organ-specific repair and regeneration using endogenous cells. *Source: Ma, X., et al., 2017. Protein Cell 8 (5), 328–348. Springer Nature.*

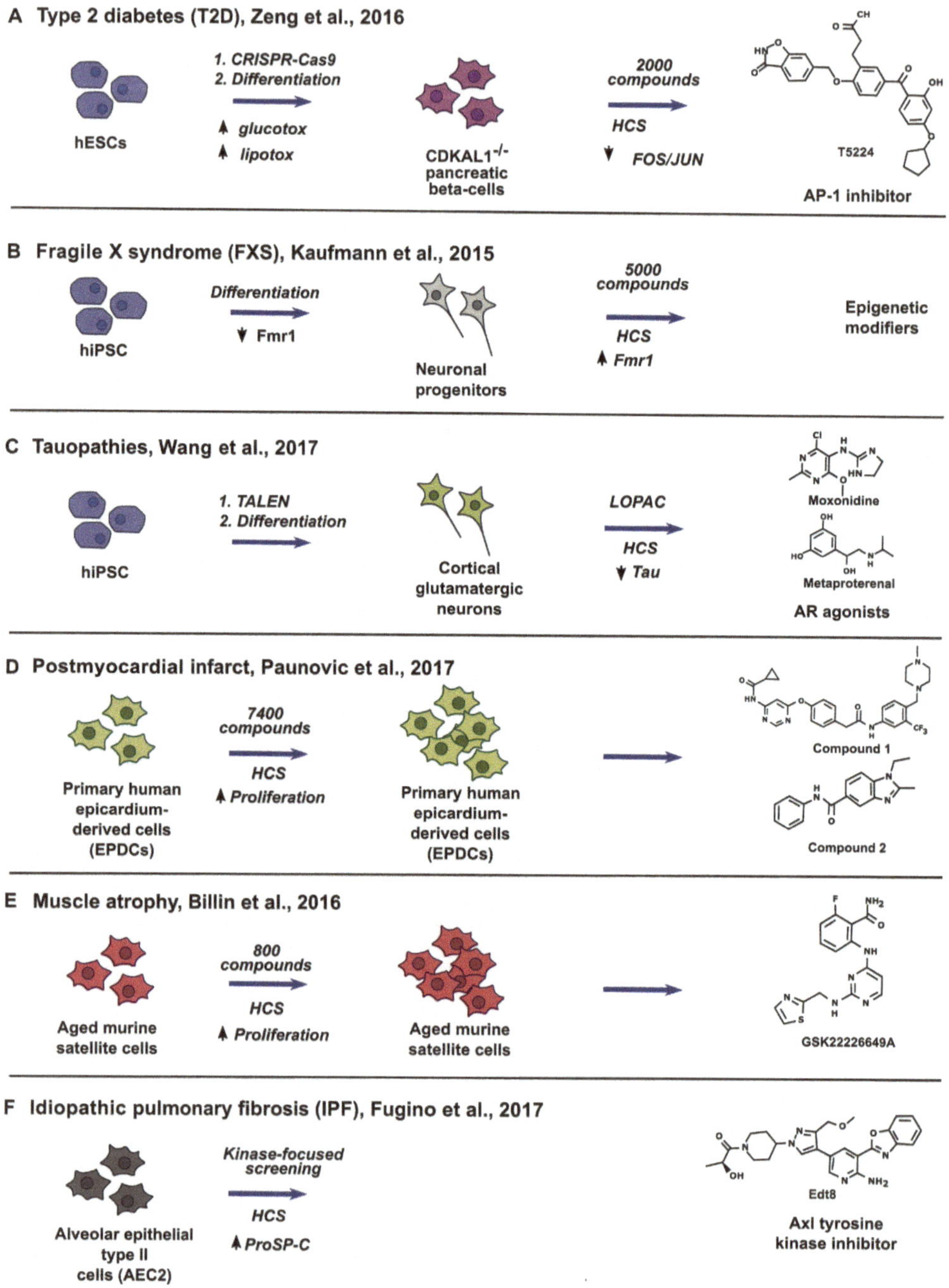

FIG. VII.20 Induced stem-based disease model could be useful for the screening of therapeutically potential molecules. Abbreviations: *AP-1*, FOS/JUN activator protein-1; *CDKAL1*, CDK5 regulatory subunit-associated protein 1-like 1; *Fmr1*, fragile X mental retardation 1; *HCS*, high-content screening; *hESC*, human embryonic stem cell; *hiPSC*, human-induced pluripotent stem cell; *LOPAC*, library of pharmacologically active compounds; *ProSP-C*, prosurfactant Protein-C; *TALEN*, transcription activator-like effector nucleases. *Source: Friese, A., 2019. Cell Chem. Biol. 26 (8), 1050–1066. Elsevier.*

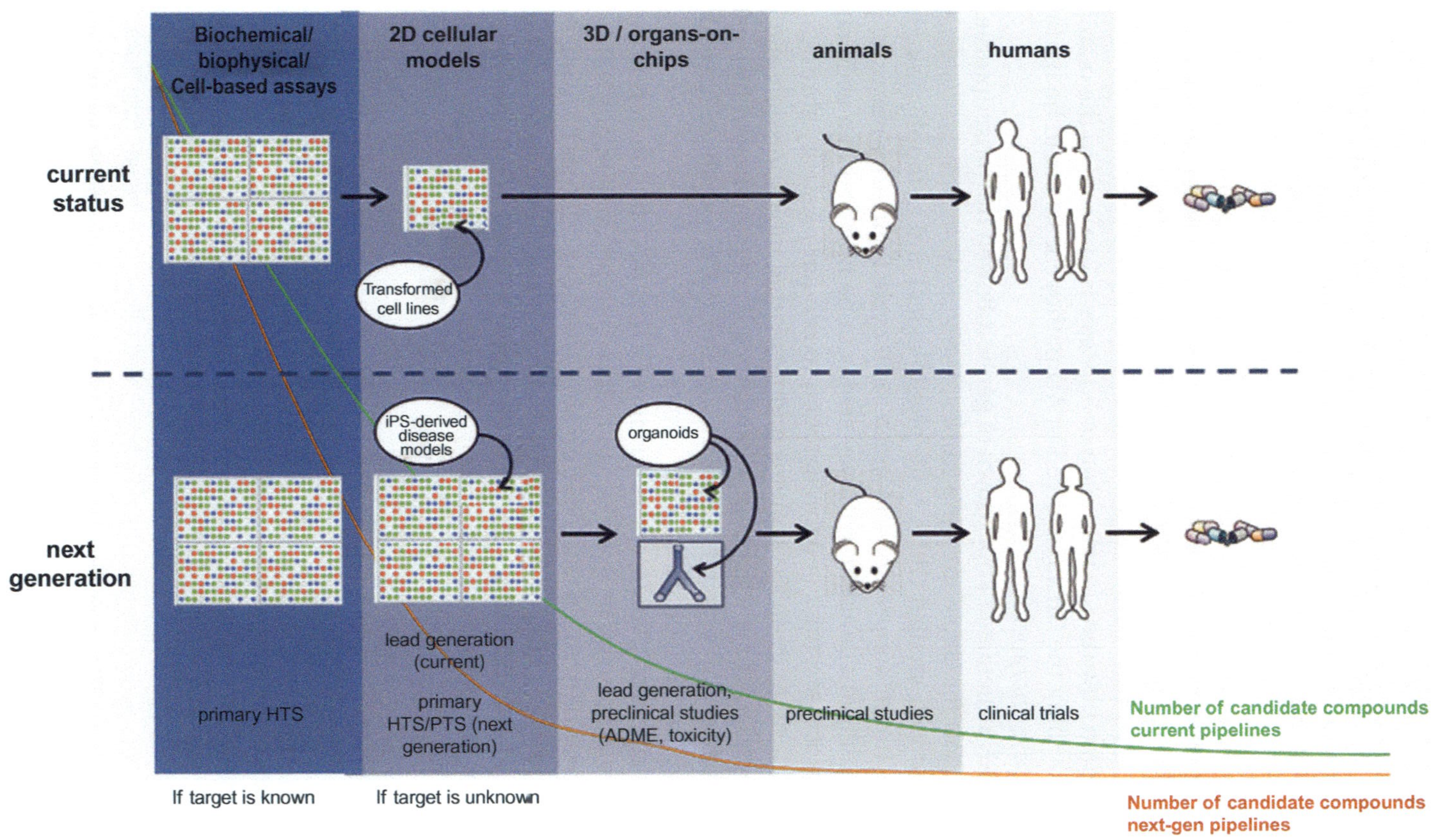

FIG. VII.21 In next-generation drug discovery and development program by employing organoids (produced by in vitro reprogramming), preclinical animal experiments may not be required. *Source: Friese, A., 2019. Cell Chem. Biol. 26 (8), 1050–1066. Elsevier.*

Chapter summary

- The principles of signal-based functioning of GRNs have been discussed in a wide range of contexts: from the simple bacteriophage λ lysis-lysogenic switch to the complex GRNs in humans comprising transcriptional regulatory networks, protein-protein interaction networks, and metabolic networks. The ultimate phenotypes of complex organisms result from functional GRNs, where the gene expression pattern is fundamentally regulated by the interactions of gene and environmental factors.
- The first hint that multiple genes may work together came from the discovery of the phenomenon, Epistasis, the nonallelic interactions in gene functions. As an example, we have presented how the phenomenon of Epistasis is involved in the manifestation of AD. Understanding the GRNs involved in AD pathogenesis could be useful for designing effective therapeutics.
- siRNA- and shRNA-mediated knockdown of genes followed by transcriptome analysis are the major tools for elucidating the GRN.
- One of the most well-studied GRNs, with physiological and pathophysiological contexts, is the tumor suppressor and master cell cycle regulatory network, Rb pathway. The multifunctional Rb pathway seems to be involved in a wide range of physiological processes, from cell proliferation to cellular plasticity, through its interactions with more than 300 proteins and regulating the expression of a large number of genes, including epigenetic modulators. Understanding of RB1 pathway is guiding the discovery of targeted cancer therapeutics. A few of these have already received FDA approval to be used in the clinics.
- Reversal of the epigenetic landscape is no longer a dream or fiction. It is proven today that the ectopic expression of 3–4 TFs can induce pluripotency in any differentiated cells. It is also possible to push the cells from one lineage to another by using these TFs. Most interestingly, this kind of cell reprogramming is also possible by a set of small molecule compounds targeting epigenetic enzymes, kinases, etc. These discoveries have created tremendous potential in drug discovery and regenerative medicines.

Further reading

Chen, P., Lee, N.V., Hu, W., Xu, M., Ferre, R.A., Lam, H., Bergqvist, S., Solowiej, J., Diehl, W., He, Y.A., Yu, X., Nagata, A., VanArsdale, T., Murray, B.W., 2016. Spectrum and degree of CDK drug interactions predicts clinical performance. Mol. Cancer Ther. 15 (10), 2273–2281.

Emmert-Streib, F., Dehmer, M., Haibe-Kains, B., 2014. Gene regulatory networks and their applications: understanding biological and medical problems in terms of networks. Front. Cell Dev. Biol. 2, 38.

Friese, A., Ursu, A., Hochheimer, A., Schöler, H.R., Waldmann, H., Bruder, J.M., 2019. The convergence of stem cell technologies and phenotypic drug discovery. Cell Chem. Biol. 26 (8), 1050–1066.

Hochedlinger, K., Plath, K., 2009. Epigenetic reprogramming and induced pluripotency. Development 136 (4), 509–523.

Kallunki, T., Barisic, M., Jäättelä, M., Liu, B., 2019. How to choose the right inducible gene expression system for mammalian studies? Cells 8 (8), 796.

Knudsen, E.S., Pruitt, S.C., Hershberger, P.A., Witkiewicz, A.K., Goodrich, D.W., 2019. Cell cycle and beyond: exploiting new RB1 controlled mechanisms for cancer therapy. Trends Cancer 5 (5), 308–324.

Liu, C.C., Kanekiyo, T., Xu, H., Bu, G., 2013. Apolipoprotein E and Alzheimer's disease: risk, mechanisms and therapy. Nat. Rev. Neurol. 9 (2), 106–118.

Ma, X., Kong, L., Zhu, S., 2017. Reprogramming cell fates by small molecules. Protein Cell 8 (5), 328–348.

Najm, R., Jones, E.A., Huang, Y., 2019. Apolipoprotein E4, inhibitory network dysfunction, and Alzheimer's disease. Mol. Neurodegener. 14 (1), 24.

Semizarov, D., Kroeger, P., Fesik, S., 2004. siRNA-mediated gene silencing: a global genome view. Nucleic Acids Res. 32 (13), 3836–3845.

Singh, A.J., Ramsey, S.A., Filtz, T.M., Kioussi, C., 2018. Differential gene regulatory networks in development and disease. Cell. Mol. Life Sci. 75 (6), 1013–1025.

Velychko, S., Adachi, K., Kim, K.P., Hou, Y., MacCarthy, C.M., Wu, G., Schöler, H.R., 2019. Excluding Oct4 from Yamanaka cocktail unleashes the developmental potential of iPSCs. Cell Stem Cell. 25 (6) 737–753.e4.

Takahashi, K., 2012. Cellular reprogramming – lowering gravity on Waddington's epigenetic landscape. J. Cell Sci. 125 (11), 2553–2560.

VIII

Chemical biology: Drug discovery targeting the functional genome

VIII.1 Proteins involved in epigenetic modifications as drug targets

Epigenetic modifications of DNA and DNA-associated histones fine-tune the regulation of gene expression and most of the biological processes. The enzymes responsible for these modifications catalyze the reactions in a concerted manner and establish signal-dependent epigenetic language. This language is read by effector modules and converted to functional outcomes by various protein complexes. These modifications are essential for cellular physiology. The dysregulation of epigenetic modifications leads to a patho-physiological condition or disease manifestation. Thus the modulation of these modifying enzymes could be an attractive therapeutic strategy. Currently, two DNA methyltransferase (DNMT) inhibitors of azanucleosides class and one histone deacetylase inhibitor have been approved by the US Food and Drug Administration (FDA) for the treatment of certain types of cancer. However, several other molecules, targeting other components of the epigenetic machinery, are under development for several diseases. In this section, we shall discuss small-molecule modulators of histone acetyltransferases, histone deacetylases, histone methyltransferases, besides, DNMT inhibitors and microRNA-based therapeutics (Fig. VIII.1).

VIII.1.1 DNA methyltransferase inhibitors for epigenetic therapy

DNA methylation is the most stable epigenetic mark involved in both transgenerational and transcellular epigenetic phenomena. Genome methylation generally occurs at the CpG dinucleotide outside of the CpG islands (see Chapter III). In eukaryotic cells, most of the gene promoters are highly enriched in CpG islands. The transcriptionally active promoters are methylation free. However, the silenced genes are placed in the heterochromatin regions where the promoter-specific CpG islands are heavily methylated. The methylation of promoter DNA prevents transcription factor binding. The hypermethylation of the genome often leads to unwanted gene silencing, which can be causally related to several diseases, including cancer. Changes in DNA methylation pattern during the process of carcinogenesis is a hallmark of many cancers and termed epimutation. In cancer cells, hypermethylation of CpG islands is often found in the promoters of tumor suppressor genes, leading to the silencing of transcription of some tumor suppressor genes. The

Chemical Biology of the Genome. https://doi.org/10.1016/B978-0-12-817644-3.00009-X

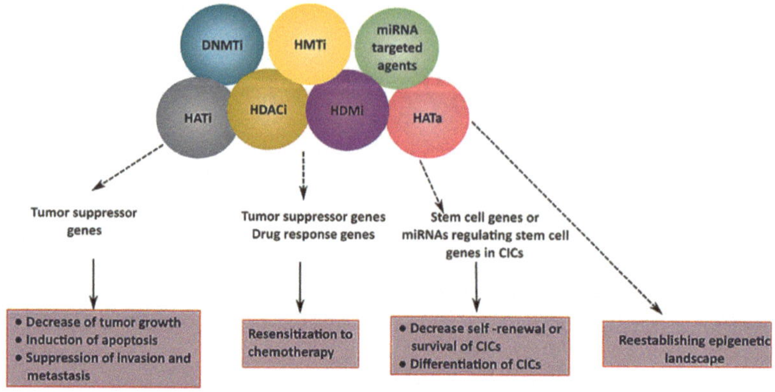

FIG. VIII.1 Epigenetic reprogramming using small-molecule modulators such as DNA methyltransferase inhibitors (DNMTi), histone methyltransferase inhibitors (HMTi), histone acetyltransferase inhibitors (HATi), histone deacetylase inhibitors (HDACi), histone demethylase inhibitors (HDMi), and histone acetyltransferase activators (HATa) and mRNA targeted small stretches of nucleotides (not) or nt conjugates are promising candidates for therapy. Some of these molecules are already approved drugs. The physiological functions of these molecules have been indicated in the boxes. *Modified from Kwon, M.J., et al., 2011. Int. J. Mol. Sci. 12(2), 983–1008. MDPI.*

hypermethylation could be caused by the overexpression of DNMTs. It was indeed found that in several cancers, including that of breast, lung, stomach, colon, and hematopoietic cells, the expression of DNMTs is deregulated. Quite rationally, it was thought that DNMTs inhibitor could be a potential class of antineoplastic therapeutics. Indeed, it was found that DNMT inhibitors reverse the methylation of tumor suppressor gene promoters and activate the expression of these genes.

There are two classes of DNMT inhibitors: nucleoside inhibitors and nonnucleoside inhibitors. The nucleoside inhibitors are often cytidine analogs, which are incorporated in the DNA during replication and form covalent suicidal complexes with the recruited DNMT enzyme. The nonnucleoside inhibitors are a diverse group of compounds that directly bind and inhibit different classes of DNMTs. We shall discuss a few inhibitors of both the classes.

VIII.1.1.1 Nucleoside inhibitors: Azacitidine, decitabine, and zebularine

Azacitidine (5-azacytidine) and decitabine (5-Aza-2′-deoxycytidine) are cytidine nucleoside analogs modified in position 5 of the pyrimidine ring (Fig. VIII.2). These compounds were used for chemotherapy against leukemia even before it was known to have any effect on DNA methylation. However, mechanistically, in human cells, these compounds are transported by nucleoside transporter protein-1. Inside the cell, these compounds are modified to their respective active triphosphate forms. The azacytidine is converted to 5-azacytidine-5′-triphosphate by uridine cytidine kinase and the decitabine to 5-aza-2′-deoxycytidine-5′-triphosphate. The azacytidine gets incorporated into both RNA and DNA (predominantly RNA), but decitabine, being a deoxyriboside analog, is only incorporated into DNA (Fig. VIII.3). During the S phase of the cell cycle, when DNA is replicating, these compounds get incorporated into the newly replicating DNA. Incorporating an abnormal base (azanucleoside) into the DNA prevents the methylation reaction because

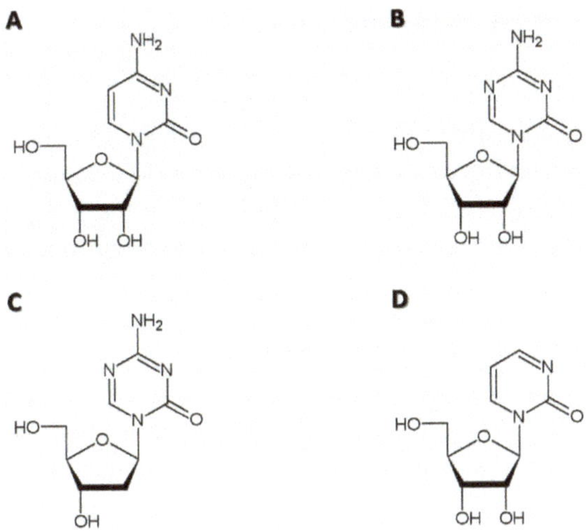

FIG. VIII.2 Chemical structures of nucleoside inhibitors. (A) Cytidine, (B) azacytidine, (C) decitabine, and (D) zebularine.

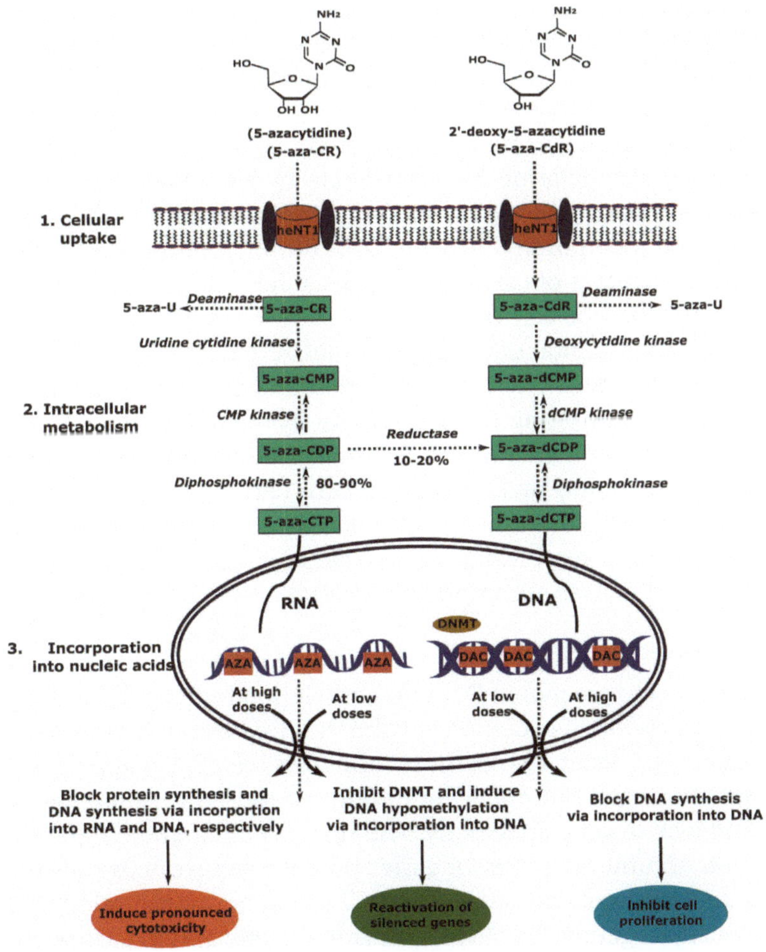

FIG. VIII.3 Molecular mechanisms of azacytidine and decitabine actions in human cells: the drugs are transported into the cell by nucleoside transporter 1 (hENT1) and processed as indicated. Source: *Agrawal, K., Das, V., et al., 2018. Pharmacol. Ther. 188, 45–79. Elsevier.*

the 5-position has a nitrogen atom instead of a carbon atom. This leads to an irreversible covalent complex formation between the DNA and DNMT because of which not only the DNMT enzyme inactivates but also induces the DNA damage signals. These events eventually reduce the level of DNA methylation because the enzyme concentration goes down, and the 5-position with a nitrogen atom cannot be methylated (Fig. VIII.4).

Both azacytidine (Vidaza; Celgene) and decitabine (5 aza 2′ deoxycytidine) (Dacogen; SuperGen) have been approved by the US FDA for the treatment of acute myeloid leukemia (AML), chronic myelomonocytic leukemia (CMML), and myelodysplastic syndrome (MDS). It is noteworthy that although azacytidine and decitabine are commonly used chemotherapeutic agents, they are limited by poor chemical stability in physiological media, poor bioavailability, and high toxicity. The enzyme cytidine deaminase (CDA) metabolizes azacytidine and decitabine into 5-azauridine. In the human liver and spleen, there is a high level of CDA, which significantly reduces the half-life of both the drugs.

Zebularine is also a nucleoside analog, but it possesses better chemical stability and bioavailability. It is highly stable at acidic to neutral pH with a half-life of ~44 h, as opposed to 41 min for azacytidine and decitabine in physiological conditions. Thus zebularine is more suitable for oral administration as a drug. Similar to azacitidine and decitabine, zebularine also forms an irreversible covalent complex of DNMT and zebularine-substituted DNA. On entering into the living cell, it gets phosphorylated by uridine-cytidine kinase and then converted into the 2′-deoxyzebularine-5′-diphosphate by ribonucleotide reductase. The 2′-deoxyzebularine-5′-diphosphate is converted to 2′-deoxyzebularine-5′ triphosphate by an unknown enzyme and incorporated into the DNA during replication of DNA. It was found that the zebularine treatment of cancer cells arrests the cells at the G2/M stage of the cell cycle and thereby reduces the cell viability.

A large body of experimental data suggests that zebularine is stable in aqueous solution, possesses selective specificity for cancer cells, and ensures minimal general toxicity. It can be used for continuous long-term treatment by the oral route, which could prevent gene remethylation. In vitro, it shows a promising effect on a wide range of cancers, including cancers of bladder, leukemia, stomach, cervix, lung, pancreas, breast, liver, colon, brain, osteosarcoma, and oral squamous cell carcinoma. Besides azacitidine, decitabine, and zebularine, several other nucleoside analogs are in the advanced stage of development (Table VIII.1), as summarized further.

VIII.1.1.2 Nonnucleoside inhibitors

Nucleoside inhibitors induce global DNA demethylation by getting incorporated into the DNA and forming an irreversible complex with DNMT. These events are highly effective in cancer therapy, and as a result, two of these inhibitors have been approved by the FDA as anticancer drugs. However, the major limitations of these drugs are cytotoxicity, mutagenic effects, and lack of gene specificity. The nonspecific demethylation not only leads to the reactivation of tumor suppressor genes but also some oncogenes that are responsible for cancer development and progression. Thus the next-generation DNMT inhibitors are being developed, targeting the DNMT enzyme directly and in a class-specific manner

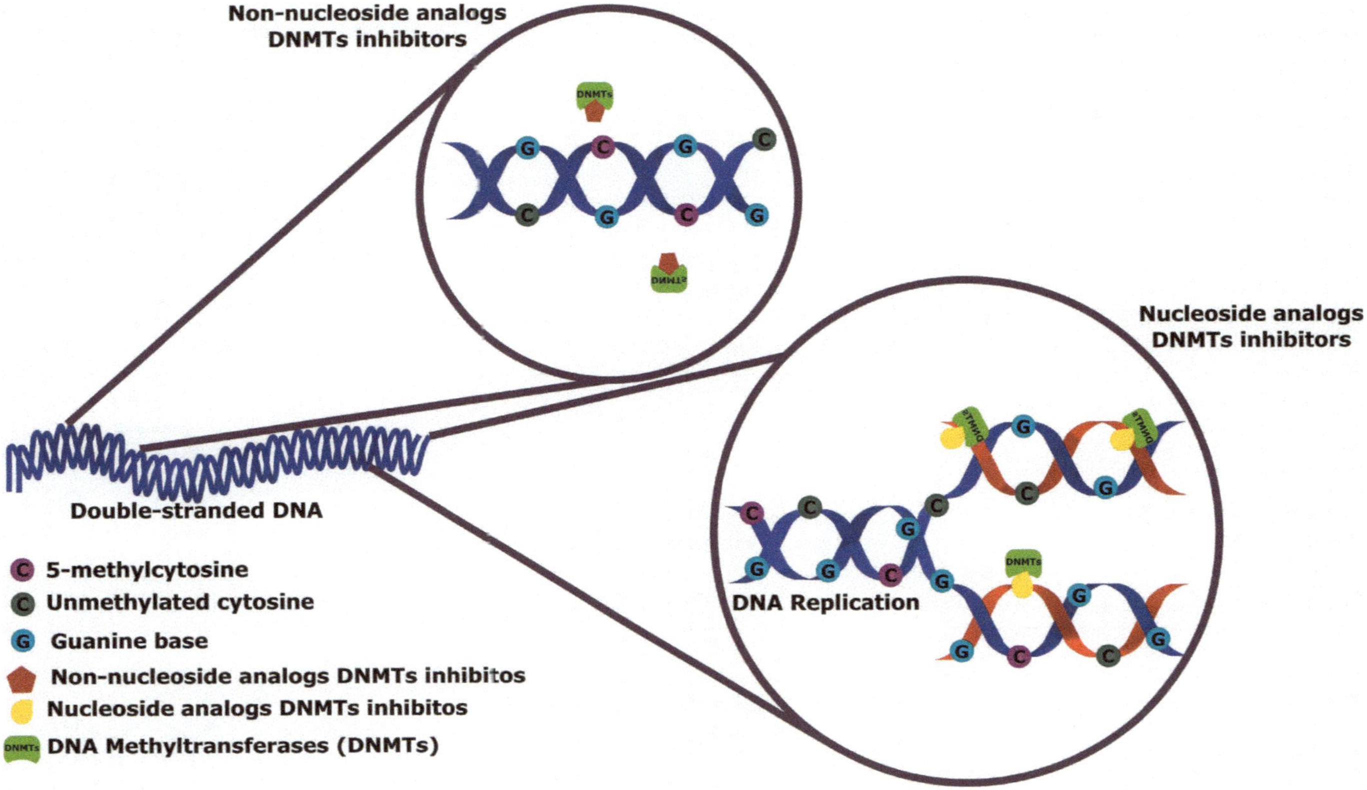

FIG. VIII.4 Mechanism of action of nucleoside and nonnucleoside DNMT inhibitors. Source: *Marques-Magalhães, Â., et al., 2018. Front. Pharmacol. 9, 366. Frontiers Media SA.*

Table VIII.1 First generation nucleosidic DNMTIs in developmental stage.

Drug	Name and structure	Characteristics	Target	Preclinical activity	Clinical phase
6-tG	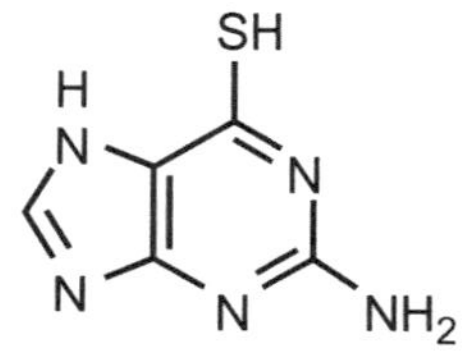	• Incorporates into DNA and RNA and inhibits DNA and RNA synthesis • Inhibits DNA methylation through proteasomal degradation of DNMT1	Leukemia, kidney	–	–
FdCyd	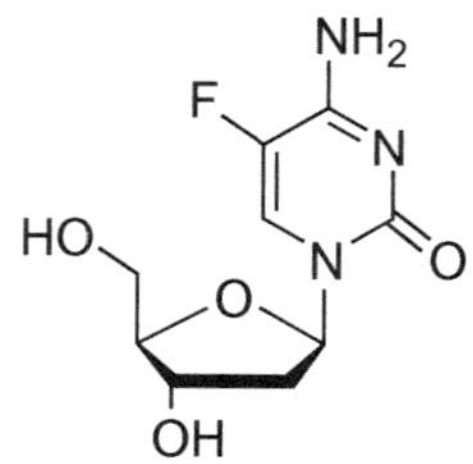	• Stable in aqueous solution • Orally bioavailable when coadministered with CDA inhibitor, THU	Cervix, melanoma, breast, colon, bladder, brain	• The combination of 6 mg/kg FdCyd + 100 mg/kg THU, i.v. dose showed no significant in vivo activity	Phase 1/2
ψϕCyd	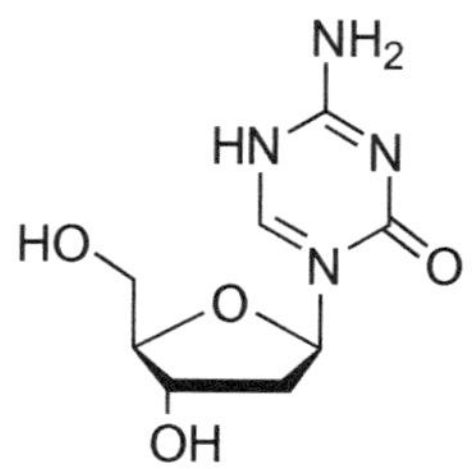	• Exceptional hydrolytic stability • Resistant to enzymatic deamination by CDA • Reduced cytotoxicity • Orally bioavailable	Leukemia	• Antileukemic activity at 60–150 mg/kg, i.p. or 100–150 mg/kg, p.o. doses in leukemia mouse models	–
DHAC	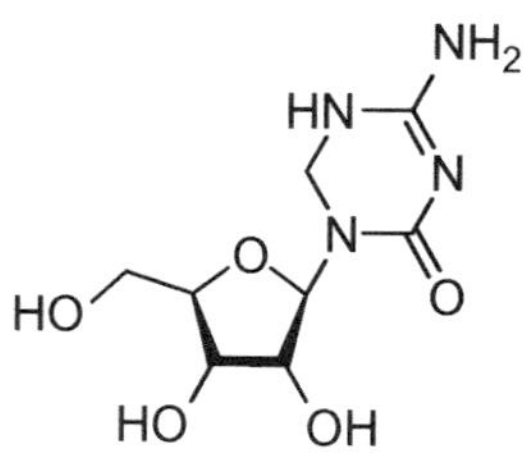	• Stable in aqueous solution • Modest hematologic toxicity profile	Lymphoma, leukemia, breast, prostate	• 25.1% and 46.3% decrease in DNA methylation at 1500 mg/kg, i.p. dose in dCK(0) and dCK(–) mouse leukemic cancer xenografts	Phase 1/2

Ara-AC	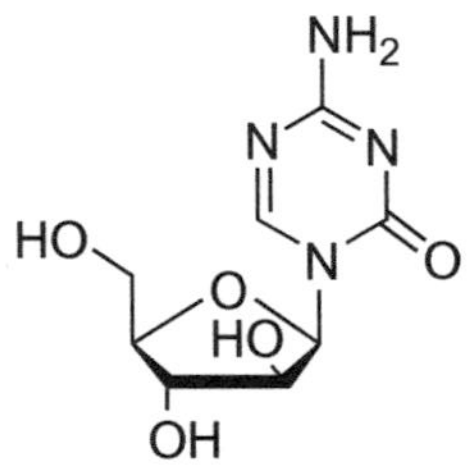	• Protected from deamination by CDA • Orally bioavailable	Colon, leukemia	• Ara-AC, administered i.p. or i.v. demonstrated	Phase 1/2
Zeb	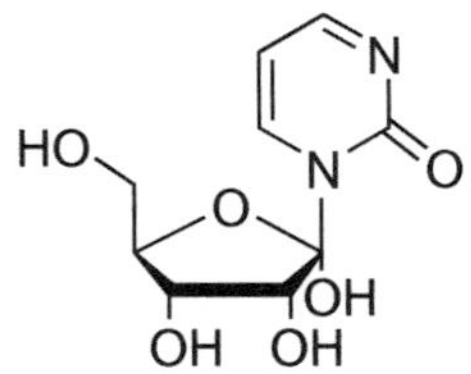	• Stable in aqueous solution • Potent inhibitor of CDA • Selective specificity for cancer cells ensures minimal general toxicity • Continuous long-term treatment possibility prevents gene remethylation • Orally bioavailable	Bladder, leukemia, stomach, cervix, lung, pancreas, breast, liver, colon, cholangiocarcinoma, brain, osteosarcoma, oral squamous cell carcinoma	• Significant reduction in tumor volume at high dose	–
DHDAC	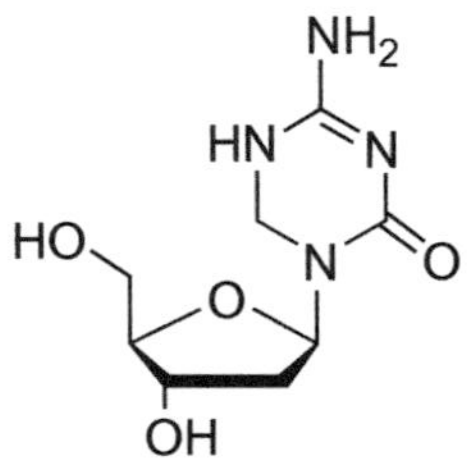	• High aqueous stability • Minimal cytotoxicity	Leukemia	–	–
TdCyd 5-aza-TdCyd	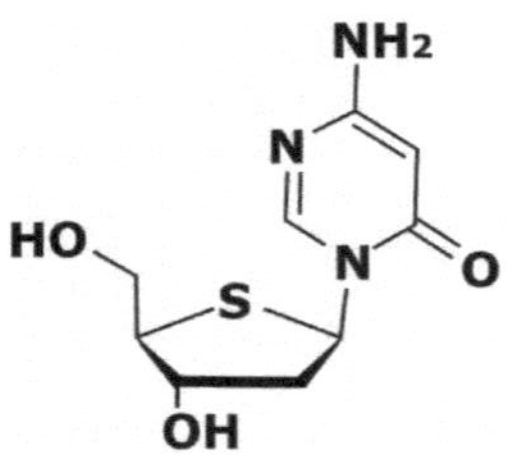 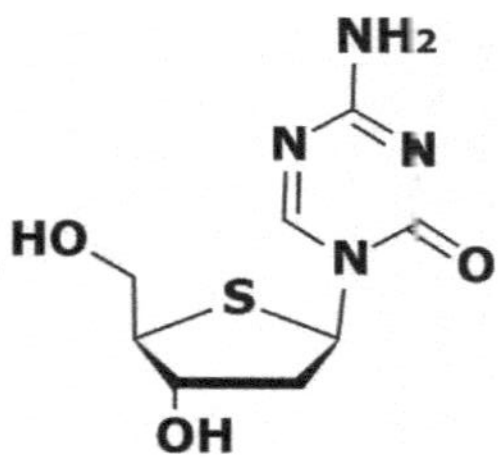	• Longer half life • Minimal off-target toxicity • Orally bioavailable	Leukemia, lung.	• TdCyd at 1.3 and 0.9 mg/kg and 5-aza-TdCyd	Phase 1/2

(Fig. VIII.4). At present, several small-molecules nonnucleoside DNMT inhibitors have been identified. These include epigallocatechin-3-gallate, hydralazine, procainamide, procaine, and RG108. Significantly, these inhibitors directly bind to the catalytic region of DNMTs but do not get incorporated into DNA.

Although conceptually, these compounds should be better compounds for therapeutic purposes, their efficiency for demethylating the gene of interest and reactivation was found to be insufficient in the cellular system. Demethylation could also be achieved by directly inhibiting the expression of a specific enzyme using antisense or siRNA. An inhibitor of DNMT1, MG98, has been developed for this purpose. It is a 20-base pair antisense oligonucleotide that binds specifically to the 3′-untranslated regions (3′UTR) of human DNMT1 mRNA and inhibits its transcription. Although MG98 was found to be partially effective in inducing demethylation of a few genes and inhibit the xenograft tumor growth in mice, it could not be developed further for therapeutic purposes due to its toxicity.

VIII.1.2 Targeting specific gene expression to counter drug resistance

Chemotherapeutic drugs are the major cause of the dramatic increase in the life expectancy of human beings and win over different infectious diseases and cancer. The bacterial, viral, and parasitic diseases are combated predominantly by the small-molecule chemical entities, besides vaccines in a few cases. The treatments of bacterial diseases with properly-known antibacterial chemical entities began with the discovery of a series of Arsenic-derived compounds discovered by Alfred Bertheim and Paul Ehrlich in 1907. In 1910 the first antibacterial drug became available commercially (marketed for human use, named "salvarsan" and now known as "arsphenamine") against "syphilis" (a sexually transmitted infection caused by the bacterium *Treponema pallidum*). The natural compound penicillin (from mold, *Penicillium chrysogenum,*) discovered by Alexander Fleming in 1928, was the first antibiotic discovered (Fig. VIII.5). However, the era of antibacterials opens with the discovery and development of sulfonamide drugs (Figs. VIII.6 and VIII.7) in

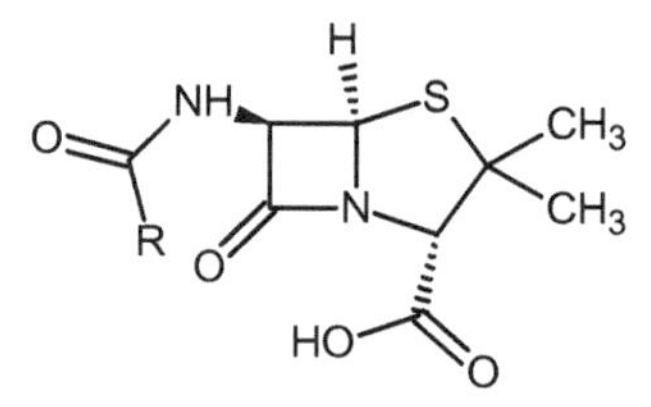

FIG. VIII.5 Chemical structure of penicillin.

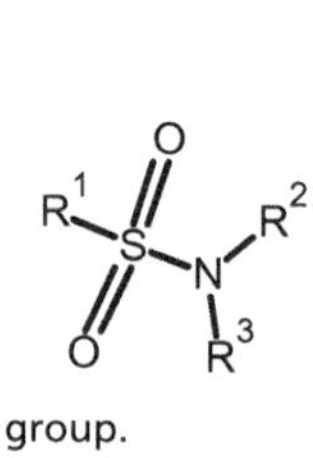

FIG. VIII.6 General structure of the sulfonamide group.

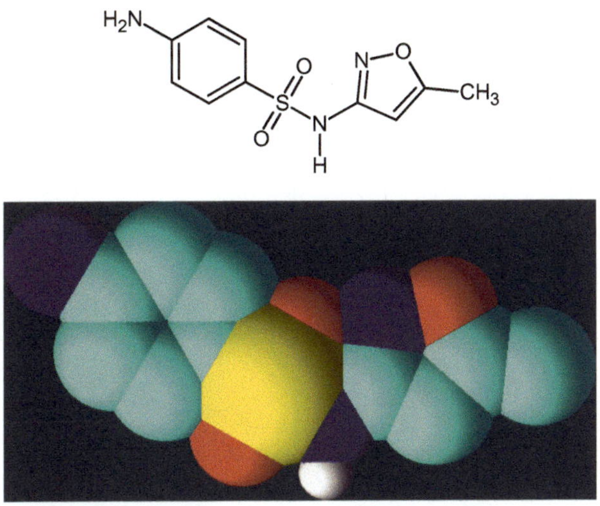

FIG. VIII.7 Chemical structure of sulfamethoxazole, a widely used antibiotic.

1932–33 by Gerhard Domagk, in 1935 for which he also got the Nobel Prize but was forced to reject it. Since then, a large number of antibacterial/antibiotics have been discovered, which belong to different classes of fungal origin compounds, glycopeptides, lipopeptides, sulfonamides, macrolides (large macrocyclic lactone), etc.

Besides bacteria, there are numerous parasites and viruses that cause infectious diseases to humans and animals. The parasitic diseases are caused by helminths, amoeba, ectoparasites, parasitic fungi, and protozoa. Broad-spectrum antiparasitic drugs may also be effective against bacterial diseases. For example, metronidazole, developed against parasites, can also be used as an antibacterial agent.

Both DNA and RNA viruses cause viral diseases. These include HIV, herpes virus, hepatitis B and C viruses, and influenza A and B viruses, and of course, the SARS-COV2 virus. Unlike most antibiotics, antiviral drugs do not destroy the target pathogen; instead, they inhibit the reproduction of the virus.

However, during the course of the long use of these drugs, antimicrobial resistance often occurs when the same microbe cannot be killed or inhibited by the given drug at the same dose or even a higher dose. The resistance may arise because of (1) natural resistance, (2) genetic mutation, or (3) horizontal gene transfer (one species of bacteria acquiring resistance from another). The extended and indiscriminate use of antibacterial drugs may induce the selection of drug-resistant bacteria. Preventing the misuse of antimicrobial drugs could be helpful in reducing the development of drug resistance. Besides the direct consumption of these drugs, human beings may also be subjected to a huge amount of antimicrobial drugs through food. For example, antibiotics are used as animal food supplements and a preventive measure to reduce the probable infection for the livestock. These animal products are a rich source of unwanted antibiotics.

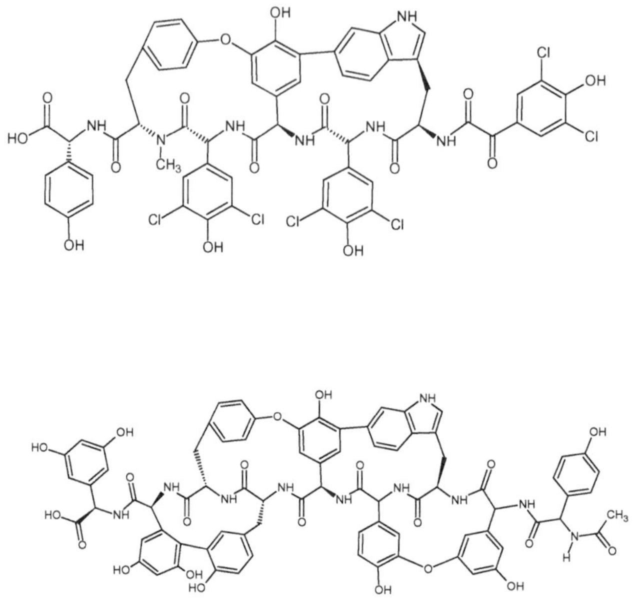

FIG. VIII.8 Chemical structures of complestatin *(upper panel)* and corbomycin *(lower panel)*.

The development of microbial drug resistance has forced scientists to search for drugs that will function by different mechanisms and find new targets through a better understanding of disease biology, especially host-pathogen interactions. In this effort, the most recent success story is the glycopeptide class of antibiotics: corbomycin and complestatin (Fig. VIII.8). These two antibiotics target the drug-resistant *Staphylococcus aureus*, a group of bacteria that can cause many serious infections. Mechanistically, these antibiotics inhibit the cleavage of the bacterial cell wall, thereby stopping the bacterial cell division.

Besides designing and exploring the discovery of new antimicrobials, the effort is also on to find out the new targets, such as host-pathogen interaction factors and epigenetic machinery of both pathogens and the host. All the pathogens (virus, bacteria, and parasites) alter the host cells' epigenetic landscape to evade the immune response. This modulation of the epigenetic state could be a unique drug target and may be useful to overcome drug resistance.

The reversible acetylation of histones by lysine acetyltransferases and deacetylases (KATs and KDACs, respectively) plays a central role in regulating gene expression associated with almost all biological phenomena, including immunological functions. The metabolites and virulence factors of the pathogenic microorganisms alter the acetylation status in the host cell by multiple routes: (1) induce degradation of KATs and thereby inhibit the KAT activity; (2) enhance the expression of KDACs; (3) induce recruitment of KDACs; and (4) inhibit KDACs. We shall discuss a few typical examples of these factors.

The enteropathogenic and enterohaemorrhagic *Escherichia coli* are significant causes of gastrointestinal diseases worldwide. These bacteria secrete a zinc metalloproteinase,

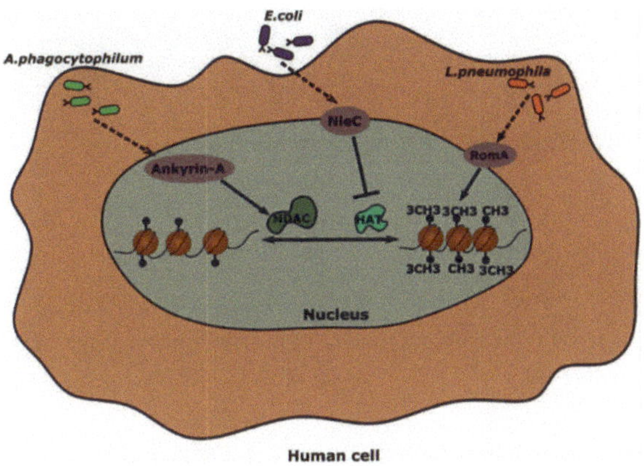

FIG. VIII.9 Infectious microorganisms modulate host cell epigenetic land scape to evade the immune response. The reversible acetylation balance is one of the key targets. For example, ankyrin A of *Anaplasma phagocytophilum* upregulates HDAC1 and promotes HDAC1 recruitment to gene promoters. On the other hand, hypoacetylation may also be created by the degradation of lysine acetyltransferase, p300, by a metalloproteinase NleC produced by *Escherichia coli* upon infection. Acetylation could also be controlled by inducing histone H3 trimethylation, by an effector molecule RomA, produced by *Legionella pneumophila. Modified from Grabiec, A.M., Potempa, J., 2018. Crit. Rev. Microbiol. 44(3), 336–350. Taylor & Francis.*

NleC, which specifically binds to the master epigenetics enzyme, KAT3B/p300, and degrades it (Fig. VIII.9). In this manner, these bacteria suppress the host inflammatory responses. Regarding induction of the expression of histone deacetylases (HDACs/KDACs) by the bacterial infection, the ideal example could be the *Pseudomonas aeruginosa*-mediated infection in the inflamed airway of patients with cystic fibrosis and burn wounds. These bacteria release a quorum-sensing molecule, 2-aminoacetophenone, which induces the expression of class I, histone deacetylase, HDAC1 in human THP-1 monocytes. The overexpression of HDAC1 leads to hypoacetylation of histone H3K18, which represses the expression of inflammatory cytokines and chemokines (e.g., TNF, IL-1b, and MCP-1), leading to the inhibition of host cell immune response. The hypoacetylation state of the epigenome to evade the immune system could also be created by the enhanced recruitment of deacetylases on a specific locus. *Anaplasma phagocytophilum* is an obligate Gram-negative bacterium that causes anaplasmosis in sheep and cattle, known as tick-borne fever and pasture fever. It survives and propagates within neutrophils and causes human granulocytic anaplasmosis too. It disrupts the neutrophil function affecting the defense-related gene expression globally. It is achieved, at least partially, by the enhanced recruitment of HDAC1 at promoters of the host defense gene through the bacteria-secreted ankyrin A (Fig. VIII.10). The enhanced recruitment of HDAC1 leads to hypoacetylation of histone H3 and thereby the suppression of the target gene expression. The pathogenic bacteria may also create a completely reverse situation for their own need. The metabolic products of these bacteria, such as short-chain fatty acids (SCFAs), which are predominant fermentation products of anaerobic bacteria, potently inhibit the host class I and II HDACs. The inhibition of HDACs induces hyperacetylations in the host

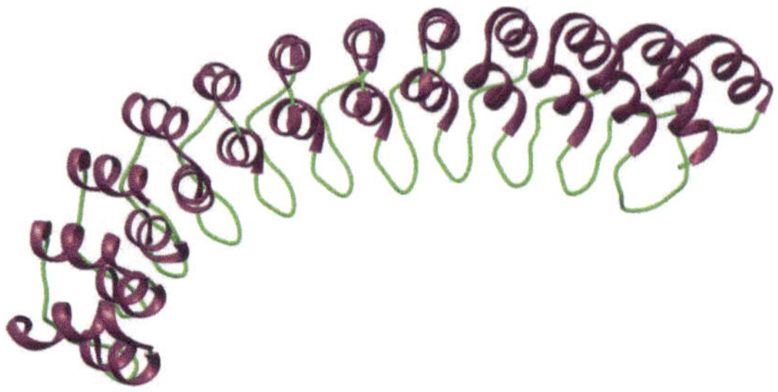

FIG. VIII.10 Crystal structure of ankyrin repeat. Ankyrin from *Anaplasma phagocytophilum* induces the expression of HDAC1.

cells and promotes the generation of regulatory T cells (Tregs). Hyperacetylation of histones H3 and H4 have also been documented and causal connection established in the bovine mastitis model (mouse) system using pathogenic *E. coli*, isolated from the natural field. The cause of hyperacetylation, however, is yet to be found (Table VIII.2).

These examples clearly establish that microbial pathogens can target both arms of the reversible histone acetylation systems of the host—KATs/HATs and KDACs/HDACs to maneuver the inflammatory response through epigenetic reprogramming. Therefore histone deacetylases, acetyltransferases, and bromodomain and extra terminal domain (BET) proteins are being considered as the new-generation targets to tackle bacterial and other microbial infections, especially drug resistance. The HDAC-specific inhibitors, rather than global HDAC inhibitors, have been found to be effective as modulators of the immune response in the case of microbial infection.

Table VIII.2 Pathogenic bacterial infection alters histone acetylation landscape in host cells.

Bacterial species	Bacterial effector	Epigenetic effect
Mycobacterium tuberculosis	Unknown	AcH4↓
Helicobacter pylori	Unknown	AcH3K23↓
Listeria monocytogenes	Listeriolysin-O	AcH3K18 and AcH4↓
	InlB	AcH3K18↓
Legionella pneumophila	RomA	AcH3K14↓
Anaplasma phagocytophilum	Ankyrin A	HDAC1↑
Pseudomonas aeruginosa	2-Aminoacetophenone	HDAC1↑
E. coli	Proteinase NleC	P300↓
Porphyromonas gingivalis	Unknown	HDAC1 and HDAC2↓
	SCFAs	AcH3, AcH4↑
E. coli	Unknown	H3K14↑, H4K8↑

↓ Reduction of histone acetylation or downregulation of protein expression?

↑ Induction of histone acetylation or upregulation of protein expression.

AcH, acetylated histone; *HDAC*, histone deacetylase; *SCFAs*, short-chain fatty acids.

Modified from Grabiec, A.M., Potempa, J., 2018. Crit. Rev. Microbiol. 44(3), 336–350. Taylor & Francis.

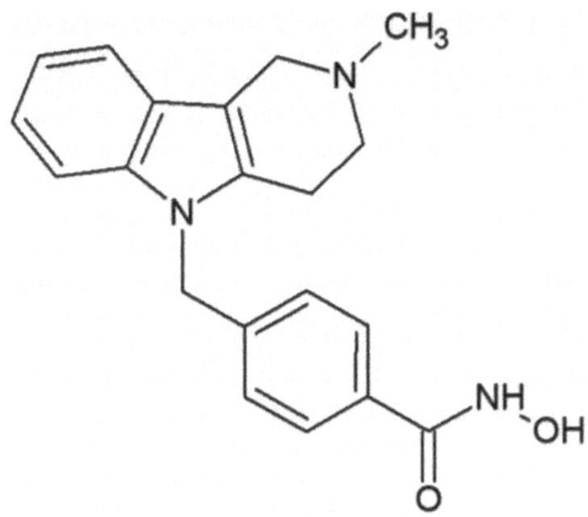

FIG. VIII.11 Chemical structure of tubastatin A.

In the case of *E. coli* infection to human macrophage cells, it was observed that the HDAC6-specific inhibitor tubastatin A (Fig. VIII.11) efficiently induced the killing of the bacteria. These results were also validated further in the lethal sepsis model system, where tubastatin A could significantly improve the survival of the animal. Treatment of tubastatin A could also restore the innate immune cell population in the sepsis model animal. However, the class I-specific inhibitor MS-275 (Fig. VIII.12) could not exhibit these functions.

However, simultaneous treatment with HDAC inhibitors promoted clearance of intracellular bacteria by enhancing the generation of mitochondrial reactive oxygen species. Interestingly, only the HDAC6-specific inhibitor tubastatin A, but not MS-275, which targets only class I HDACs, improved bacterial killing by macrophages, indicating a unique and nonredundant role for HDAC6 in this process. There is another similar example in this context, for *Bacillus anthracis* infection. It was found that HDAC8-specific inhibitor, PCI-34051, could significantly prevent the lethal effect of anthrax toxin. Specific inhibition of HDAC8 at least could partially restore the immune response of the host on *B. anthracis* infection. On the contrary, when microbial infection creates the hyperacetylation state due to inhibition of HDACs or activation of HATs (?), the inhibitors of HATs or BET protein modulators could be used to reverse the cellular response.

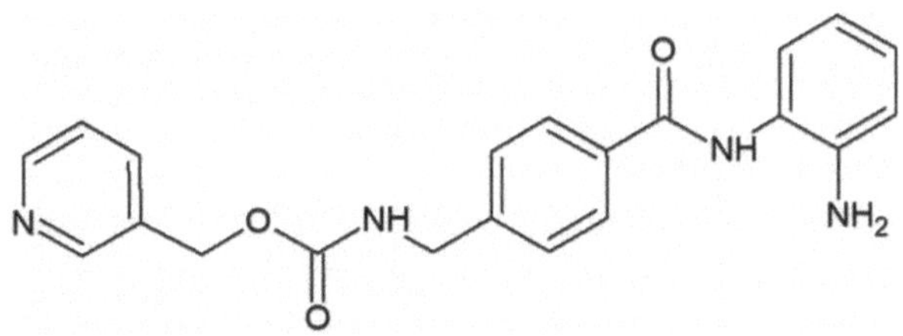

FIG. VIII.12 Chemical structure of MS-275.

Epigenetic machinery is not the monopoly of eukaryotic organisms. All microorganisms, including bacteria, possess well-orchestrated epigenetic tools. They use it not only to establish the pathogenicity but also to develop drug resistance. We shall discuss here only the bacterial epigenetic system in the context of drug resistance, although other microorganisms, especially protozoan parasites, own highly evolved epigenetic systems.

Although the bacterial genome is not packed in orderly nucleosomal structures, besides DNA and RNA methylation, reversible protein acetylation and methylation are also involved in regulating bacterial gene expression. There are two types of DNA methyltransferases (MTases): restriction-modification system (RM system) (e.g., EcoRV, CfrBI, etc.) and orphan methyltransferases. The orphan methyltransferases are of two kinds—adenine methyltransferases (Dam, CcrM, Yhdj, etc.) and cytosine methyltransferases (Dcm, VchM, etc.). Physiologically, prokaryotic DNA MTases are involved in multiple key functions, including antiviral defense, cell cycle regulation, DNA replication and repair, and transcriptional regulation.

The RM system and orphan methyltransferases play a significant role in the modulation of gene expression, either interfering in the RNA polymerase binding to the promoter or recruitment of transcriptional cofactors. The binding of RNA polymerase is inhibited when the promoters of transposon Tn10, genes of tryptophan operon (trpR and trpS), are methylated. However, in bacteria, methylation can also promote gene expression. A transcriptional cofactor of sigma 70 from *Caulobacter crescentus* stably binds to the promoter region only when the nearby adenine is methylated. There are several examples where methylation of DNA prevents the binding of the repressor—for example, the binding of repressor OxyR in *E. coli* involved in biofilm formation. Several genes involved in the horizontal gene transfer are regulated by the methylation of DNA. Because horizontal gene transfer is directly involved in the development of drug resistance, it is evident that the bacterial epigenetic system plays a critical role in the development of drug resistance.

Exposure to a sublethal dose of antibiotics or other antibacterial leads to adaptive resistance due to genetic and epigenetic changes in the bacterial genome. For example, when *E. coli* bacteria are subjected to a low concentration of antibiotics from the β-lactam and quinolone classes, methylation of the DNA is enhanced, which helps the development of antibiotic resistance. Methylation of cytosine induces the cytosine to thymine mutation because the frequency of deamination of m5C is several-fold higher than that of cytosine (Fig. VIII.13). These methylation-mediated genetic changes may help to develop drug resistance. Therefore to combat drug resistance, the epigenetic machinery of both the host and the microorganisms are the most promising upcoming targets.

VIII.1.2.1 Drug resistance in cancer

The challenge of drug resistance in cancer is quite similar to microbial infectious diseases. The initial regime of cancer therapy was single-agent chemotherapy. The development of drug resistance against single-agent chemotherapy was addressed by the invention of combinatorial therapy, where multiple therapeutic agents with nonoverlapping

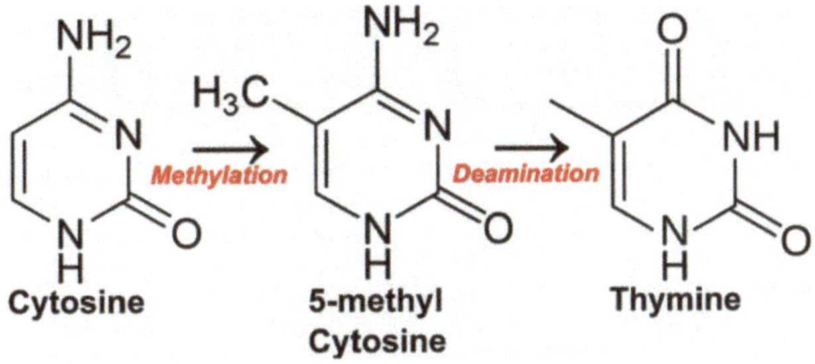

FIG. VIII.13 Mechanism of cytosine to thymine mutation.

mechanisms of functions are administrated. However, after 50 years of administration, polychemotherapy has also become less effective. It is evident that surgery, polychemotherapy, and radiation therapy failed to stop cancerous growth and metastasis. Meanwhile, our understanding of the disease biology of cancer has advanced substantially. The major hallmarks of cancers (Fig. VIII.14) were identified. The gene networks behind the operation of these hallmarks are the pathological determinants of the cancers and, thus, become the effective targets of cancer therapy. These targets include tyrosine kinases and nuclear receptors. In this context, massive success was achieved by the antagonists of the estrogen receptor (ER) and androgen receptor (AR) and inhibitors of BCR-ABL, HER2, and EGFR. However, the development of resistance to these drugs also becomes just a matter of time. The next cancer therapeutics discovered is immunotherapy: Anti-CTLA4 and anti-PD-1/PD-L1 monoclonal antibodies, which target the adaptive immune system. As expected, the resistance also developed against immunotherapy.

At this juncture, one will be interested to know how the resistance develops. The drug resistance is impregnated in the inherent nature of cancer itself. There are multiple factors that contribute to the drug resistance of cancer, which include (1) immune system and tumor microenvironment; (2) tumor burden; (3) tumor growth; (4) tumor heterogeneity; (5) physical barrier; (6) undruggable genome; and (7) therapeutic pressure (Fig. VIII.15). In the advancement of our understanding of these factors, along with the elucidation of the functional mechanisms of cancer hallmarks, all these factors have been explored for designing therapeutics (Table VIII.3).

Targeted cancer therapy created a revolution in cancer care. For example, selective estrogen receptor modulators are used in (ERa)-positive (ER^+) breast cancer treatment. ER^+ breast cancer accounts for almost 70% of all types of breast cancer. Overexpression of ER induces tumor growth. Tamoxifen (TAM), an antineoplastic nonsteroidal selective estrogen receptor modulator (SERM), is predominantly used to treat a patient with ER^+ breast cancer. It is a 1,2-diphenylethylene moiety-containing compound. Tamoxifen binds to the estrogen receptor and thereby prevents the binding of estradiol. Tamoxifen-bound receptor cannot bind to the cognate gene promoter, which leads to the inhibition of estrogen response (Fig. VIII.16). The treatment regime with TAM dramatically reduced the recurrent events and increased patient survivability. However,

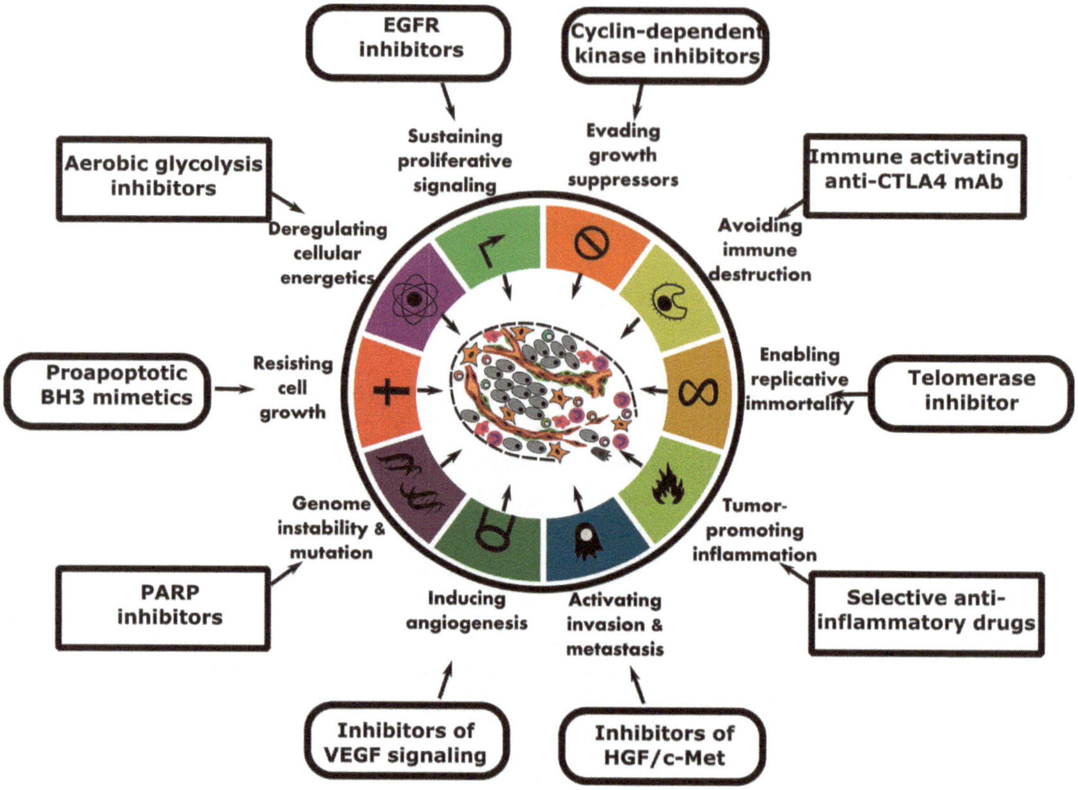

FIG. VIII.14 Cancer cell hallmarks are the target of drugs: During the development of malignancy, cancer cells acquire important characteristics to avail tremendous advantages and abilities, which are proliferative signaling, evading growth suppressors, resisting cell death, enabling replicative immortality, inducing angiogenesis, and activating invasion and metastasis. Behind each of these hall marks, there are highly orchestrated gene network regulated genetically and epigenetically. Promising cancer therapeutics targeted to the hall marks is indicated. Source: *Hanahan, D., Weinberg, R.A., 2011. Cell 144(5), 646–674. Elsevier.*

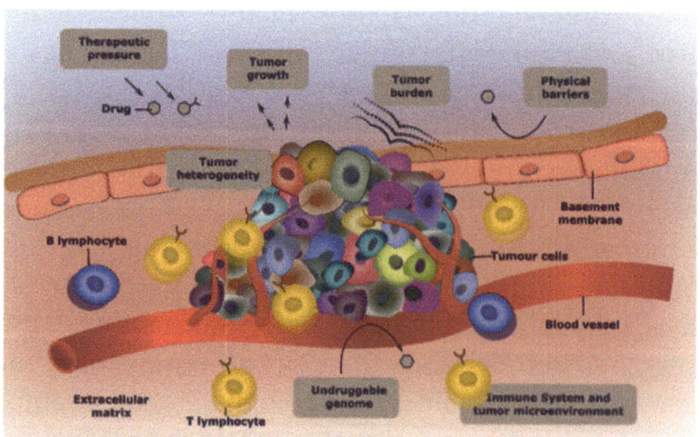

FIG. VIII.15 Biologically cancerous tumor develops in a physical and chemical environment that is resistant to the therapeutics. The factors that contribute to the drug resistance of the malignant tumor growth are shown here. Source: *Vasan, N., et al., 2019. Nature 575(7782), 299–309. Springer Nature.*

Table VIII.3 Measures to overcome the drug resistance in cancer.

	Tumor growth	Tumor burden	Tumor heterogeneity	Physical barriers	Immune system and tumor microenvironment	Undruggable genome	Therapeutic pressure
How to overcome resistance	• Dynamic monitoring • Functional-imaging studies • Dose density and more complete killing of cells	• Radiotherapy • Surgery • Neoadjuvant chemotherapy	• Early detection • Combination therapy • Targeting tumor neoantigen	• Local therapies • Functional-imaging studies • Small molecules engineered to penetrate sanctuary sites	• Enhancing tumor recognition by the immune system • Antiangiogenic therapy • Cellular therapies	• Transcription factor inhibitors • Allele-specific inhibitors • Restoring the function of tumor suppressors	• Next-generation TKIs that overcome resistance mutations • Allosteric inhibitor • Antibody-drug conjugates

Source: Vasan, N., et al., 2019. Nature 575(7782), 299–309. Springer Nature.

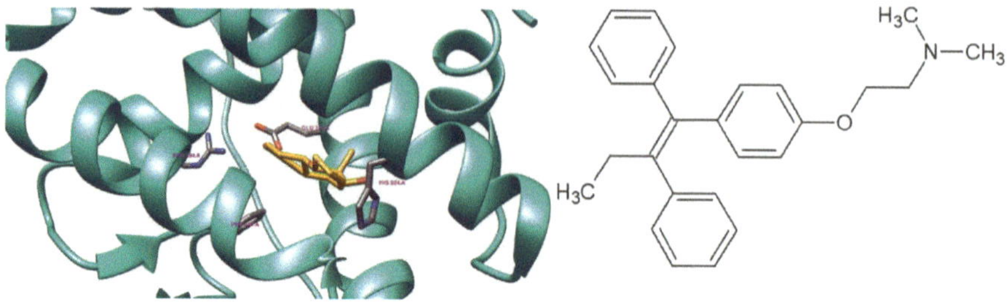

FIG. VIII.16 Tamoxifen, an antineoplastic nonsteroidal selective estrogen receptor modulator (SERM). *(Left panel)* Tamoxifen; *(Right panel)* ER-bound tamoxifen.

unfortunately, over the period, approximately 30%–40% of patients treated with TAM as adjuvant therapy showed deadly uncontrollable metastatic symptoms. These TAM-resistant patients generally do not survive. Several mechanisms have been proposed for the development of TAM resistance. Recent evidence has established an epigenetic link to TAM resistance.

It has been shown that the expression of Spalt-like transcription factor 2 (SALL2) is downregulated in several cancers, including breast cancer, ovarian cancer, and esophageal squamous cell carcinoma. Interestingly, reduced expression of SALL2 was also found in TAM-resistant patients where tumor growth is not dependent on estrogen. The expression of SALL2, however, could be restored using a DNMT inhibitor. Furthermore, the treatment with a DNMT inhibitor could resensitize TAM-resistant breast cancer to TAM therapy, both in the orthotopic breast cancer model system and a few breast cancer patients (Fig. VIII.17).

VIII.1.3 Histone lysine and arginine methylation as the target of therapeutics

The deregulation of lysine and arginine methyltransferases and the corresponding demethylases are causally associated with many diseases and the aging process. These diseases include cancer, neurological disorders, metabolic disorders, and kidney diseases. The methylation of K and R is also critical for infections by microorganisms and organ developments. The corresponding genes of these enzymes may undergo amplifications, deletions, mutations, and abnormal expression due to deregulated promoter function. Epigenetic reprogramming using small-molecule modulators is, thus, now being considered as a major upcoming therapeutic target. These enzymes, besides acting on histones, methylate large numbers of nonhistone substrates with functional implications, including transcription regulation, translational regulation, DNA repair, and RNA splicing.

VIII.1.3.1 Lysine methyltransferases

Based on the protein motif and recruitments, different lysine methyltransferases methylate different residues of histones: predominantly, histone H3 and histone H4 at lysine 20 (Fig. VIII.18) with distinct functional implications. Deregulation of these enzymes

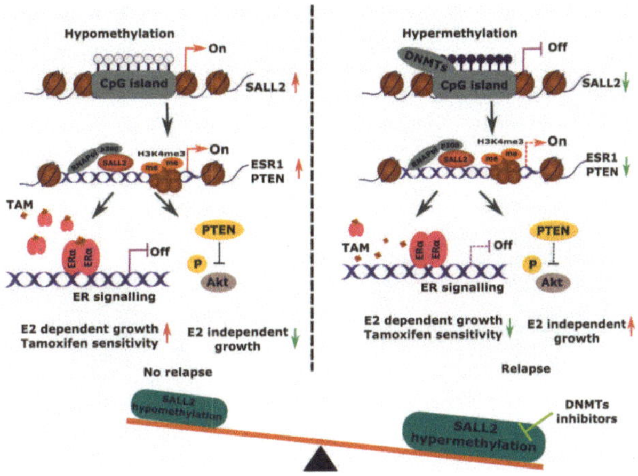

FIG. VIII.17 In TAM-resistant cells, SALL2 gene promoter is hypermethylated, and thereby, there is no expression SALL2 *(right)*. SALL2 is essential for the ER-dependent transcription. Because these cells are not ER dependent, they are not sensitive to TAM. Treatment of DNMT inhibitor, 5-Aza-dC, restored the SALL2 expression and thereby significantly increased the antitumor effect of TAM. *TAM*, tamoxifen. Source: *Ye, L., et al., 2019. EMBO Mol. Med. 11(12), e10638. John Wiley and Sons.*

may cause several diseases or could be directly causally related. Several small-molecule modulators of these enzymes have been discovered, some of which are already approved drugs and some under different stages of development. To appreciate the tremendous progress of this field, we shall discuss a few.

VIII.1.3.2 H3K79 methyltransferase DOT1L and its inhibitor

DOT1L (disruptor of telomeric silencing 1-like) is a unique lysine methyltransferase, which methylates lysine in the globular domain of the core histone H3 (H3K79) instead of lysine residues in the histone tails. Furthermore, unlike most of the lysine methyltransferase, which possesses the SET domain in the catalytic site, its catalytic domain is more similar to that of protein arginine methyltransferases (PRMTs). DOT1L methylate lysine 79 of histone H3 as a part of the active transcription complex and plays a key role in sustaining the active state of the chromatin. Physiologically, DOT1L is essential for proper embryonic development. It is critical for maintaining normal hematopoiesis. The function of DOT1L was also implicated in heart and kidney functions.

The histone methylation activity of DOT1L is linked to several pathologies. In cancer, DOT1L plays a unique role in the leukemogenesis of mixed-lineage leukemia (MLL)-rearranged oncogenes. MLL/KMT2A is also a lysine methyltransferase, specific for histone H3K4 methylation. The gene of MLL gets fused with different transcription factor (AF4, AF9, AF10, and ENL) genes, without its catalytic domain, due to chromosomal translocation. The catalytically inactive fused MLL protein functions as an oncoprotein in an active transcription complex recruits DOT1L, which methylates H3K79, and activates the genes for leukemogenesis, such as, HoxA9, HoxA7, and Meis1. The overexpression of these genes causes leukemia (Fig. VIII.19).

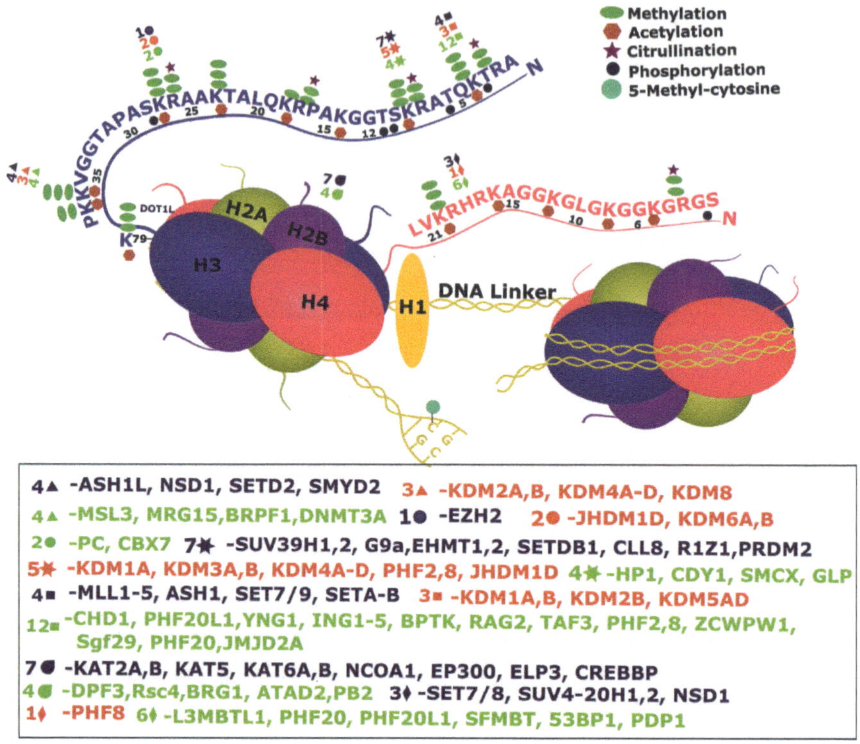

FIG. VIII.18 Nucleosome structure and principal modification sites on H3, H4, and DNA. The reported writers (blue), erasers (red), and readers (green) for these modifications are also depicted. *Reproduced from Morera, L., Lübbert, M., Jung, M., 2016. Targeting histone methyltransferases and demethylases in clinical trials for cancer therapy. Clin. Epigenet. 8, 57. https://doi.org/10.1186/s13148-016-0223-4.*

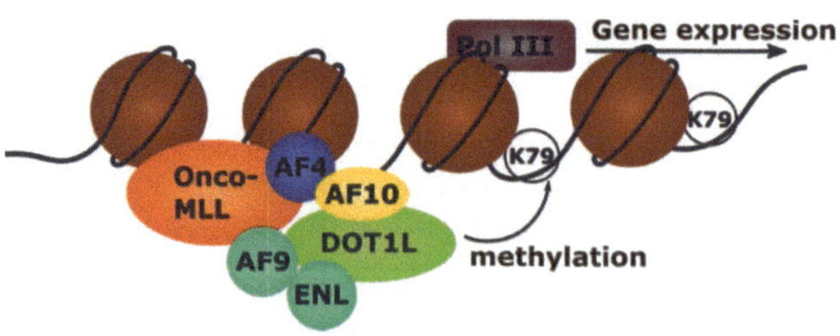

FIG. VIII.19 The catalytically inactive AF4, AF9, AF10, or ENL transcription factor-fused onco-MLL protein form an active complex that can recruit DOT1L. Owing to the methyltransferase activity of DOT1L, histone H3K79 gets methylated and causes overexpression of genes related to leukemogenesis. *Modified from Anglin, J.L., et al., 2013. J. Med. Chem. 55(22), 8972–8983. American Chemical Society.*

The histone methyltransferase activity of DOT1L, thus, appeared to be an important drug target for MLL-rearranged leukemia. This subtype of leukemia shows an inferior prognosis. Almost ~75% of acute leukemia in infants is of this class. Small-molecule inhibitors of DOT1L are being developed with a great deal of optimism. So far, most of

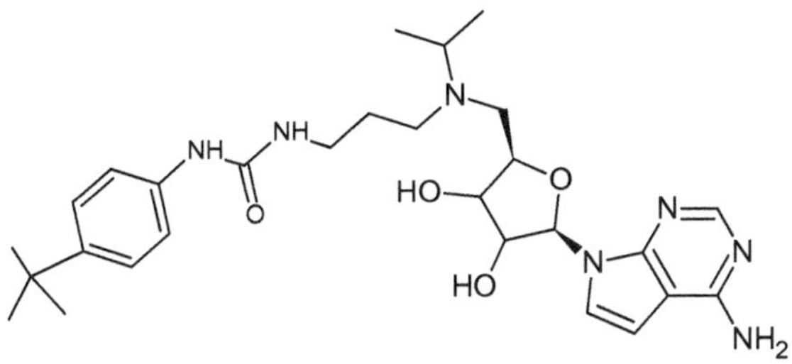

FIG. VIII.20 EPZ004777: a potent, selective inhibitor of DOT1L with an IC_{50} of 0.4 nM.

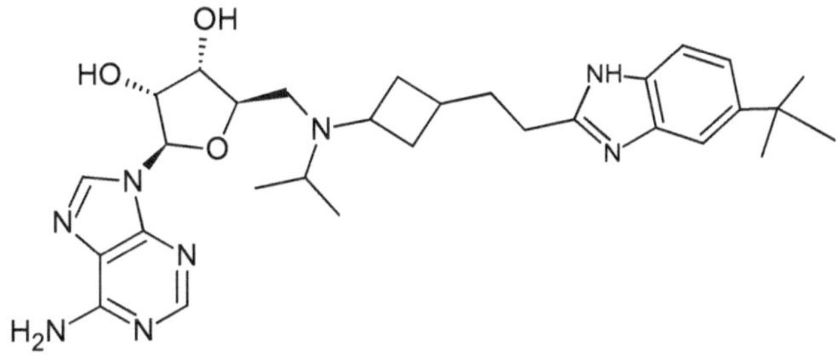

FIG. VIII.21 Pinometostat (EPZ-5676): a potent inhibitor of DOT1L with a IC_{50} of 80 pM.

the inhibitors disclosed are adenosine or its derivatives and competitively inhibit the binding of SAM. One of the first specific inhibitors of DOT1L discovered is EPZ004777 (Fig. VIII.20). It could dramatically reduce the expression of aberrantly regulated MLL1/KMT2A-target genes and thereby inhibits the leukemogenesis in mice. Another therapeutically potential DOT1L inhibitor is pinometostat (EPZ-5676) (Fig. VIII.21). It has reached the next level of drug development, beyond cultured cells and animal model systems. In phase 1 clinical trial, the safety and efficacy in adult patients with an advanced stage of leukemia caused by MLL gene rearrangement are being studied.

VIII.1.3.3 H3K4 methylation by MLL (including demethylase LSD1 inhibitor)

Histone H3K4me3 is a signature of transcriptionally active chromatin. The H3K4 residue can be methylated by the MLL in all three forms (mono-, di-, and trimethylation) (Fig. VIII.18). In the chromosomal translocation, one copy of the MLL gene may be rearranged and lead to the generation of an oncoprotein. As described earlier, the other copy of the wild type enzyme methylates the H3K4 residue as usual. However, the activity of the wild-type MLL was found to be essential for the malignancy. Therefore the MLL histone methyltransferase activity could also be an important target for anticancer therapy. The MLL activity-specific inhibitor is yet to be discovered. The pure MLL protein activity is very low. The protein complex of MLL-WDR5, ASH2L, and RbBP5 is fully active. Among these, the interaction of WDR5 and MLL is very critical for the methyltransferase activity of MLL. In the MM-401, the compound can disrupt the interaction between MLL-WDR5 and thereby directly inhibit the MLL activity (Fig. VIII.22).

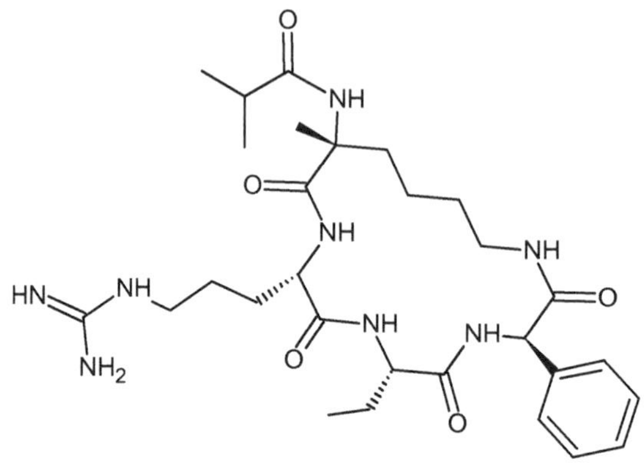

FIG. VIII.22 MM-401 inhibits the interaction between MLL-WRD5 and there by inhibits the MLL activity.

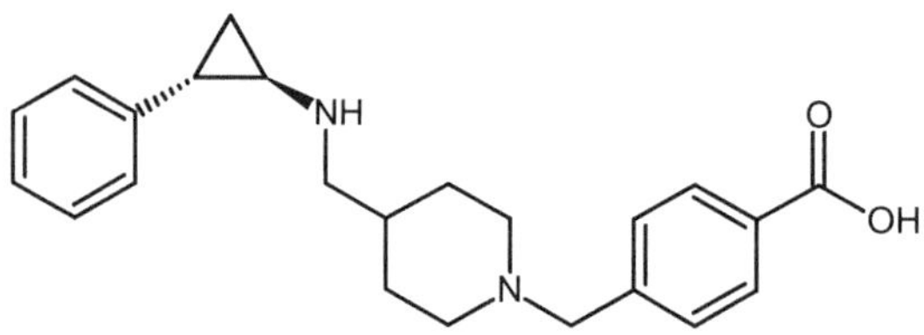

FIG. VIII.23 GSK2879552, orally active LSD1 inhibitor.

The activity of MLL is controlled in vivo by the opposite action (demethylation) of an eraser, the lysine-specific demethylase 1 (LSD1). Primarily, LSD1 demethylates H3K4me1 and me2. To counteract and balance the transcriptional activity, LSD1 has been placed in the MLL transcription complex. LSD1 is overexpressed in lung, breast, and prostate cancer. Inhibitors of LSD1 were highly effective in reducing the tumor burden in these cancers, especially lung cancer. Among the different small-molecule inhibitors discovered, GSK2879552 (Fig. VIII.23) seems to be very promising because it is in a clinical trial now for small-cell lung carcinoma (SCLC). It is an orally active, selective, and irreversible inhibitor of LSD1/KDM1A. GSK2879552 binds to and inhibits the demethylase activity of LSD1. In a tumor, cellular inhibition of the demethylase enhances H3K4 methylation, which in turn induces the expression of tumor suppressor genes.

Next, we shall discuss an interesting inhibitor of LSD1, Vafidemstat (ORY-2001) (Fig. VIII.24), which inhibits both LSD1 demethylase activity and the mitochondrial membrane protein monoamine oxidase B (MAO-B). It is being developed for the treatment of neurodegenerative diseases such as Alzheimer's disease. This small-molecule inhibitor is orally bioavailable and capable of crossing the blood-brain barrier (BBB). Presently, it is in phase II clinical trials.

VIII.1.3.4 H3K27 methyltransferase EZH2: EZH2 selective degrader

Enhancer of zeste homolog 2 (EZH2) is a histone H3K27me3 methyltransferase (KMT6) encoded by the EZH2 gene. EZH2 is the functional enzymatic component of the polycomb

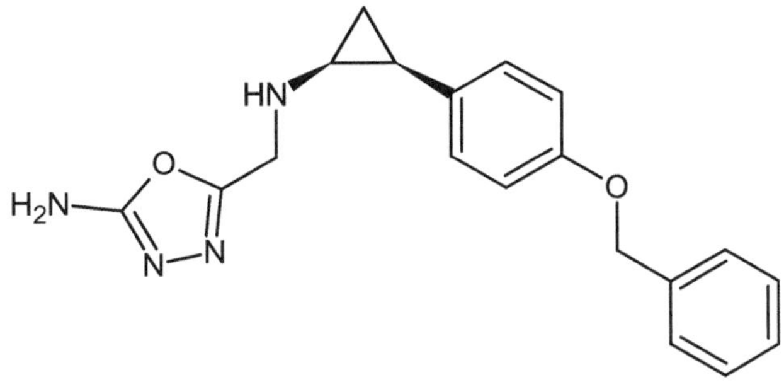

FIG. VIII.24 Vafidemstat (ORY-2001) is a dual lysine-specific histone demethylase (LSD1)/MAO-B inhibitor.

repressive complex 2 (PRC2) and plays a key role in the maintenance of the repressive heterochromatin state of the genome. The EZH2 is often overexpressed or mutated in solid tumors: breast, prostate, lung, skin, and colon cancer. The overexpression of EZH2 is correlated with tumor initiation, progression, stem cell self-renewal, migration, and angiogenesis. The invasiveness of breast and prostate cancer cells was found to be caused by the expression of EZH2. Thus the EZH2 activity is one of the most important epigenetic targets of antineoplastic therapeutics.

The most successful effort targeting EZH2 was the discovery of tazemetostat (Fig. VIII.25), an orally available, small-molecule selective and *S*-adenosylmethionine (SAM)-competitive inhibitor of histone methyltransferase EZH2. Tazemetostat (brand name: Tazverik) has recently been approved by the FDA, United States, for adult patients with relapsed or refractory follicular lymphoma. It is being used for the treatment of metastatic epithelioid sarcoma.

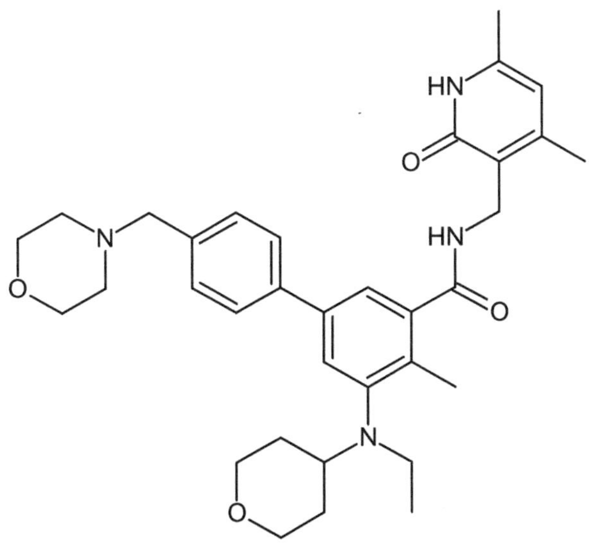

FIG. VIII.25 FDA-approved EZH2 inhibitor, Tazemetostat (Tazverik).

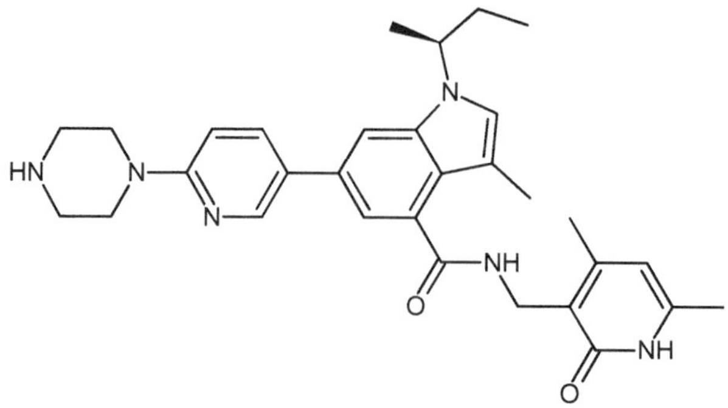

FIG. VIII.26 GSK2816126, a very high-affinity and selective EZH2 inhibitor.

GSK2816126 (GSK126) (Fig. VIII.26) is an inhibitor of EZH2, which targets the mutant (at Tyr641 and Ala677) found in certain cancers, rather than the wild-type enzyme. It also competes with the methyl group donor SAM for binding to EZH2. GSK126 potently inhibits the H3K27me3 activity of the mutant enzyme and functions as an antineoplastic molecule targeting large B-cell lymphoma. However, it is not orally bioavailable.

The CPI-1205 selectively inhibits the activity of both wild-type and mutated forms of EZH2, and it is bioavailable on oral administration (Fig. VIII.27). Being a trifluoro compound, it has tremendous potential to be a drug. This inhibitor has entered the clinical trial now for EZH2-mutated non-Hodgkin lymphoma. Drug discovery to the clinic is a long process and requires years of follow-up through the regulatory pipeline (Fig. VIII.28).

VIII.1.3.5 EZH2 inhibitor to overcome breast cancer drug resistance

Drug resistance is one of the most challenging barriers to remission for patients with cancer. In the case of metastatic breast cancer treatment, taxane (chemotherapeutic agents, including paclitaxel and docetaxel) is being successfully used in both neoadjuvant and adjuvant settings. Although taxane treatment is one of the most effective regimes, taxane-resistant triple-negative breast cancer (TNBC) has often been detected in recent years.

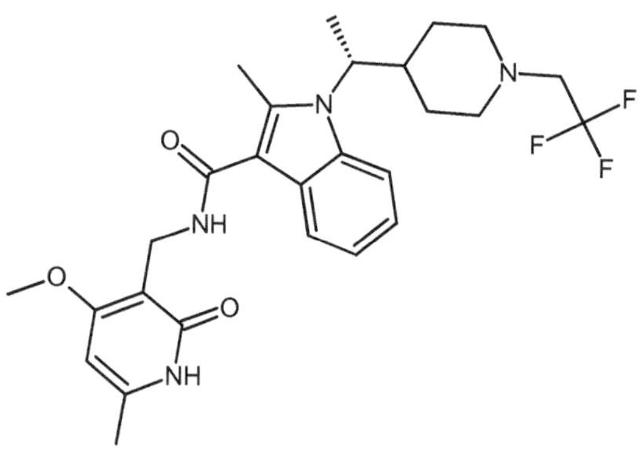

FIG. VIII.27 CPI-1205, a potent and selective histone methyltransferase EZH2 inhibitor.

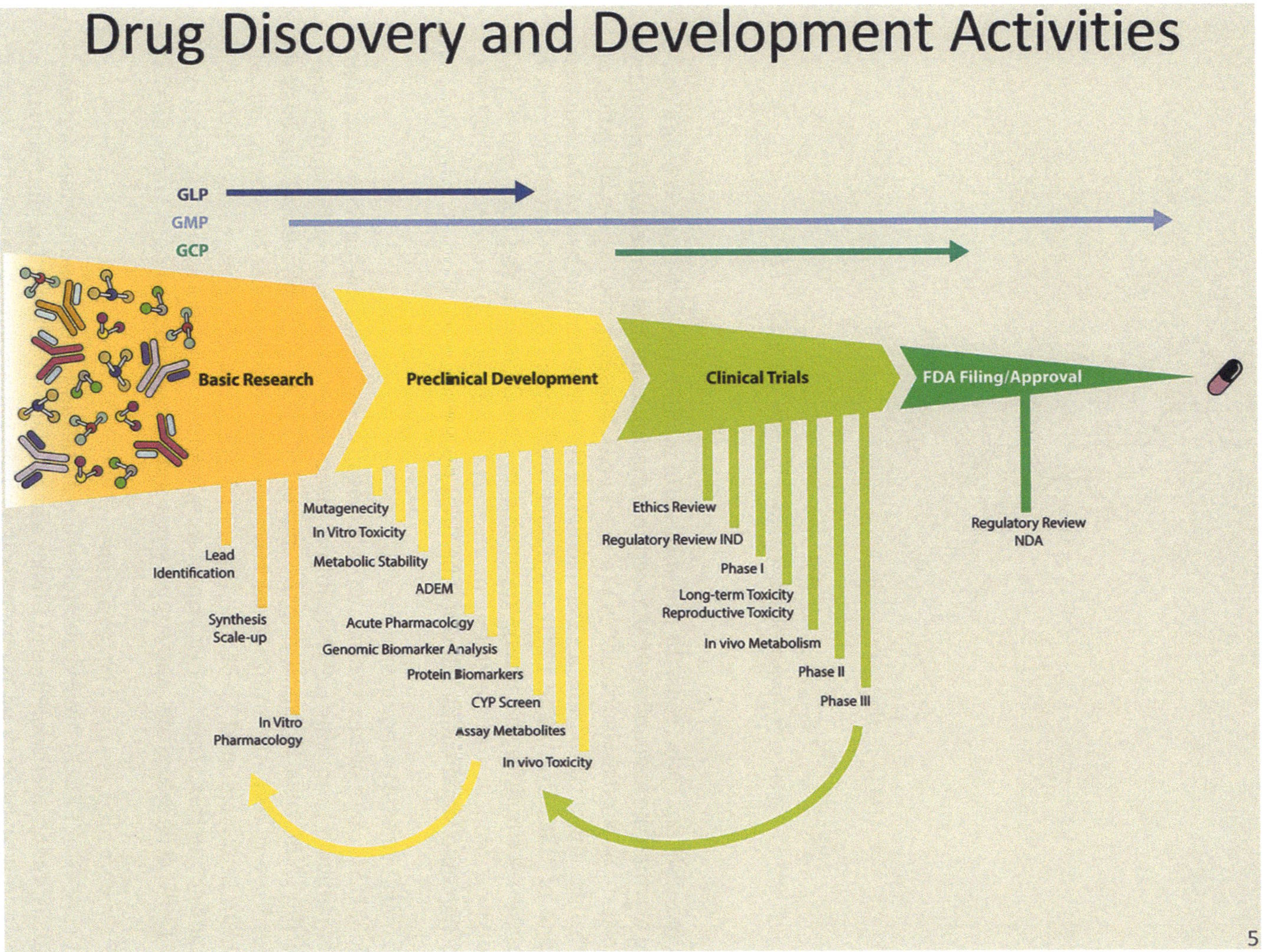

FIG. VIII.28 The path of drug discovery. A schematic of the activities involved in the drug discovery and development process. At the left are shown icons depicting small molecules (NMEs) and biological molecules (NBEs) being considered for development. At the top are the time lines for quality assurance guides governing the process; they are good laboratory practice (GLP), good manufacturing practice (GMP), and good clinical practice (GCP). Specific activities in the stages of development are shown at the bottom; they include studies of absorption, distribution, excretion, and metabolism (ADEM), screening for activity at cytochrome P450 (CYP) liver enzymes, and regulatory filings for investigational new drug (IND) and new drug application (NDA). *NBE*, new biological entity; *NME*, new molecular entity. *Reproduceed from Mohs, R.C., Greig, N.H., 2017. Drug discovery and development: role of basic biological research. Alzheimers Dement (NY)3 (4), 651–657.*

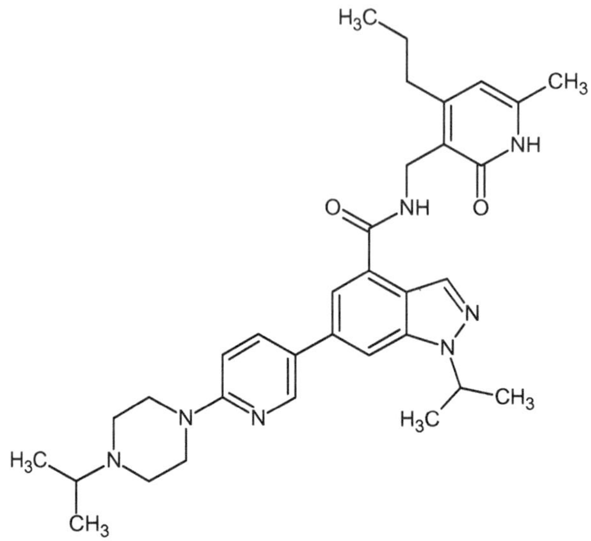

FIG. VIII.29 UNC1999 is a potent, orally bioavailable and selective inhibitor of EZH2 and EZH1.

In taxane-resistant TNBC cells, the metabolism of methionine is often altered, as a consequence of which the availability of SAM reduces dramatically. The decreased availability of SAM leads to hypomethylation of DNA in these cells. However, the hypomethylation of DNA in the loci enriched in transposable element (TE) is maintained by EZH2-mediated methylation of H3K27me3. The EZH2-mediated methylation of these large chromatin domains was essential for the taxane resistance in TNBC cells. In the cells and TNBC patient-derived models, the treatment of an EZH2 inhibitor (for example, UNC1999) (Fig. VIII.29) could significantly alleviate the drug resistance.

VIII.1.3.6 EZH2 selective degrader

Although EZH2 is often overexpressed in triple-negative breast cancer patients with poor prognosis, clinically promising EZH2 inhibitors discussed earlier are not sufficiently effective to kill the TNBC cells. By employing the concept of PROTACs methodology (discussed in Chapter II), very recently, an EZH2 selective degrader, MS1943, has been synthesized (Fig. VIII.30). In the cultured TNBC cell lines, MS1943 could efficiently reduce the level of EZH2 and so the H3K27me3. Most importantly, MS1943 showed a significant cytotoxic effect on TNBC cells selectively, and suppresses tumor growth in a xenograft mice model.

VIII.1.3.7 Protein arginine methyltransferases

Different classes of PRMTs methylate arginine residues in histone and nonhistone proteins; initially, mono-methylation and finally, dimethylation patterns. The dimethylation could be symmetric (type II) or asymmetric (type I) in nature, depending on the class of PRMTs (Fig. VIII.31). Arginine methylation is an important epigenetic modification that

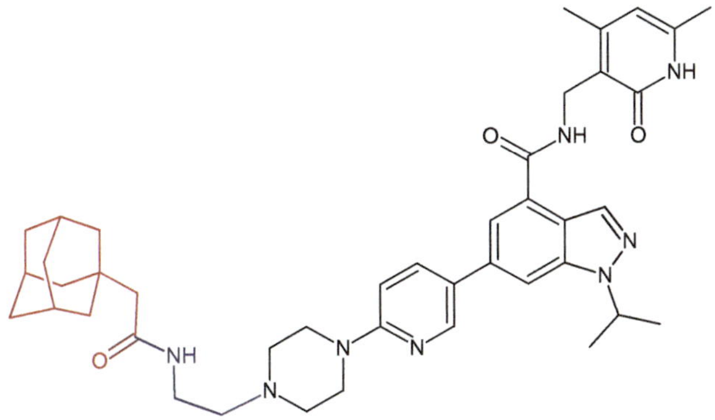

FIG. VIII.30 Chemical structure of selective EZH2 degrader with IC_{50} of 120 nM, which is orally bioavailable.

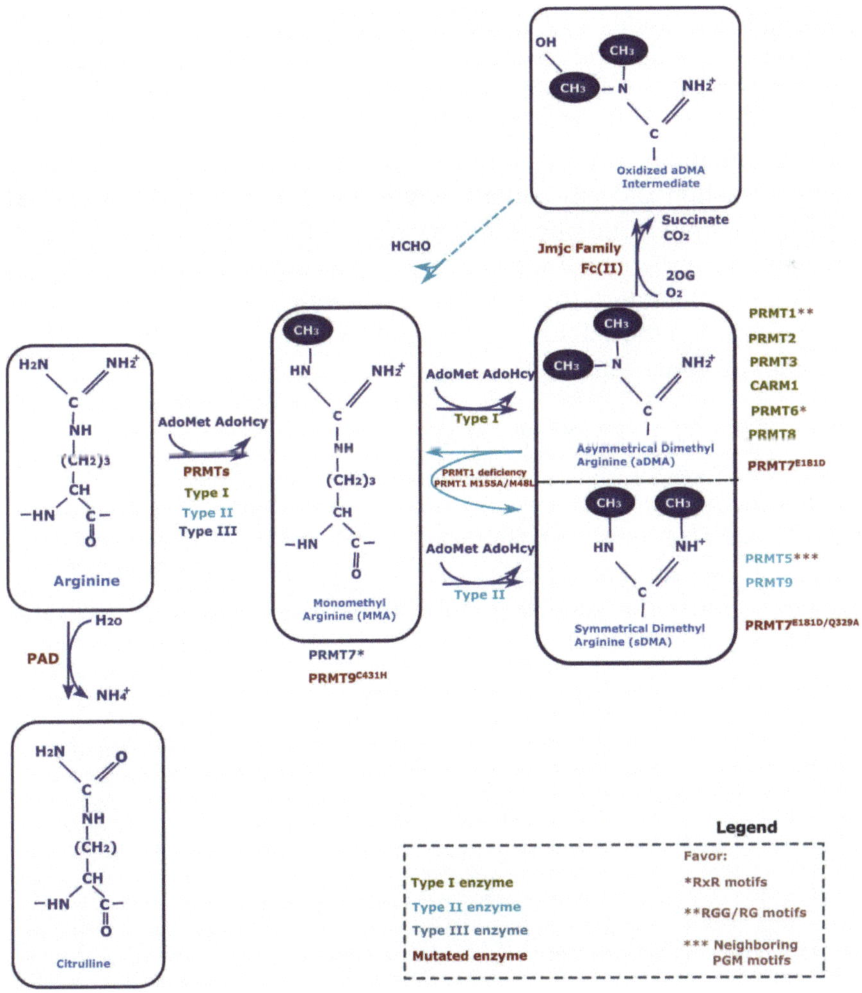

FIG. VIII.31 Different classes of PRMTs. Source: *Blanc, R.S., et al., 2017. Mol. Cell 65(1), 8–24. Elsevier.*

plays a crucial role in pre-mRNA splicing, DNA damage signaling, mRNA translation, cell signaling, and cell fate decision. Diseases such as cancer and metabolic, neurodegenerative, and muscular disorders are often associated with the dysregulation of PRMTs activity. Small-molecule inhibitors of PRMTs are, thus, at the center stage of the new-generation drug discovery regime.

VIII.1.3.8 Type I PRMT and inhibitors

One of the broad-spectrum inhibitors discovered for targeting class I PRMTs is MS023 (Fig. VIII.32). It is a noncompetitive inhibitor. Although it can inhibit the arginine methylation activity of PRMT1, PRMT3, PRMT4, PRMT6, and PRMT8, it does not affect the activity of protein lysine methyltransferases and DNA MTases even at 25 μM concentration. It inhibits the asymmetric dimethylation of H4R3, mediated by the abovementioned PRMTs with IC_{50} <100 nM. The binding of MS203 in the active site pocket of PRMT6 seems to be highly specific (Fig. VIII.33). It was found to be nontoxic to cells even after 96 h of treatment, suggesting its suitability as a therapeutic molecule.

The first specific inhibitor of CARM1/PRMT4 discovered was a natural compound, ellagic acid (TBBD), which selectively inhibits methylation at arginine 17 of histone H3 (H3R17) but not H3R26 methylation. It binds to the enzyme-substrate (histone) complex preferentially at the signature motif, "KAPRK," where the proline residue (Pro-16) plays a critical role in interaction and subsequent enzyme inhibition (Fig. VIII.34). TBBD shows a promising effect on some metabolic disorders where PRMT4 activity is causally related to disease pathology. Recently, a high-affinity selective inhibitor of PRMT4, TP-064, has been reported (Fig. VIII.35). It inhibits the PRMT4 activity with an IC_{50} of <10 nM. This inhibitor was found to be active in the cellular system, and its treatment could inhibit the proliferation of multiple myeloma cells.

VIII.1.3.9 Type II PRMTs and their inhibitors

Among the different type II PRMTs, PRMT5 is the most well-studied arginine methyltransferase. PRMT5 links the histone methylation and DNA methylation directly. The symmetric methylation of histone H4 arginine 3 (H4R3me2s) by the protein arginine methyltransferase PRMT5 induces the recruitment of DNMT3A in a locus that methylates the CpG island and transcriptionally silences the gene (Fig. VIII.36). This is also a unique example where

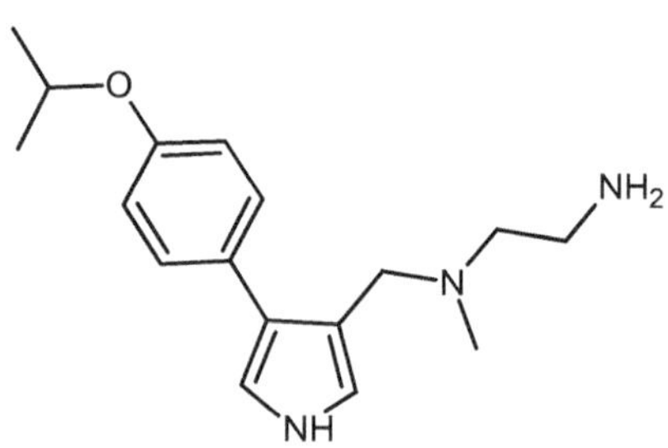

FIG. VIII.32 MS023: a potent and nonselective inhibitor of type-1 PRMTs.

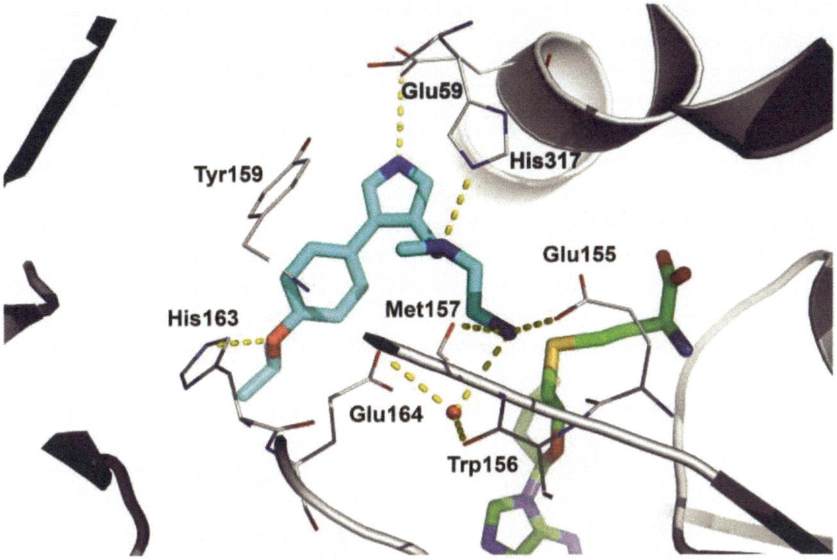

FIG. VIII.33 The X-ray crystal structure of a ternary complex of MS023 *(cyan)*, SAH *(green)*, and PRMT6 *(gray)*. Key interactions are shown in *yellow dotted lines*. Source: *Eram, M.S., et al., 2016. ACS Chem. Biol. 11(3), 772–781. American Chemical Society.*

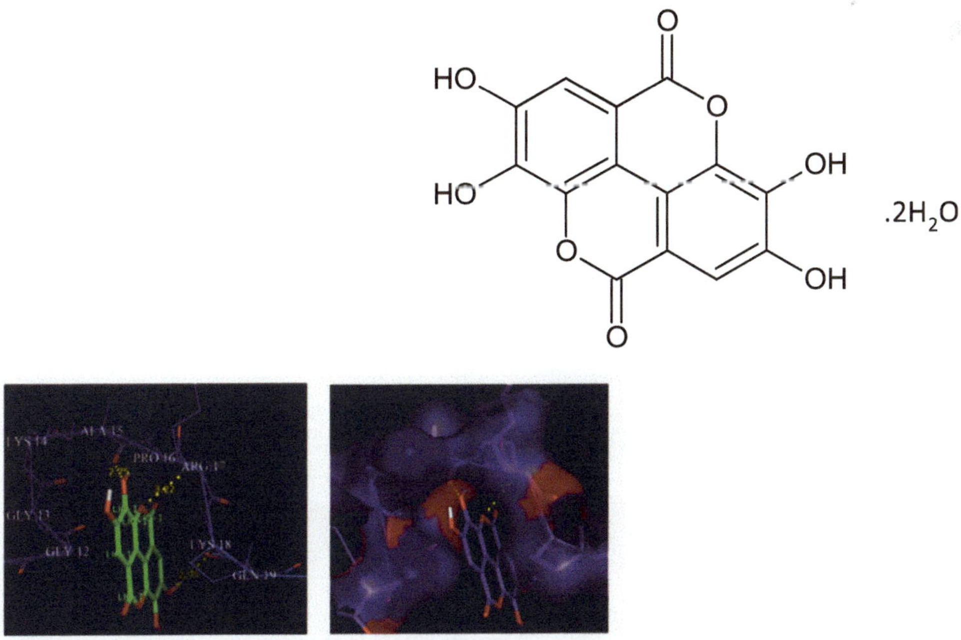

FIG. VIII.34 PRMT4/CARM1 selective inhibitor TBBD (ellagic acid). The docked conformation of TBBD near the KAPRK motif of histone H3 on the CARM1 active site. Source: *Selvi, B.R., et al., 2010. J. Biol. Chem. 285(10), 7143–7152. American Society for Biochemistry and Molecular Biology.*

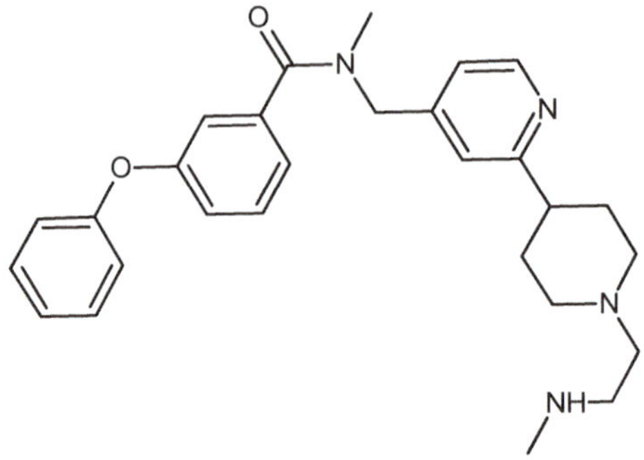

FIG. VIII.35 TP-064, a potent inhibitor of protein arginine methyltransferase 4 (PRMT4) with IC_{50} <10nM.

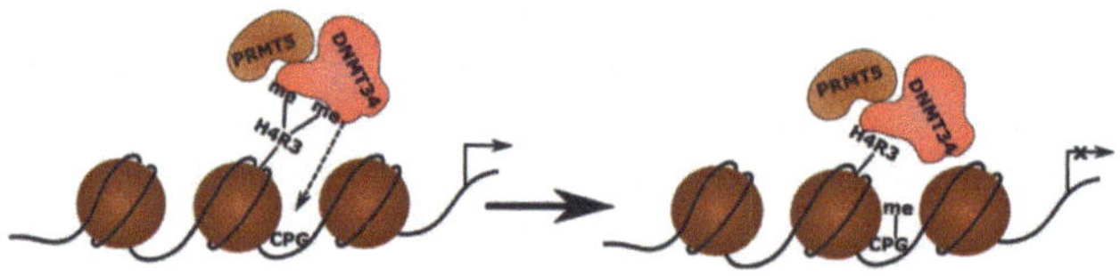

FIG. VIII.36 Model of PRMT5-induced silencing of gene expression. Symmetric dimethylation of H4R3 by PRMT5 induces direct binding of DNMT3A, resulting in methylation of adjacent CpG dinucleotides and gene silencing.

DNMT3A serves as both a reader (recognizes H4R3me2s) and writer (methylates cytosine of CpG island) in the dynamic epigenetic regulation of genome organization. Besides histone H3, PRMT5 also methylates a large number of nonhistone proteins with functional consequences such as mRNA translation, splicing, and DNA repair.

Tumor suppressor protein p53 is a bona fide substrate of PRMT5. Methylation of p53 by PRMT5 regulates its gene specificity (Fig. VIII.37). The arginine-methylated p53 activates the cell cycle arrest-specific genes. It is found that the silencing of PRMT5 indeed induces p53-dependent apoptosis. PRMT5 is overexpressed in several cancers. Small-molecule inhibitors of PRMT5 have, thus, now received earnest focus. One of these molecules,

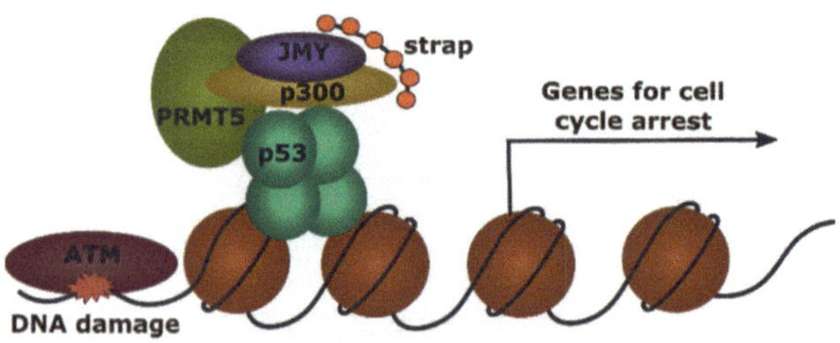

FIG. VIII.37 Methylation of p53 by PRMT5 induces the expression of cell cycle arrest-specific genes. On DNA damage signal, a protein complex comprising Strap/JMY/p300 recruits PRMT5 in the promoter, which then methylates p53. The methylated p53 activates the expression of cell cycle arrest-related genes. Source: *Durant, S.T., et al., 2009. Cell Cycle 8(6), 801–802. Taylor & Francis.*

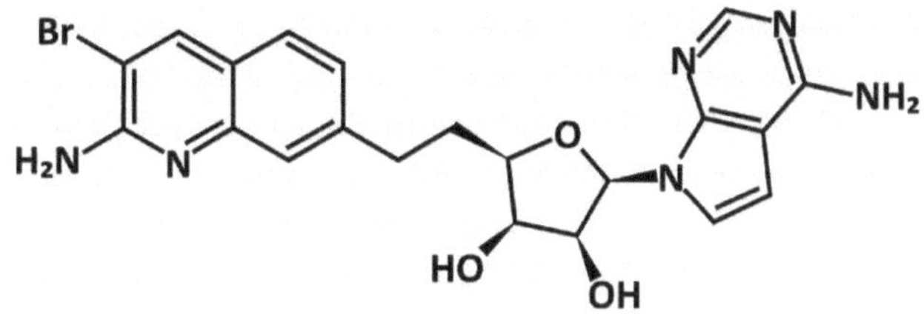

FIG. VIII.38 JNJ-64619178, a highly selective PRMT5 inhibitor.

JNJ-64619178 (Fig. VIII.38), is now under clinical trial for relapsed/refractory B-cell non-Hodgkin lymphoma. It is a selective, orally active, and pseudo-irreversible inhibitor of PRMT5 with an IC_{50} of 0.14 nM.

VIII.1.4 Small-molecule modulators (inhibitors and activator) of reversible histone acetylation

Reversible acetylation by lysine acetyltransferases and deacetylases is one of the most dynamic epigenetic modifications. Grossly, acetylation of histones and nonhistone proteins is associated with a relatively open and transcriptionally active genome. Although the role of acetylation in trans-generational epigenetics is not appreciated much, the case of transcellular epigenetics and prompt response to any signal depends on acetylation of histones and nonhistone proteins by a large number of different classes of KATs. Thus KATs play a central role in both physiology and pathophysiology. The dysregulation of KATs is associated with several diseases, including cancer, neurological, inflammatory, respiratory disorders, and cardiovascular diseases. Hyperacetylation of histones due to overexpression or enhanced autoacetylation of KATs is the signature of oral cancer, some types of breast cancer, and rhabdoid sarcoma. In oral cancer, the autoacetylation of p300 is dramatically enhanced, which actives its KAT activity several folds. In myeloid leukemia (MLL), due to gene fusion, p300/CPB forms chimeric proteins with other classes of KATs, MOZ and MOF, and lysine methyltransferase MLL. Gene mutation of p300/CBP is the cause of developmental disorders such as Rubinstein-Taybi syndrome, which is due to the lack of KAT activity of the mutant protein. Similar to most KATs, p300/CBP acetylates a huge repertoire of nonhistone substrates, including NF-kB and p53. Acetylation of NF-kB not only enhances its DNA-binding ability but also decreases affinity for the inhibitory protein IkB and thereby favors the inflammatory pathway of gene expression. Recent pieces of evidence suggest a critical role of p300/CBP in neurodegenerative disorders (e.g., Alzheimer's disease) and spinal cord injury (SCI). Besides p300/CBP, the GNAT family KATs are also dysregulated in diseases. Gcn5 is overexpressed in lung and colon cancer, which upregulates cell cycle genes and oncogenes (e.g., Myc). The aberrant expression of GNAT family KATs is also very common in progressive type 2 diabetes. Gcn5 and PCAF acetylate PGC-1α (peroxisome proliferator-activated receptor γ coactivator-1α). PGC1a acetylation induces its degradation. Because PGC1a is a coactivator of hepatic gluconeogenesis related genes,

PCAF/Gcn5 KAT activity directly affect the type II diabetes pathogenesis. To elucidate the role of KATs in life and death, many small-molecule modulators (inhibitors and activators) have been developed to better understand the molecular mechanisms of KATs' function and to design therapeutically potential molecules as well.

VIII.1.4.1 Inhibitors of lysine acetyltransferases

The first successful attempt was to synthesize bisubstrate analogs mimicking the ternary complex of the cofactor acetyl-CoA and the substrate lysine. A covalently linked acetyl-CoA and lysine unexpectedly shows highly specific and inhibitory activity toward the full-length recombinant p300. However, when acetyl-CoA was conjugated with histone H3-derived, 20 residues N-terminal peptide, the inhibitor turned out to be specific for the KAT activity of PCAF (Fig. VIII.39). The other peptide constructs of similar length did not show any inhibitory activity. The bisubstrate concept was later on used by many groups to synthesize enzyme inhibitors.

Natural products are the richest source of drug molecules. Over the decades, a great number of drugs and biological tools have been isolated from natural sources. The diversity and chemical space of natural products are the greatest advantages for designing drug molecules. However, the specificity of natural products toward any enzyme is the most challenging issue. The synthesis of more selective compounds through the semisynthetic route is a solution.

Most of the natural product-based KAT inhibitors are polyphenolic in nature and, thus, are not specific, except embelin, which is specific for GNAT family KATs (Table VIII.3). However, the first natural product identified as a KAT inhibitor is anacardic acid, isolated from cashew nut shell liquid. It is a broad-spectrum KAT inhibitor. Garcinol, polyisoprenylated benzophenone from *Garcinia indica*, is also shown to have similar activity. Using

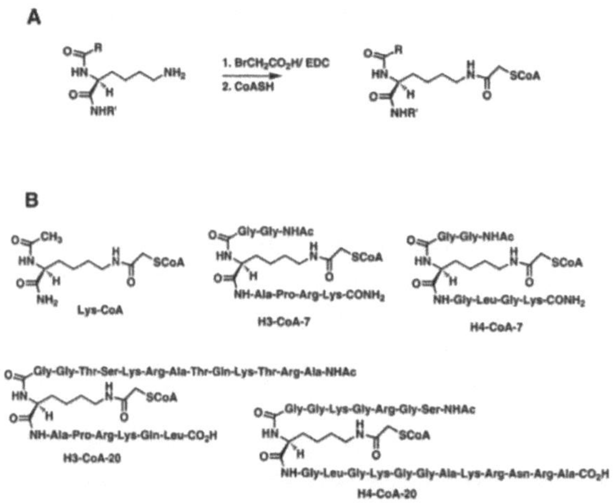

FIG. VIII.39 Design and synthesis of bisusbatrate analog KAT inhibitors: (A) scheme for CoA coupling to the lysine e-NH_2 groups. (B) Peptide CoA conjugates. Source: *Lau, O.D., et al., 2000. Mol. Cell 5(3), 589–595. Elsevier.*

garcinol as synthon, a semisynthetic compound LTK14 was synthesized, which is specific for p300 KAT activity and could potently inhibit the HIV viral proliferation in the HeLa cell model system. In an orthotopic breast cancer model, garcinol, in combination with subtoxic doses of cisplatin, was also shown to reduce the tumor burden very effectively. Curcumin-based semisynthetic, water-soluble broad spectrum KAT inhibitor CTK7A was found to be a potent anticancer molecule in a xenograft animal model system of oral cancer. CTK7A was also shown to be significantly bioavailable and effective in reducing prostate tumor growth. Although natural product-based KAT inhibitors have offered a significant number of scaffolds to develop KATi-based therapeutics, none of these compounds has yet proceeded further in the drug discovery pipeline.

Recently, using a virtual screening approach, a high affinity, selective synthetic inhibitor of KATs has been synthesized. One of the most successful efforts was the discovery of pyrazolone-containing inhibitor, C646. On screening ~500,000 commercially available compounds virtually docked on the bisubstrate-binding pocket, followed by synthesis and assay of a few, C646 was identified. It is a competitive inhibitor of p300 with a Ki of 400 nM (Fig. VIII.40).

Unlike the bisubstrate analog inhibitors, C646 was highly permeable to cells and exhibited potent antineoplastic activity. It could inhibit the cell growth of lung cancer cell lines and induce apoptosis on administration in AML1-ETO positive leukemia cells. One of the major disadvantages of C646 was its nonspecific reactivity with different cellular cysteine-containing proteins. This conjugation causes intrinsic fluorescence of the compound. This shortcoming was addressed by replacing the furan group with a phenyl ring (Fig. VIII.41) in C646.

The most recent success in the area of p300/CBP inhibitors for drug development is the discovery of a highly potent, selective, drug-like competitive inhibitor, A-485 (Fig. VIII.42). Its antineoplastic activity was established in the cellular and animal model of prostate cancer and several hematological malignancies. More recently, it has been shown that in mice, A-485 inhibit lipogenesis and hepatic gluconeogenesis, targeting the acetylation-dependent gene expression. Presumably, p300/CBP specific inhibition could be useful to treat the metabolic disorder as well.

Besides p300/CBP, potent synthetic inhibitors for MYST class of KATs, MOZ/MORF inhibitors WM8014, and WM1119 have also been reported. Although many potent and interesting small-molecule inhibitors of KATs have been discovered, and our understanding of disease biology clearly indicates the need, none of the KAT modulators have yet seen the light of the day in the field of drug discovery.

VIII.1.4.2 Small-molecule activators of the lysine acetyltransferase p300/CBP

Using broad-spectrum natural KAT inhibitor, anacardic acid, as synthon, *N*-(4-chloro-3--trifluoromethyl-phenyl)-2-ethoxy-6-pentadecyl-benzamide (CTPB) was synthesized (Fig. VIII.43). CTPB could specifically activate the KAT activity of p300, presumably by inducing the autoacetylation of p300 allosterically. Because CTPB is not very well permeable to cells, many derivatives were synthesized with limited success. One of the active

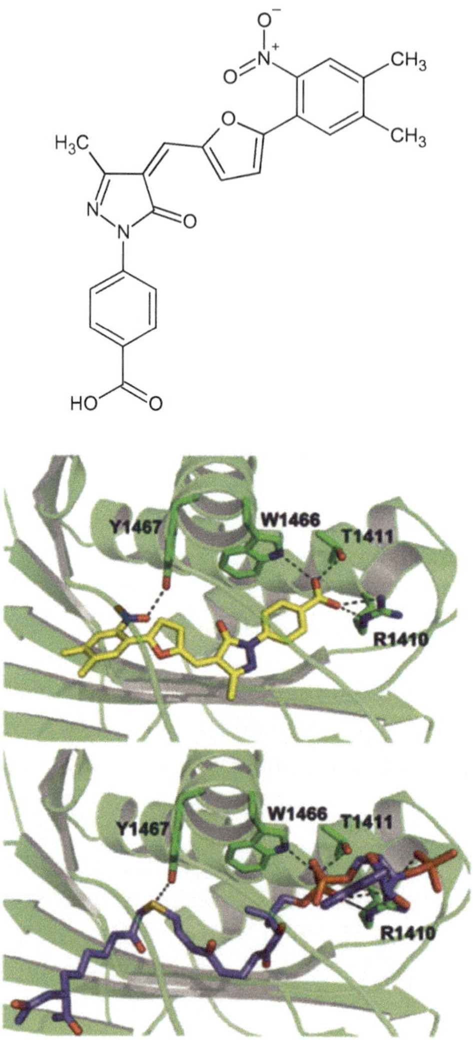

FIG. VIII.40 *Top*, chemical structure of C646. *Middle*, model of C646 bound to p300 HAT active site (- -). *Bottom*, the X-ray crystal structure of p300 HAT complexed with Lys-CoA (- -). Source: *Bowers, E.M., et al., 2010. Chem. Biol. 17(5), 471–482. Elsevier.*

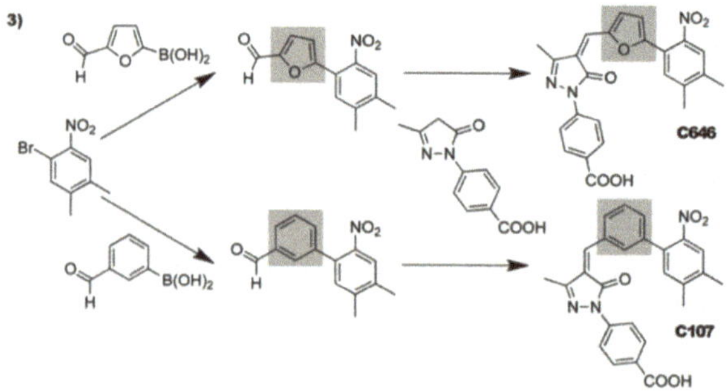

FIG. VIII.41 Synthesis of fluorescent inactive derivative of C646 and C107. Source: *Dancy, B.M. et al., 2012. Chembiochem 13(14), 2113–2121. John Wiley and Sons.*

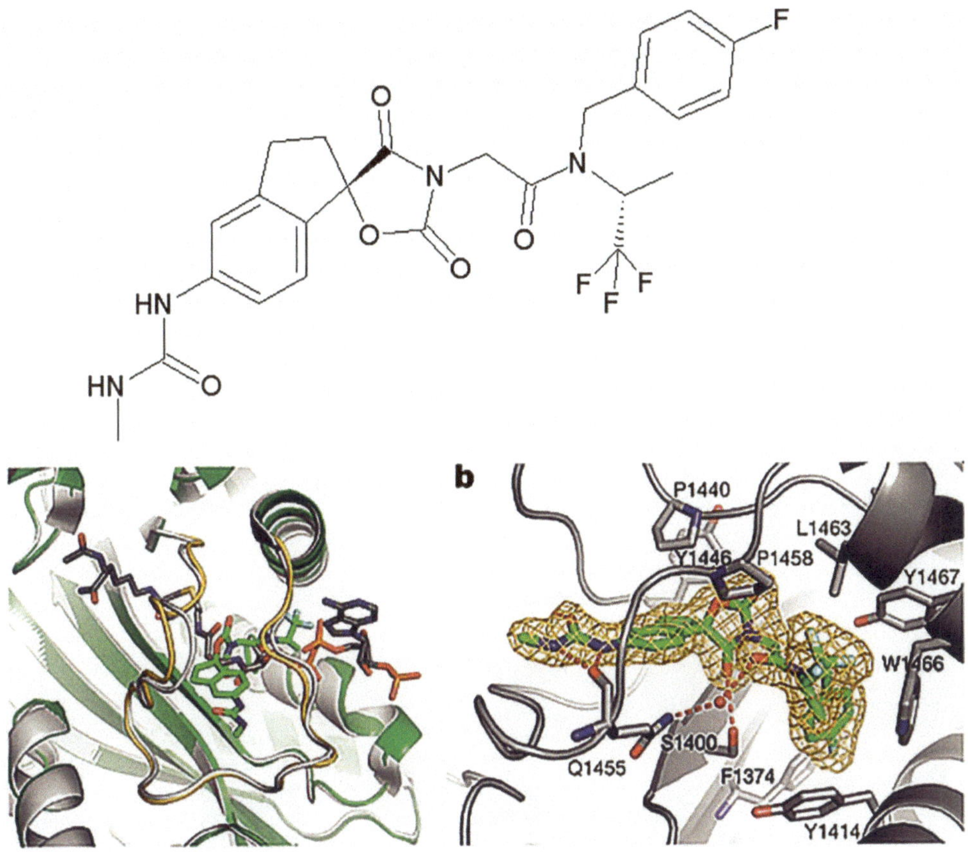

FIG. VIII.42 *Top*, chemical structure of A-485. *Bottom*, crystal structure of A-485 bound to Δp300 HAT. (A) Superposition of the Δp300 HAT *(dark green)* in complex with the A-485 complex *(bright green)*, with Δp300 HAT *(gray)* in complex with Lys-CoA (PDB ID: 3BIY). (B) Enlarged view of the A-485 binding site. Key interacting residues are shown with hydrogen bonds indicated by *red dashes* (some residues are omitted for viewing clarity). The 2Fo-Fc electron density map calculated at 1.95 Å for A-485 is contoured at 1σ. Source: *Lasko, L.M., et al., 2017. Nature 550 (7674), 128–132. Springer Nature.*

derivatives, TTK21, was conjugated with cell and organ permeable glucose-derived carbon nanosphere (CSP). The CSP-TTK21 could cross the cell membrane and the BBB and activate the p300/CBP activity in the mice brain. The conjugated molecule could induce histone hyperacetylation in the hippocampus of mice brain, thereby inducing neurogenesis and long-term spatial memory formation. It has been reported that indeed the specific activation of p300/CBP could result in almost complete recovery of memory in the Alzheimer's disease model (Fig. VIII.44). This activation could also dramatically lead to the repair of spinal injury in mice and rats.

VIII.1.4.3 Lysine deacetylases, the successful targets of epigenetic therapy

Besides DNMTs, the most successful targets for epigenetic drug discovery are the lysine deacetylases (KDAC/HDACs). Many small-molecule inhibitors have been designed,

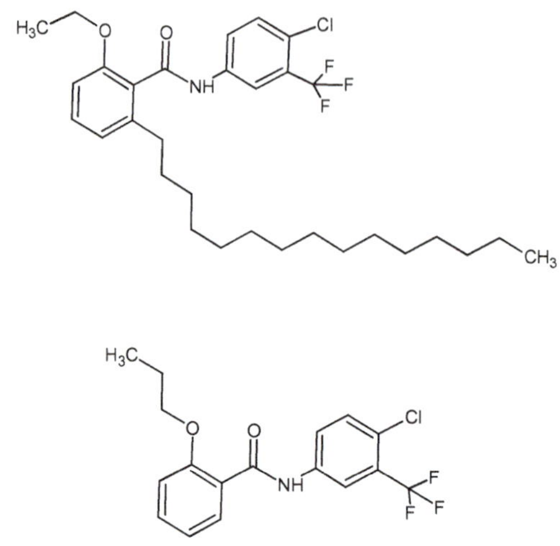

FIG. VIII.43 The first-known small-molecule activator of p300/CBP, CTPB, and its derivative TTK21.

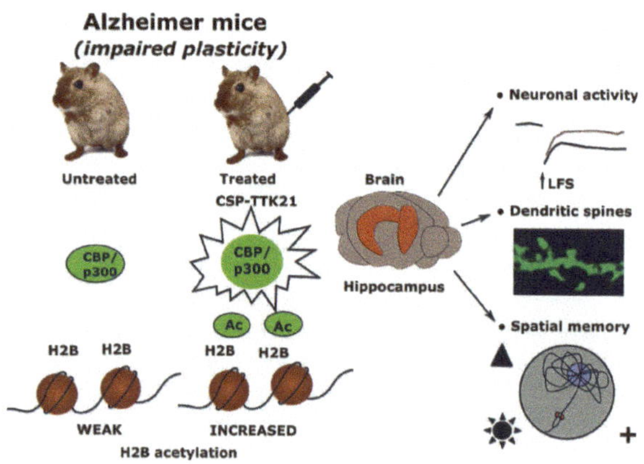

FIG. VIII.44 Activation of p300 HAT activity with CSP-TTK21 treatment restores plasticity in the hippocampus and spatial memory of Alzheimer's disease mice model: Histone H2B tetra-acetylation is induced by the activator, resulting in the normal neural activity, formation of dendritic spines, and, ultimately, complete recovery of memory. *From Chatterjee, S., et al., 2018. EMBO Mol. Med. 10(11), e8587. John Wiley and Sons.*

synthesized, and tested for biological activities targeting classes I, II, and IV HDACs. Several of these molecules have been shown to possess antineoplastic activity against many cancers. A few of these inhibitors have already got a place in the clinic upon FDA approval, and some are in the pipeline (Table VIII.4).

All HDACs are present as highly stable functional protein complexes in the cellular systems, responsible for very specific signal-dependent functions. By employing sophisticated chemoproteomics and chemical phylogenetic approaches, the selectivity and affinity of these inhibitors have been established. The results establish the physiological specificity of HDACi (Fig. VIII.45). Mechanistically, inhibition of HDACs creates a

Table VIII.4 The FDA approved and different phases of clinical trials of HDAC inhibitors.

Compound	Organization	Indication	Status
Romidepsin	Bristol Myers Squibb	Cutaneous T-cell lymphoma	FDA approved
		Peripheral T-cell lymphoma	FDA approved
		HIV, +3BNC117	Phase 2, no. 03041012
Vorinostat	Merck	Cutaneous T-cell lymphoma	FDA approved
		Epilepsy	Phase 2, no. 03894826
		Cushing disease	Phase 2, no. 04339751
		Breast cancer, +pembrolizumab, tamoxifen	Phase 2, no. 04190056
Tefinostat	GlaxoSmithKline	Hepatocellular carcinoma	Phase 1/2, no. 02759601
Ricolinostat	Regenacy	Diabetic neuropathic pain	Phase 2, no. 03176472
Belinostat	Onxeo, Spectrum	Peripheral T-cell lymphoma	FDA approved
Panobinostat	Novartis	Multiple myeloma	FDA, EMA approved
		Graft-versus-host disease	Phase 3, no. 04326764
		Myelodysplastic syndrome, +azacitidine	Phase 2, no. 03151304
Pracinostat	Helsinn, Menarini	Acute myeloid leukemia, +azacitidine	Phase 3, no. 03151408
		myelodysplastic syndrome, +azacitidine	Phase 2, no. 03151304
Givinostate	Italfarmaco	Duchenne muscular dystrophy	Phase 2/3, no. 03373968
		Polycythemia vera, +hydroxyurea	Phase 2, no. 00928707
Abexinostat	Xynomics	Follicular lymphoma	Phase 2, no. 03934567
		Renal carcinoma, +pazopanib	Phase 3, no. 03592472
Entinostat	Syndax	Breast cancer, +exemestane	Phase 3, no. 02115282 and no. 03538171
		Renal carcinoma, +IL-2	Phase 2, no. 03501381
Tucidinostat	Chipscreen	Peripheral T-cell lymphoma	CFDA approved
		B-cell lymphoma	Phase 3, no. 04231448
		Breast cancer, +exemestane	NMPA approved
		Soft-tissue sarcoma, +toripalimab	Phase 2, no. 04025931
Domatinostat	4SC	GI cancers, +avelumab	Phase 2, no. 03812796

[a] For combination trials, the other drugs involved are indicated with "+." The clinical trial is identified by the ClinicalTrials.gov NCT number. Reproduced from Ho, T.C.S., et al., 2020. J. Med. Chem. 63(21), 12460–12484. https://doi.org/10.1021/acs.jmedchem.0c00830.[a]

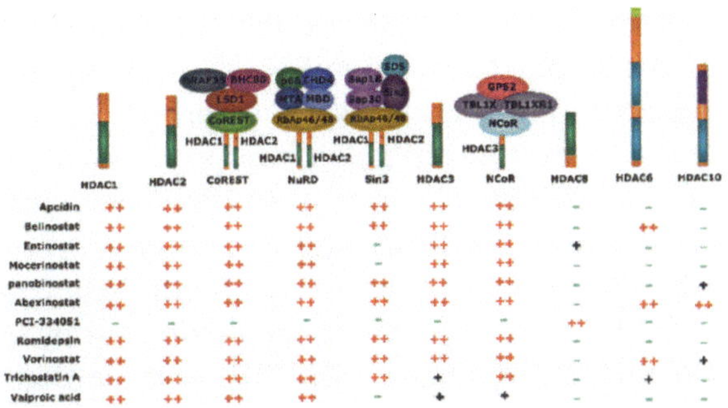

FIG. VIII.45 The specificity of HDACis for HDACs in the associated protein complex. *BHC80*, BRAF35-HDAC complex protein; *CHD4*, chromodomain helicase DNA binding protein 4; *CoREST*, REST corepressor 1; *LSD1*, lysine-specific demethylase 1; *MTA*, metastasis-associated protein; *MBD*, methyl-CpG binding domain protein; *NuRD*, nucleosome remodeling and deacetylase; *RbAp46*, retinoblastoma-binding protein p46; *Sap30*, Sin3A-associated protein, 30 kDa; *SDS*, serine dehydratase; *TBL1X*, transducin β-like 1X-linked; *TBL1XR1*, transducin β-like 1 X-linked receptor 1; *GPS2*, G protein pathway suppressor 2. Source: *West, A.C., et al., 2014. J. Clin. Invest. 124(1), 30–39. The American Society.*

hyperacetylated state of the genome. Many nonhistone proteins (such as transcription factors, coactivators and nonhistone chromatin proteins) also become hyperacetylated. Several repressed genes get activated. The final results of these could be the induction of apoptosis, senescence, differentiation, cell cycle arrest, immunogenicity, immune suppression, and angiogenesis (Fig. VIII.46). These diverse effects of HDACi are also the causes of general toxicity.

VIII.1.4.4 Dual HDAC therapeutics

As we have discussed so far, five HDACi have been approved by the FDA as antineoplastic therapeutics. However, owing to poor isoform specificity, drug efficacy and the consequent side effects become the major limitation of these classes of molecules. To overcome these difficulties, researchers have come up with the development of bifunctional HDACi, where single molecules contain an HDAC inhibitor conjugated with another moiety that targets a different function. In one example, the latter fragment may target ubiquitination machinery for inducing proteolysis (Fig. VIII.47).

VIII.2 MicroRNA-based therapeutics

VIII.2.1 miRNA biogenesis

MicroRNAs (miRNAs) are small noncoding RNAs, which are transcribed by RNA polymerase II. There are two classes of miRNAs: canonical and noncanonical, based on the processing of the pri-miRNA. In the canonical pathway of miRNA biogenesis, pri-miRNA is cleaved by Drosha (a RNase) and DGCR8 (the regulatory subunit) protein complex, generating 60–70 nucleotides (nts) long pre-miRNA in the nucleus. The pre-miRNA is then exported to the cytoplasm by Exportin 5 (Expo5)-Ran-GTP complex. In the cytoplasm, the pre-miRNA is further processed by another RNase, Dicer, to produce a miRNA duplex intermediate of 22 base pairs. In the next step, Argonaute (AGO) protein binds to the duplex miRNA precursor and produces a single-standard 22-nt mature miRNA. Some of the Dicer-binding proteins, such as TRBP and PACT, may facilitate the miRNA duplex formation (Fig. VIII.48). In the biogenesis of miRNA through the noncanonical pathway, all the protein machinery mentioned above is not essential.

MicroRNA (miRNA) genes are transcribed as primary miRNAs (pri-miRNAs) by RNA polymerase II (Pol II) in the nucleus. The long pri-miRNAs are cleaved by the microprocessor complex, which includes DROSHA and DiGeorge syndrome critical region 8 (DGCR8), to produce the 60–70-nucleotide precursor miRNAs (pre-miRNAs). The pre-miRNAs are then exported from the nucleus to the cytoplasm by exportin 5 (XPO5) and further processed by DICER1, a ribonuclease III (RIII) enzyme that produces the mature miRNAs. One strand of the mature miRNA (the guide strand) is loaded into the miRNA-induced silencing complex (miRISC), which contains DICER1 and Argonaute (AGO) proteins, directs the miRISC to target mRNAs by sequence complementary binding and mediates gene expression suppression by targeted mRNA degradation and translational

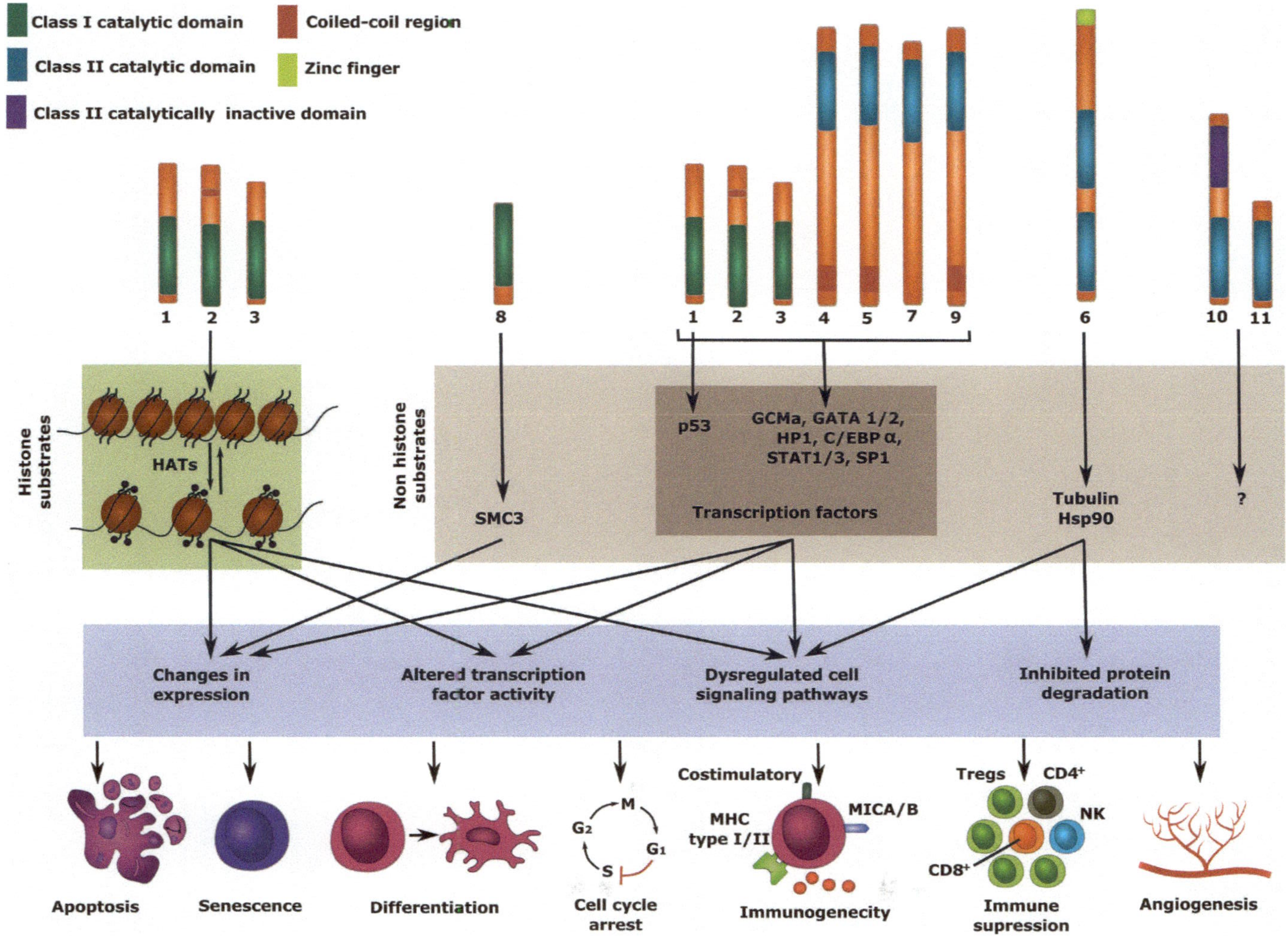

FIG. VIII.46 The cellular effect of HADC inhibitors. *SMC3*, structural maintenance of chromosomes 3; *GCMa*, glial cells missing homolog 1; *HAT*, histone acetyltransferase; *HP1*, heterochromatin protein 1. Source: *West, A.C., et al., 2014. J. Clin. Invest. 124(1), 30–39. The American Society.*

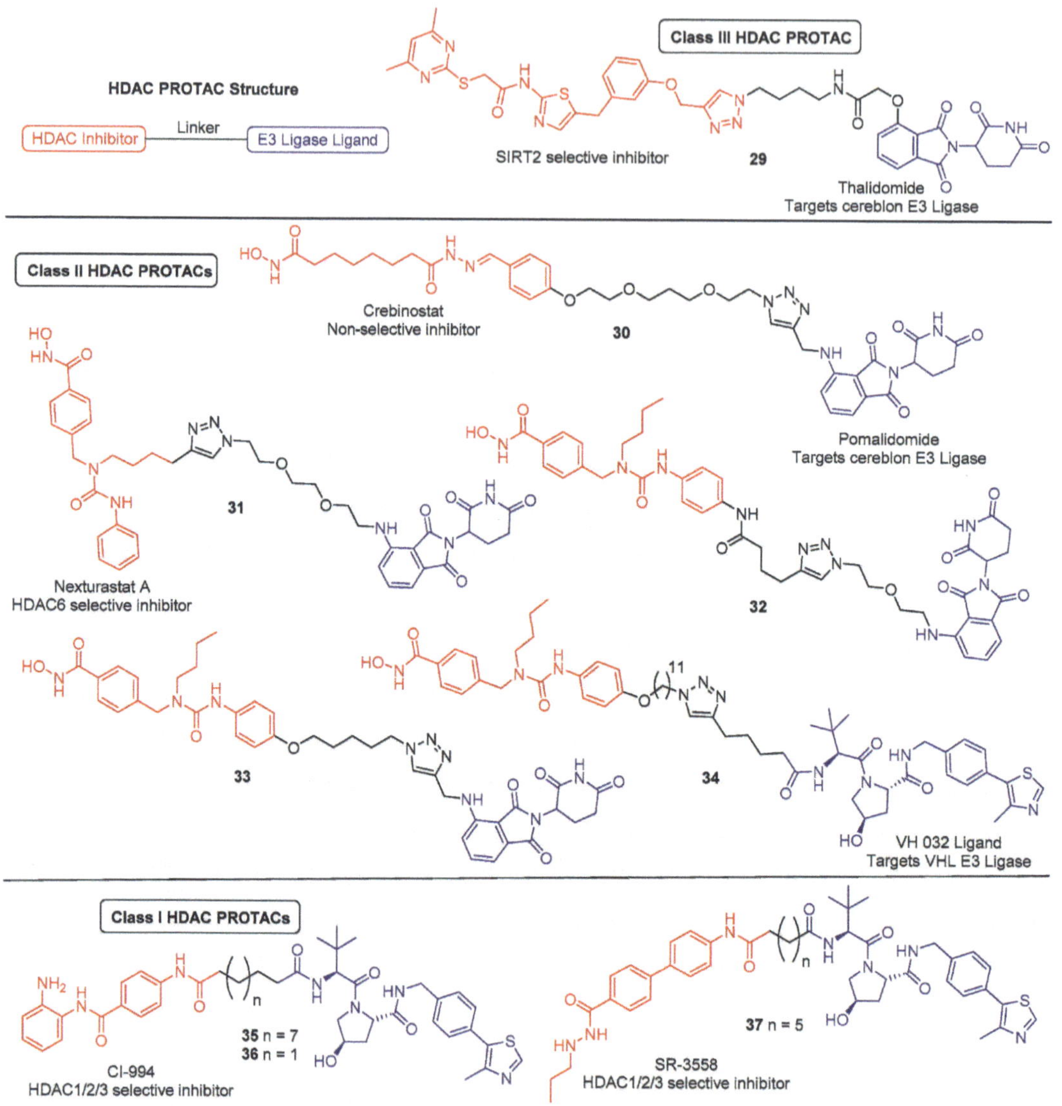

FIG. VIII.47 Dual HADCi: HDAC inhibitor *(red)*, small-molecule E3 ligase ligand *(blue)*, and linker *(black)*. *Reproduced from Smalley, J.P., Cowley, S.M., Hodgkinson, J.T., 2020. Bifunctional HDAC therapeutics: one drug to rule them all?* Molecules *25 (19), 4394 (published by MDPI).*

repression in processing bodies (P-bodies). *TRBP,* transactivation-responsive RNA-binding protein.

VIII.2.2 The physiological role of miRNA

The expression and biogenesis of miRNA are very tightly regulated both temporally and spatially. Epigenetic modifications, especially DNA methylation, directly control the

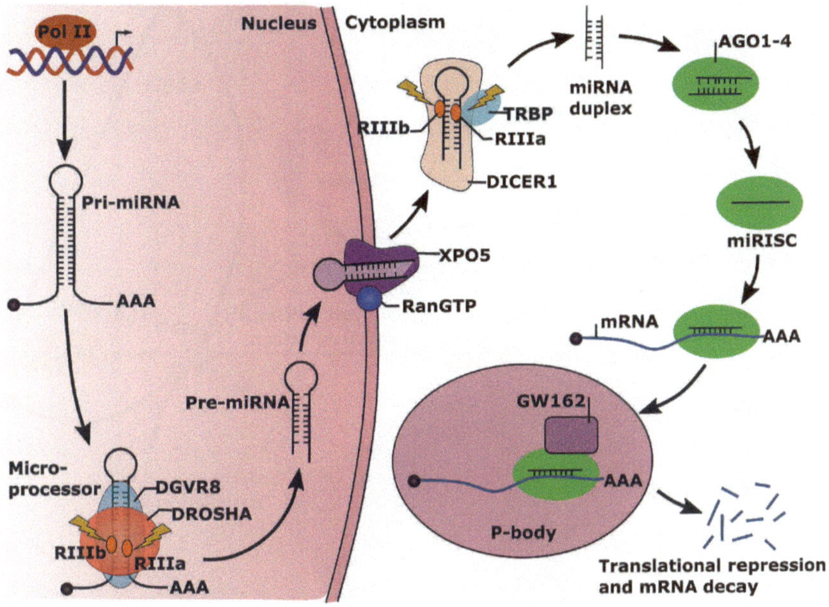

FIG. VIII.48 Overview of miRNA biogenesis pathway. *Reproduced from Lin, S., Gregory, R.I., 2015. MicroRNA biogenesis pathways in cancer. Nat. Rev. Cancer 15 (6), 321–333. https://doi.org/10.1038/nrc3932.*

expression of miRNA. The biogenesis of miRNA could be regulated by Drosha processing, Dicer processing, RNA editing, RNA methylation, uridylation, adenylation, AGO modification, and RNA decay. The detailed mechanisms have not been discussed here. Appropriate references provided at the end of this chapter may be consulted. Although the major role of miRNA is RNA silencing (gene silencing), either by the degradation of mRNA or blocking the translation, it may be involved in cell-cell communication and, in rare cases, gene activation (Fig. VIII.49).

Recent evidence suggests that miRNA plays a central role in most biological processes, including the evolution of organisms. The dysregulation of miRNA is almost a signature of cancer in humans. Besides cancer, aberrant expression profiles of miRNAs are also predominant in metabolic and neurodevelopmental and neurodegenerative disorders. The functional mechanisms of altered expression of miRNA are being elucidated. The transcriptional repression of miRNAs by hypermethylation of their corresponding promoter loci could be one of the most significant causes.

VIII.2.3 miRNA as a therapeutic target

The aberrant expression of miRNAs is now considered an effective drug target, either alone or in combination with other targets. In cancer progression, miRNAs play key regulatory roles in cell proliferation, apoptosis, migration, and invasion. The activity of miRNA could be modulated by several means: antisense oligonucleotide construction,

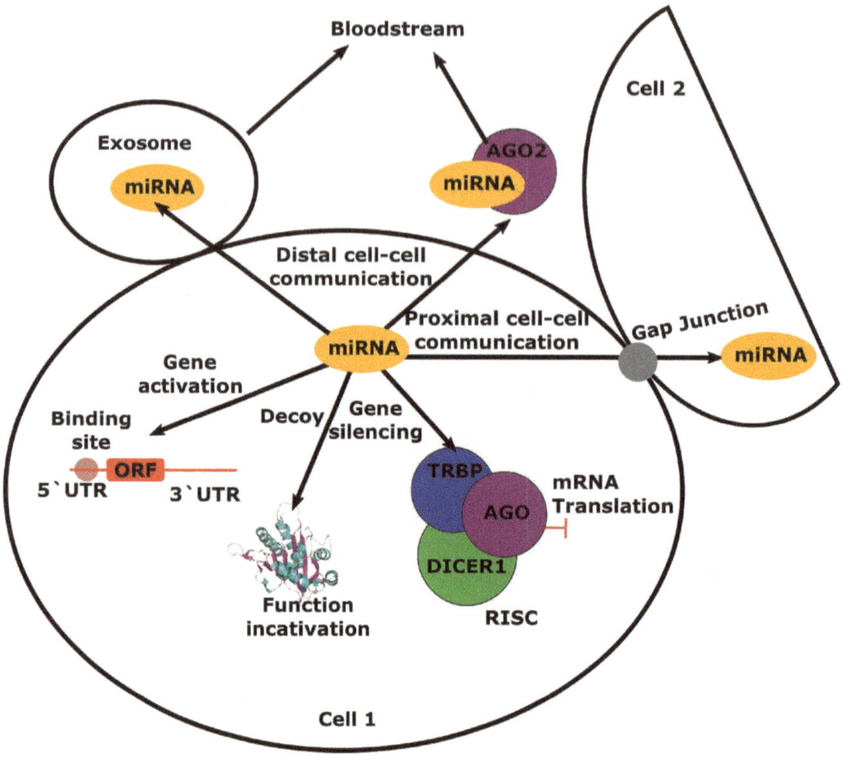

FIG. VIII.49 miRNAs are not only involved in gene silencing but also activation and more significantly cell-to-cell communication through gap junction and exosomes. *AGO*, Argonaute protein; *ORF*, open reading frame; *RISC*, RNA-induced silencing complex; *TRBP*, TAR-binding protein; *UTR*, untranslated region. Source: *Lopez-Serra, P., et al., 2012. Oncogene 31(13), 1609–1622. Springer Nature.*

such as small RNA zippers, miRNases (miRNA-targeted artificial ribonucleases), miRNA sponges, miRNA masks, anti-miRNA oligonucleotides, and synthetic miRNA mimics (Fig. VIII.50). These methods could be applied either solo or in combination, depending on cancer and stage. Collective data from cellular and animal models, as well as a clinical trial, are highly promising.

Cellular signaling of all the hallmarks of cancer is at least partially regulated by different miRNAs. A microRNA that is associated with cancer is termed an Oncomir. Oncomirs are causally related to the fundamental process of carcinogenesis, malignant transformation, and metastasis. These oncomirs could be oncogenic or a tumor suppressor. For therapeutic purposes, the oncogenic oncomirs could be targeted by anti-miRNA, whereas the miRNA mimics could enhance the tumor suppressor activity. miRNA-based therapeutics may be applied with only mixtures of anti-mirs/miRNA mimics and in combination with chemotherapeutic agents, as shown in Fig. VIII.51.

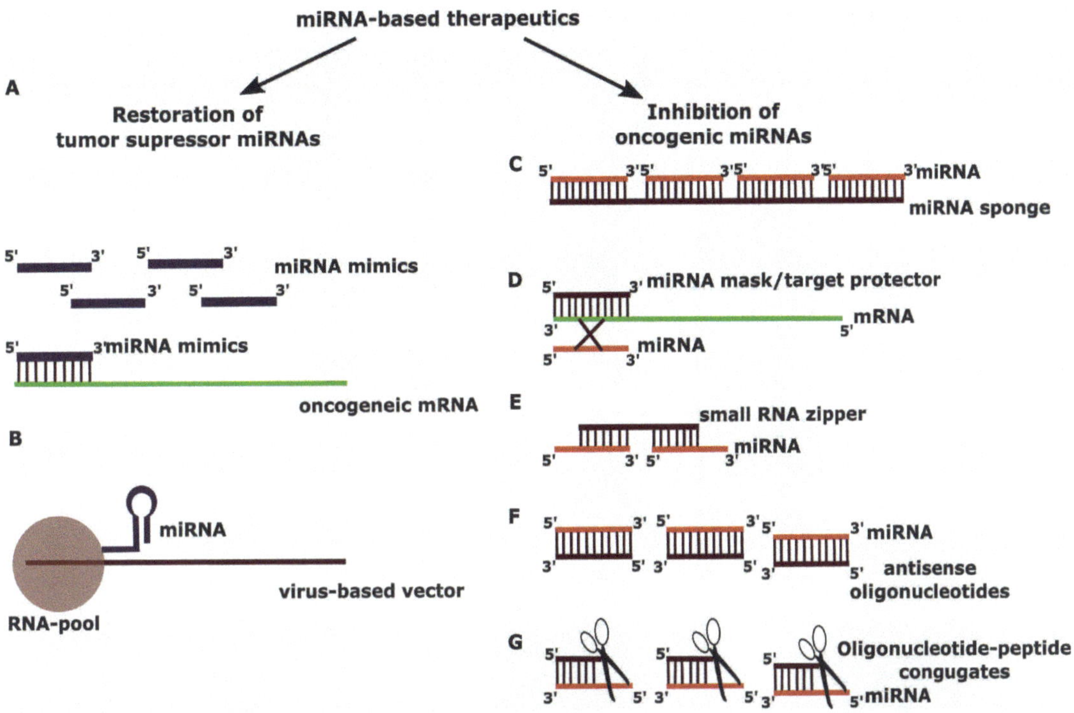

FIG. VIII.50 Strategies for oligonucleotide-based miRNA-targeted therapeutics. (A) To restore the concentration of tumor suppressor miRNA synthetic miRNA; (B) these miRNAs could also be expressed endogenously in the cell using viral vector, which will be transcribed by the cellular RNA polymerase. The oncogenic miRNAs (oncomirs) maybe inhibited by: (C) miRNA sponges; (D) miRNA masks (target protectors); (E) small RNA zippers; (F) antisense oligonucleotides; and (G) oligonucleotide-peptide conjugates (miRNases). Source: *Miroshnichenko, S., et al., 2019. Front. Pharmacol. 10, 488. Frontiers Media S.A.*

The delivery of miRNA-based therapeutics is one of the major challenges for using this type of drugs in clinics. Targeted delivery with the assistance of nanovehicles has been successfully demonstrated in the animal model system. Cancer cell-derived extracellular vesicles (TEVs) were used for this purpose (Fig. VIII.52). In a doxorubicin (DOX)-resistant subcutaneous breast tumor model, anti-miR-21, which blocks the function of an oncogenic miR-21 overexpressed in the cancer cells was effectively delivered using this system.

Besides cancer, extensive research work is being done to understand miRNA's role in various other diseases. It was found that dysregulation of miRNA expression is predominant in metabolic disorders (e.g., diabetes and liver disorders), neurodegenerative diseases, and in stroke, traumatic brain injury (TBI), and SCI. A few miRNA-based therapeutics are already in a clinical trial (Table VIII.5), and several others are in the pipeline.

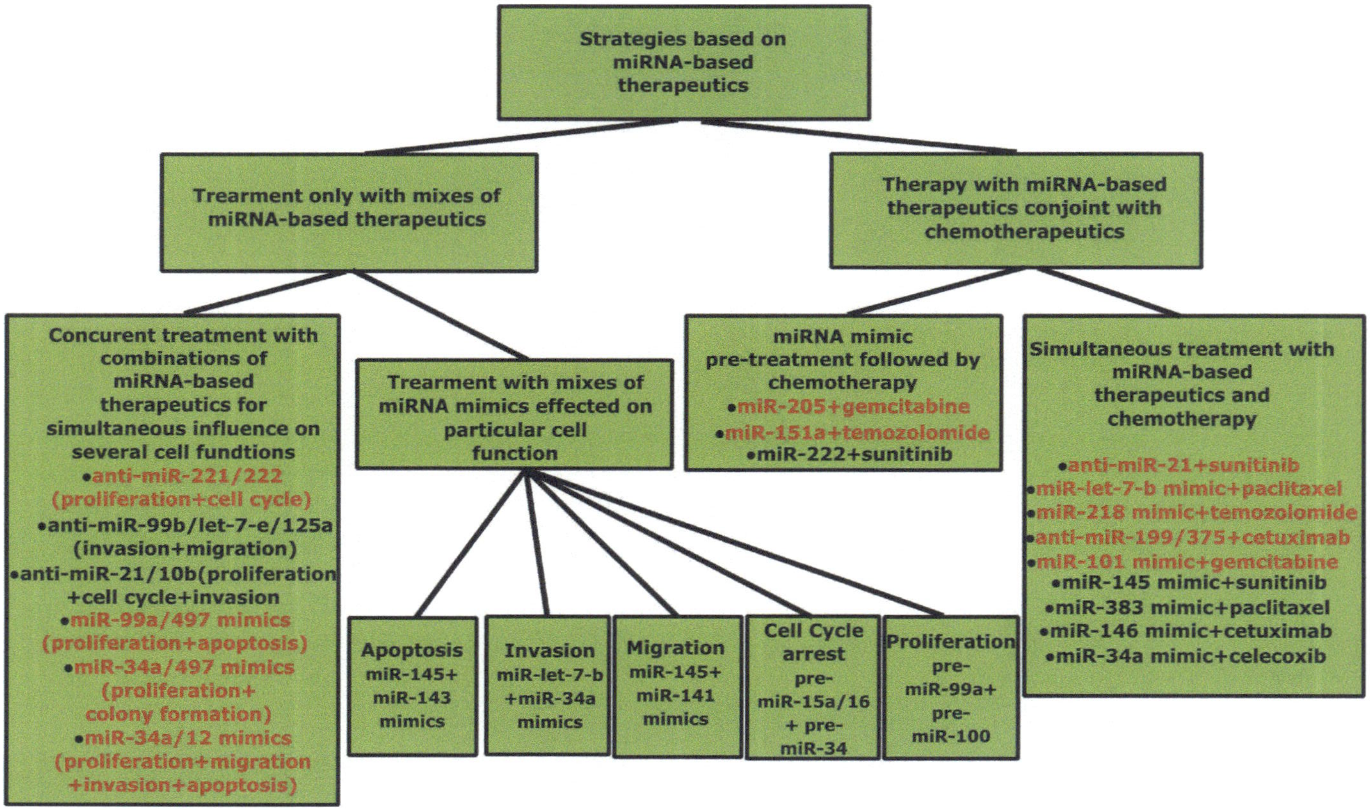

FIG. VIII.51 Potential combinatorial anticancer therapeutics based on dysregulated miRNAs: In this figure, different anti-miRNA and miRNA mimics have been shown with their possible targets of cancer hallmarks. The combinations of miRNA-based therapeutics and chemotherapeutics are indicated. Highly efficient antineoplastic strategies are indicated in *green*. Source: *Miroshnichenko, S., et al., 2019. Front. Pharmacol. 10, 488. Frontiers Media S.A.*

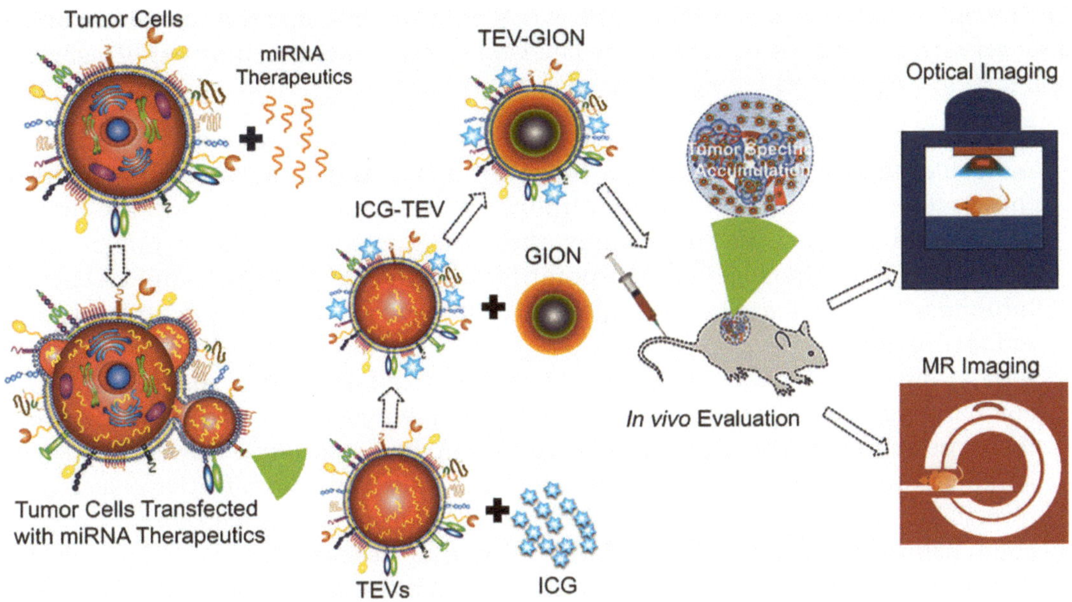

FIG. VIII.52 Nanovehicle-mediated delivery of miRNA therapeutics: anti-miRNA was packed in cancer cell-derived extracellular vesicles (TEVs), which were conjugated with functionalized gold-iron oxide nanoparticles (GIONs). The GIONs provided, this nanoplatform, two additional features: phototherapy treatments as well as the magnetic resonance imaging of cancer. The nanovehicle was also labeled with fluorescent dye, indocyanine green (ICG) to study the biodistribution of drug-containing nanoparticle. Source*: Bose, R.J.C., et al., 2018. ACS Nano 12(11), 10817–10832. American Chemical Society.*

Table VIII.5 miRNA-based therapeutics in clinical trials.

Disease	Targeted miRNA	Therapeutics	Chemical nature	Therapeutic implications
Liver disease	miR-122↓	Miravirsen	miRNA inhibitor	Antihepatitis C virus
Liver disease	miR-122↓	RG-101	miRNA inhibitor	Antihepatitis C virus
Diabetes and liver disease	miR-103/107↓	RG-125/AZD 4076	miRNA inhibitor	Antitype 2 diabetes and nonalcoholic fatty liver disease
Cancer	miR-34↑	MRX34	miRNA mimic	Antihepatocellular carcinoma
Cancer	MiR-16↑	TargomiRs	miRNA mimic	Antirecurrent thoracic cancer

(↑) Indicates upregulation and (↓) indicates downregulation by the therapeutics.

Chapter summary

In this chapter, we have discussed how the different types of epigenetic machinery are being targeted to develop therapeutics. Besides developing new drugs, these enzymes and noncoding RNAs have also been exploited to combat drug resistance. In this chapter, we put forward the recent developments of FDA-approved drugs, the potential therapeutic molecules under different phases of clinical trials, and the candidates under discovery stages, focusing on the dysregulated epigenetic system. A summary of each section is as follows:

- DNA methyltransferase inhibitors are the first set of epigenetic modulator molecules to be approved as drugs; however, at the time of the approval, targets of these molecules were not known. Although these nucleoside analogs are being used in the clinics quite successfully, these molecules are highly cytotoxic because of their nonspecific incorporation into the DNA/RNA. Nonnucleoside inhibitors of DNMT are being exploited at present for their specificity and reduced toxicity.
- Besides directly using them as therapeutics, DNMT inhibitors may also be used to combat drug resistance because dysregulation of methylation of the genome is one of the most frequently encountered coincidences of drug resistance. However, antimicrobial drug resistance also exploits other epigenetic modifications, such as reversible acetylation. Thus HDAC inhibitors could also be candidates for combinatorial therapeutics against drug-resistant microbes.
- Lysine (K) and Arginine (R) methylation are often dysregulated in multiple diseases, including cancer. A large number of K, R methyltransferase-specific inhibitors have been discovered. Some of these inhibitors have also received FDA approval for use in the clinics. A substantial number of inhibitors are in the pipeline of drug discovery. Besides K/R methyltransferases, histone demethylases are also a hot target of drug development. Inhibitors of demethylases are also in the developmental phases.
- Next to DNMT inhibitors, the most successful area is the HDAC inhibitor research. A few molecules of this class have been approved as therapeutics, and some are in the active phase of development.
- Although extensive research has been done in lysine acetyltransferases (KAT/HAT), several natural, semisynthetic, and synthetic inhibitors of KATs have been discovered, none of these molecules have succeeded as therapeutics. Recently, a small-molecule activator of KAT3 (p300/CBP) has been discovered and seems to have a good potential for drug development.
- miRNA-based therapeutics seem to be highly promising. Because miRNAs are one of the key regulators of gene expression, programming of miRNA dysregulation using anti-miRNA and miRNA mimics show a highly optimistic outcome. These classes of molecules are also in the pipeline of development. miRNA-based therapeutics could also be used as combinatorial therapeutics with other chemotherapeutic drugs.

Further reading

Agrawal, K., Das, V., Vyas, P., Hajdúch, M., 2018. Nucleosidic DNA demethylating epigenetic drugs—a comprehensive review from discovery to clinic. Pharmacol. Ther. 188, 45–79.

Bandyopadhaya, A., Tsurumi, A., Maura, D., Jeffrey, K.L., Rahme, L.G., 2016. A quorum-sensing signal promotes host tolerance training through HDAC1-mediated epigenetic reprogramming. Nat. Microbiol.. 1, 16174.

Bose, R.J.C., Uday Kumar, S., Zeng, Y., Afjei, R., Robinson, E., Lau, K., Bermudez, A., Habte, F., Pitteri, S.J., Sinclair, R., Willmann, J.K., Massoud, T.F., Gambhir, S.S., Paulmurugan, R., 2018. Tumor cell-derived extracellular vesicle-coated nanocarriers: an efficient theranostic platform for the cancer-specific delivery of anti-miR-21 and imaging agents. ACS Nano 12 (11), 10817–10832.

Chatterjee, S., Cassel, R., Schneider-Anthony, A., Merienne, K., Cosquer, B., Tzeplaeff, L., Halder Sinha, S., Kumar, M., Chaturbedy, P., Eswaramoorthy, M., Le Gras, S., Keime, C., Bousiges, O., Dutar, P., Petsophonsakul, P., Rampon, C., Cassel, J.C., Buée, L., Blum, D., Kundu, T.K., Boutillier, A.L., 2018. Reinstating plasticity and memory in a tauopathy mouse model with an acetyltransferase activator. EMBO Mol. Med.. 10(11), e8587.

Culp, E.J., Waglechner, N., Wang, W., Fiebig-Comyn, A.A., Hsu, Y.P., Koteva, K., Sychantha, D., Coombes, B.K., Van Nieuwenhze, M.S., Brun, Y.V., Wright, G.D., 2020. Evolution-guided discovery of antibiotics that inhibit peptidoglycan remodelling. Nature 578 (7796), 582–587.

Dancy, B.M., Crump, N.T., Peterson, D.J., Mukherjee, C., Bowers, E.M., Ahn, Y.H., Yoshida, M., Zhang, J., Mahadevan, L.C., Meyers, D.J., Boeke, J.D., Cole, P.A., 2012. Live-cell studies of p300/CBP histone acetyltransferase activity and inhibition. ChemBioChem 13 (14), 2113–2121.

Deblois, G., Tonekaboni, S.A.M., Grillo, G., Martinez, C., Kao, Y.I., Tai, F., Ettayebi, I., Fortier, A.M., Savage, P., Fedor, A.N., Liu, X., Guilhamon, P., Lima-Fernandes, E., Murison, A., Kuasne, H., Ba-Alawi, W., Cescon, D.W., Arrowsmith, C.H., De Carvalho, D.D., Haibe-Kains, B., Locasale, J.W., Park, M., Lupien, M., 2020. Epigenetic switch-induced viral mimicry evasion in chemotherapy-resistant breast cancer. Cancer Discov. 10 (9), 1312–1329.

Durant, S.T., Cho, E.C., La Thangue, N.B., 2009. p53 methylation—the argument is clear. Cell Cycle 8 (6), 801–802.

Fiorentino, F., Mai, A., Rotili, D., 2020. Lysine acetyltransferase inhibitors from natural sources. Front. Pharmacol. 11, 1243.

Grabiec, A.M., Potempa, J., 2018. Epigenetic regulation in bacterial infections: targeting histone deacetylases. Crit. Rev. Microbiol. 44 (3), 336–350.

Graves, P., Zeng, Y., 2012. Biogenesis of mammalian microRNAs: a global view. Genomics Proteomics Bioinformatics 10 (5), 239–245.

Lasko, L.M., Jakob, C.G., Edalji, R.P., Qiu, W., Montgomery, D., Digiammarino, E.L., Hansen, T.M., Risi, R.M., Frey, R., Manaves, V., Shaw, B., Algire, M., Hessler, P., Lam, L.T., Uziel, T., Faivre, E., Ferguson, D., Buchanan, F.G., Martin, R.L., Torrent, M., Chiang, G.G., Karukurichi, K., Langston, J.W., Weinert, B.T., Choudhary, C., de Vries, P., Van Drie, J.H., McElligott, D., Kesicki, E., Marmorstein, R., Sun, C., Cole, P.A., Rosenberg, S.H., Michaelides, M.R., Lai, A., Bromberg, K.D., 2017. Discovery of a selective catalytic p300/CBP inhibitor that targets lineage-specific tumours. Nature 550 (7674), 128–132.

Lau, O.D., Kundu, T.K., Soccio, R.E., Ait-Si-Ali, S., Khalil, E.M., Vassilev, A., Wolffe, A.P., Nakatani, Y., Roeder, R.G., Cole, P.A., 2000. HATs off: selective synthetic inhibitors of the histone acetyltransferases p300 and PCAF. Mol. Cell 5 (3), 589–595.

Ma, A., Stratikopoulos, E., Park, K.S., Wei, J., Martin, T.C., Yang, X., Schwarz, M., Leshchenko, V., Rialdi, A., Dale, B., Lagana, A., Guccione, E., Parekh, S., Parsons, R., Jin, J., 2020. Discovery of a first-in-class EZH2 selective degrader. Nat. Chem. Biol. 16 (2), 214–222.

Marques-Magalhães, Â., Graça, I., Henrique, R., Jerónimo, C., 2018. Targeting DNA methyltranferases in urological tumors. Front. Pharmacol. 9, 366.

Miroshnichenko, S., Patutina, O., 2019. Enhanced inhibition of tumorigenesis using combinations of miRNA-targeted therapeutics. Front. Pharmacol. 10, 488.

Modak, R., Das Mitra, S., Krishnamoorthy, P., Bhat, A., Banerjee, A., Gowsica, B.R., Bhuvana, M., Dhanikachalam, V., Natesan, K., Shome, R., Shome, B.R., Kundu, T.K., 2012. Histone H3K14 and

H4K8 hyperacetylation is associated with *Escherichia coli*-induced mastitis in mice. Epigenetics 7 (5), 492–501.

Nakayama, K., Szewczyk, M.M., Dela Sena, C., Wu, H., Dong, A., Zeng, H., Li, F., de Freitas, R.F., Eram, M.S., Schapira, M., Baba, Y., Kunitomo, M., Cary, D.R., Tawada, M., Ohashi, A., Imaeda, Y., Saikatendu, K.S., Grimshaw, C.E., Vedadi, M., Arrowsmith, C.H., Barsyte-Lovejoy, D., Kiba, A., Tomita, D., Brown, P.J., 2018. TP-064, a potent and selective small molecule inhibitor of PRMT4 for multiple myeloma. Oncotarget 9 (26), 18480–18493.

Rennoll-Bankert, K.E., Garcia-Garcia, J.C., Sinclair, S.H., Dumler, J.S., 2015. Chromatin-bound bacterial effector ankyrin A recruits histone deacetylase 1 and modifies host gene expression. Cell. Microbiol. 17 (11), 1640–1652.

Shames, S.R., Bhavsar, A.P., Croxen, M.A., Law, R.J., Mak, S.H., Deng, W., Li, Y., Bidshari, R., de Hoog, C.L., Foster, L.J., Finlay, B.B., 2011. The pathogenic *Escherichia coli* type III secreted protease NleC degrades the host acetyltransferase p300. Cell. Microbiol. 13 (10), 1542–1557.

Simon, R.P., Robaa, D., Alhalabi, Z., Sippl, W., Jung, M., 2016. KATching-up on small molecule modulators of lysine acetyltransferases. J. Med. Chem. 59 (4), 1249–1270.

Sun, P., Liu, D.Z., Jickling, G.C., Sharp, F.R., Yin, K.J., 2018. MicroRNA-based therapeutics in central nervous system injuries. J. Cereb. Blood Flow Metab. 38 (7), 1125–1148.

Vasan, N., Baselga, J., Hyman, D.M., 2019. A view on drug resistance in cancer. Nature 575 (7782), 299–309.

Ye, L., Lin, C., Wang, X., Li, Q., Li, Y., Wang, M., Zhao, Z., Wu, X., Shi, D., Xiao, Y., Ren, L., Jian, Y., Yang, M., Ou, R., Deng, G., Ouyang, Y., Chen, X., Li, J., Song, L., 2019. Epigenetic silencing of SALL2 confers tamoxifen resistance in breast cancer. EMBO Mol. Med.. 11(12), e10638.

Zhao, Q., Rank, G., Tan, Y.T., Li, H., Moritz, R.L., Simpson, R.J., Cerruti, L., Curtis, D.J., Patel, D.J., Allis, C.D., Cunningham, J.M., Jane, S.M., 2009. PRMT5-mediated methylation of histone H4R3 recruits DNMT3A, coupling histone and DNA methylation in gene silencing. Nat. Struct. Mol. Biol. 16 (3), 304–311.

Index

Note: Page numbers followed by *f* indicate figures, *t* indicate tables, and *b* indicate boxes.

D

E

H

Kolibri 360 GmbH
Ferdinand-Jühlke-Straße 7
99095 Erfurt, Deutschland
produktsicherheit@kolibri360.de